# Springer Handbook of Auditory Research

Volume 76

**The ASA Press**

ASA Press, which represents a collaboration between the Acoustical Society of America and Springer Nature, is dedicated to encouraging the publication of important new books as well as the distribution of classic titles in acoustics. These titles, published under a dual ASA Press/Springer imprint, are intended to reflect the full range of research in acoustics. ASA Press titles can include all types of books that Springer publishes and may appear in any appropriate Springer book series.



ASA PRESS

Brigitte Schulte-Fortkamp • André Fiebig
Joseph A. Sisneros • Arthur N. Popper
Richard R. Fay
Editors

# Soundscapes: Humans and Their Acoustic Environment

ASA PRESS

Springer

*Editors*
Brigitte Schulte-Fortkamp
HEAD-Genuit Foundation
Herzogenrath, Germany

André Fiebig
Department of Engineering Acoustics
Technical University Berlin
Berlin, Germany

Joseph A. Sisneros
Department of Psychology
University of Washington
Seattle, WA, USA

Arthur N. Popper
Department of Biology
University of Maryland
College Park, MD, USA

Richard R. Fay
Department of Psychology
Loyola University Chicago
Chicago, IL, USA

ISSN 0947-2657 ISSN 2197-1897 (electronic)
Springer Handbook of Auditory Research
ISBN 978-3-031-22778-3 ISBN 978-3-031-22779-0 (eBook)
https://doi.org/10.1007/978-3-031-22779-0



This Springer imprint is published by the registered company Springer Nature Switzerland AG
The registered company address is: Gewerbestrasse 11, 6330 Cham, Switzerland

# The Acoustical Society of America

On 27 December 1928, a group of scientists and engineers met at Bell Telephone Laboratories in New York City to discuss organizing a society dedicated to the field of acoustics. Plans developed rapidly, and the Acoustical Society of America (ASA) held its first meeting on 10–11 May 1929 with a charter membership of about 450. Today, ASA has a worldwide membership of about 7000.

The scope of this new society incorporated a broad range of technical areas that continues to be reflected in ASA's present-day endeavors. Today, ASA serves the interests of its members and the acoustics community in all branches of acoustics, both theoretical and applied. To achieve this goal, ASA has established Technical Committees charged with keeping abreast of the developments and needs of membership in specialized fields, as well as identifying new ones as they develop.

The Technical Committees include acoustical oceanography, animal bioacoustics, architectural acoustics, biomedical acoustics, engineering acoustics, musical acoustics, noise, physical acoustics, psychological and physiological acoustics, signal processing in acoustics, speech communication, structural acoustics and vibration, and underwater acoustics. This diversity is one of the society's unique and strongest assets since it so strongly fosters and encourages cross-disciplinary learning, collaboration, and interactions.

ASA publications and meetings incorporate the diversity of these Technical Committees. In particular, publications play a major role in the society. *The Journal of the Acoustical Society of America* (JASA) includes contributed papers and patent reviews. *JASA Express Letters* (JASA-EL) and *Proceedings of Meetings on Acoustics* (POMA) are online, open-access publications, offering rapid publication. *Acoustics Today*, published quarterly, is a popular open-access magazine. Other key features of ASA's publishing program include books, reprints of classic acoustics texts, and videos. ASA's biannual meetings offer opportunities for attendees to share information, with strong support throughout the career continuum, from students to retirees. Meetings incorporate many opportunities for professional and social interactions, and attendees find the personal contacts a rewarding experience. These experiences result in building a robust network of fellow scientists and engineers, many of whom become lifelong friends and colleagues.

From the society's inception, members recognized the importance of developing acoustical standards with a focus on terminology, measurement procedures, and criteria for determining the effects of noise and vibration. The ASA Standards Program serves as the Secretariat for four American National Standards Institute Committees and provides administrative support for several international standards committees.

Throughout its history to present day, ASA's strength resides in attracting the interest and commitment of scholars devoted to promoting the knowledge and practical applications of acoustics. The unselfish activity of these individuals in the development of the society is largely responsible for ASA's growth and present stature.

*In Memoriam and Honor*

*Murray R. Schafer*
*(July 18, 1933 – August 14, 2021)*
*who laid the foundation of all soundscape work what was to come*

# Series Preface

## Springer Handbook of Auditory Research

The following preface is the one that we published in Volume I of the *Springer Handbook of Auditory Research* back in 1992. As anyone reading the original preface, or the many users of the series, will note, we have far exceeded our original expectation of eight volumes. Indeed, with books published to date and those in the pipeline, we are now set for over 75 volumes in SHAR. Once volume 77 is completed, we are turning the series over to new Series Editors who will carry on with additional volumes.

We are very proud that there seems to be consensus, at least among our friends and colleagues, that SHAR has become an important and influential part of the auditory literature. While we have worked hard to develop and maintain the quality and value of SHAR, the real value of the books is very much because of the numerous authors who have given their time to write outstanding chapters and to our many co-editors who have provided the intellectual leadership to the individual volumes. We have worked with a remarkable and wonderful group of people, many of whom have become great personal friends of both of us. We also continue to work with a spectacular group of editors at Springer. Indeed, several of our past editors have moved on in the publishing world to become senior executives.

But the truth is that the series would and could not be possible without the support of our families, and we want to take this opportunity to dedicate all of our SHAR books to them. Our wives, Catherine Fay and Helen Popper, and our children, Michelle Popper Levit, Melissa Popper Levinsohn, Christian Fay, and Amanda Fay Sierra, have been immensely patient as we developed and worked on this series. We thank them and state, without doubt, that this series could not have happened without them. We also dedicate the future of SHAR to our next generation of (potential) auditory researchers – our grandchildren – Ethan and Sophie Levinsohn, Emma Levit, Nathaniel, Evan, and Stella Fay, and Sebastian Sierra.

# Preface 1992

The Springer Handbook of Auditory Research presents a series of comprehensive and synthetic reviews of the fundamental topics in modern auditory research. The volumes are aimed at all individuals with interests in hearing research including advanced graduate students, post-doctoral researchers, and clinical investigators. The volumes are intended to introduce new investigators to important aspects of hearing science and to help established investigators to better understand the fundamental theories and data in fields of hearing that they may not normally follow closely.

Each volume presents a particular topic comprehensively, and each serves as a synthetic overview and guide to the literature. As such, the chapters present neither exhaustive data reviews nor original research that has not yet appeared in peer-reviewed journals. The volumes focus on topics that have developed a solid data and conceptual foundation rather than on those for which a literature is only beginning to develop. New research areas will be covered on a timely basis in the series as they begin to mature.

Each volume in the series consists of a few substantial chapters on a particular topic. In some cases, the topics will be ones of traditional interest for which there is a substantial body of data and theory, such as auditory neuroanatomy (Vol. 1) and neurophysiology (Vol. 2). Other volumes in the series deal with topics that have begun to mature more recently, such as development, plasticity, and computational models of neural processing. In many cases, the series editors are joined by a co-editor having special expertise in the topic of the volume.

Richard R. Fay (Deceased), Chicago, IL, USA
Arthur N. Popper, College Park, MD, USA

*SHAR logo by Mark B. Weinberg, Potomac, Maryland, used with permission*

# Volume Preface

Soundscape is playing an emergent role in society when and wherever people gather, which simply means everywhere. Therefore, multiple disciplines have adopted the soundscape concept to study the impact of anthropogenic, biophonic, and geophonic sounds on humans and animals in all kinds of environments, in various contexts, and from different perspectives. With it, new disciplines have emerged to bridge intersectoral barriers and overcome disciplinary borders. The multidisciplinary conceptual approach to the study of soundscape has gained ground in the fields of urban (including park) noise, community noise, and environmental noise control, as well as in the context of parks and wilderness.

Following this idea, this SHAR volume is an introduction to the field of soundscape research and application. The chapters provide discussions of how soundscape research can enhance the quality of life about the acoustic environment for both humans and non-humans. The book brings together broad ideas on soundscape research to enhance our understanding and provide insight into the major considerations for how soundscape is studied and how it impacts humans.

Brigitte Schulte-Fortkamp and André Fiebig show in Chap. 1, "Soundscape: The Development of a New Discipline," how the soundscape approach has been adopted by multiple disciplines leading to the evolution of an entire new discipline. The need to understand sound perception and the challenges to accurately measure perception are discussed in Chap. 2 by André Fiebig: "Soundscape—a Construct of Human Perception." This is followed in Chap. 3, "Soundscape: The Holistic Understanding of Acoustic Environment," where Brigitte Schulte-Fortkamp and Pamela Jordan introduce soundscape as a construct of human perception and introduce the paradigm shift in noise control.

Continuing with the importance of soundscape for quality of life, Bennett M. Brooks considers how soundscape techniques can be applied to urban planning for communities in Chap. 4, "Soundscape and Urban Planning." Then, in Chap. 5, "Architectural Soundscapes," Gary W. Siebein and Keely W. Siebein focus on the transformative steps that can be taken to translate soundscape data and analyses into the physical form of a building for which sound is conceived as a generator of form.

The importance of psychoacoustic data for a comprehensive evaluation of acoustic environments is highlighted in Chap. 6. "Psychoacoustics in Soundscape Research," by Klaus Genuit, Brigitte Schulte-Fortkamp, and André Fiebig. Methodology becomes important in Chap. 7, "Measurements and Techniques in Soundscape Research," where Giovanni Brambilla and André Fiebig describe various methods and techniques used in soundscape investigations to measure the main aspects of soundscapes that include perception, acoustic environment, and context. Dick Botteldooren, Bert De Coensel, Francesco Aletta, and Jian Kang discuss the need for consolidating data from mixed methods in Chap. 8, "Triangulation as a Tool in Soundscape Research," and show how triangulating information has become an essential practice in soundscape studies. Indeed, the application of triangulation has important implications for soundscape data collection and also for theory development.

In Chap. 9, "Soundscape and Health," Peter Lercher and Angel M. Dzhambov review the theory, practice, and assessment of current research that investigates how acoustic environments affect the quality of life and health. More specific applications of health-related soundscape studies follow in Chap. 10, "Soundscape in Hospitals" by Ilene Busch-Vishniac and Erica Ryherd, in which they discuss how soundscapes can impact patient recovery and staff resilience.

In the final chapter, André Fiebig and Brigitte Schulte-Fortkamp discuss "How to Put Soundscape into Practice." They focus on the fact that there are still challenges in transferring the soundscape concept with its inherent holistic demand and its interdisciplinary foundation to real-world application. However, they show that significant progress is already being made.

Beyond doubt, the soundscape concept of assessing acoustic environments through the "lens" of perception will gain in significance, whether in research or practice.

It is evident that the acoustic environment we experience is much more than just sound level. We aim for experiences of soundscapes that can be described as pleasant, vibrant, and exciting, or that are restorative and relaxing, depending on the function of the place.

Thus, let us develop the tools and methods to achieve this goal and put those into practice.

Brigitte Schulte-Fortkamp, Herzogenrath, Germany
André Fiebig, Berlin, Germany
Joseph A. Sisneros, Seattle, WA, USA
Richard R. Fay, Chicago, IL, USA
Arthur N. Popper, College Park, MD, USA

# Contents

# Contributors

**Francesco Aletta** Institute for Environmental Design and Engineering, The Bartlett, University College London, London, UK

**Dick Botteldooren** Department of Information Technology, Ghent University, Ghent, Belgium

**Giovanni Brambilla** Department of Earth and Environmental Sciences (DISAT), University of Milano-Bicocca, Milano, Italy

**Bennett M. Brooks** Brooks Acoustics Corporation, Pompano Beach, FL, USA

**Ilene Busch-Vishniac** BeoGrin Consulting, Baltimore, MD, USA

**Bert De Coensel** Department of Information Technology, Ghent University, Ghent, Belgium

**Angel M. Dzhambov** Department of Hygiene, Faculty of Public Health, Medical University of Plovdiv, Plovdiv, Bulgaria

**André Fiebig** Department of Engineering Acoustics, Technische Universität Berlin, Berlin, Germany

**Klaus Genuit** HEAD Acoustics GmbH, Herzogenrath, Germany

**Pamela Jordan** University of Amsterdam, Amsterdam, The Netherlands

**Jian Kang** Institute for Environmental Design and Engineering, The Bartlett, University College London, London, UK

**Peter Lercher** Institute for Highway Engineering and Transport Planning, Graz University of Technology, Graz, Germany

**Erica Ryherd** Durham School of Architectural Engineering and Construction, University of Nebraska, Omaha, NE, USA

**Brigitte Schulte-Fortkamp** HEAD-Genuit Foundation, Herzogenrath, Germany

**Gary W. Siebein** Siebein Associates, Inc., Gainesville, FL, USA

**Keely M. Siebein** Siebein Associates, Inc., Gainesville, FL, USA

# Chapter 1
# Soundscape: The Development of a New Discipline

**Brigitte Schulte-Fortkamp and André Fiebig**

**Abstract** The concept of soundscape as a paradigm shift for understanding, measuring, and analyzing environmental sound is more than 50 years old. Many disciplines have adapted the soundscape concept and approach to study the impact of sound on humans (and animals) more holistically with perception gaining increased significance. In the beginning there were inconsistent applications of soundscape concepts, ambiguous soundscape definitions, and multiple understandings of the appropriate approach to study soundscapes, which impeded progress. Consequently, the need for a certain level of consensus across disciplines and professions was recognized and terms, methods, and analyses were internationally standardized, leading to more consistent research endeavors. With international standards and established procedures, the field of soundscape research continues to move beyond the use of simple standards and technical specifications. More than ever, soundscape research is performed to understand more deeply the impact of sound on humans in specific contexts. With this approach, humans are acknowledged to be more than passive receivers of their acoustic environments; rather, humans interact with their environments as both creators and receivers of the soundscapes. This perspective on interrelationships between person, activity, and place has led to substantial research efforts that continue to yield valuable insights into understanding soundscape.

**Keywords** Soundscape · Health · Social aspects · Community noise · Urban planning

---

B. Schulte-Fortkamp (✉)
HEAD-Genuit Foundation, Herzogenrath, Germany
e-mail: bschulte_f@web.de

A. Fiebig
Department of Engineering Acoustics, Technische Universität Berlin, Berlin, Germany
e-mail: Andre.Fiebig@tu-berlin.de



B. Schulte-Fortkamp et al. (eds.), *Soundscapes: Humans and Their Acoustic Environment*, Springer Handbook of Auditory Research 76,
https://doi.org/10.1007/978-3-031-22779-0_1

## 1.1 Introduction

The soundscape is a relatively recent concept applied to the methods used to evaluate acoustic environments. For decades, noise control was based on measurements that were intended to reduce the burden of noise but were never really successful. Noise reduction was done at the sources and through regulations about road traffic, train, and airplane noise based on noise levels. Schultz (1978) stated that people were not satisfied with those measurements because they still felt burdened by the reduced noise.

### *1.1.1 What Is Soundscape?*

Soundscape refers to the perceptual construct of the (acoustic) environment; the acoustic environment represents sound at certain locations as described by physics. Because the soundscape describes the perception of the acoustic environment, efforts continue to focus on developing reliable methods that quantify the perception of complex environments for humans and animals in more detail.

### *1.1.2 Purpose of this Volume*

Soundscape is playing an emergent role when and wherever society gathers, which simply means everywhere. Therefore, multiple disciplines have adopted the soundscape concept to study the impact of anthropogenic sounds (or animal sounds) in all kinds of environments, in various contexts, and from different points of view. Even new disciplines have emerged to bridge intersectoral barriers and overcome disciplinary borders. In particular, the concept of soundscape has gained ground in the fields of urban sound, community noise, and environmental noise control (To et al. 2018).

This book is an introduction to the field of soundscape research. The chapters provide discussions of how soundscape research can enhance the quality of life about the acoustic environment for humans and non-humans as well. The chapters bring together broad ideas on soundscape to enhance understanding and provide insight into the major considerations for how soundscape is studied and how it impacts humans. Each chapter refers to soundscape within the definition "acoustic environment as perceived or experienced and/or understood by a person or people, in context" (ISO 12913-1 2014), including an introduction to the variety of soundscapes in different areas. The concept of soundscape includes all the sounds in one's environment, but the focus of research relies primarily on evaluating human perceptions and the interrelationships between persons, activities, and places in both space and time. Understanding soundscape will provide a greater appreciation for the diversity of acoustic environments and their effects on people.

Soundscape is not only related to humans but to animals as well. The quality of the soundscape is as important for their health and well-being as for humans. (e.g., Slabbekoorn 2018; Derryberry et al. 2020). While this book focuses on the human soundscape, many of the ideas and principles that are discussed are equally valid for both terrestrial and aquatic animals, and so those interested in other animal groups will benefit from the research perspectives provided in these chapters.

### *1.1.3 Soundscape Research Priorities*

The concept of soundscape integrates individuals into the process of assessing and changing acoustic environments based on the perceptions and responses of affected individuals. In addition, the study of soundscape changes the research priorities: perception is assessed first and then acoustical measurements are made if needed for a deeper understanding of variables in the environment. Basically, a soundscape is clearly distinguished from the acoustic environment.

The environments that are subject to soundscape studies vary widely, from natural areas (e.g., woods, oceans) to urban areas (small cities, major cities). Although traditional environmental and community noise assessments still rely heavily on sound pressure level (SPL) indicators to quantify noise exposure, predict noise annoyance, and derive noise effects, it is increasingly apparent that a sound level-oriented perspective cannot capture several relevant aspects of noise and its perception in specific contexts. Thus, methods have been developed to measure and analyze human perception of sound in context. The discipline of environmental psychology evaluates the perception following common principles and rules that possess an inherent logic. Explorations of those principles and rules continue to be subjects of active research.

Schafer (1977) introduced his concept of soundscape from a musical point of view, and this new way of dealing with and understanding environmental noise was adopted by many others. In the 1990s, Schafer's concept was increasingly discussed and extended scientifically. That scientific discourse of the soundscape concept led to the development of advanced methods and tools to systematically collect soundscape-related data that resulted in international standardization efforts, which ultimately led to a broader dissemination of the soundscape concept.

Human perception of noise is influenced by several factors. In complex environments, humans usually recognize patterns out of sensations by using mediational processes to put together diverse sensations. The recognition of patterns constitutes perception and represents the perceptual construction of the external world. In general, the apparent link between a noise stimulus that causes a physiological reaction (an auditory sensation) that is cognitively processed as a perception (a hearing event) can be altered by numerous effects. For example, stimulation of another sensory system concurrently can modify the percept of sound in remarkable ways. The assigned meaning to sound can influence human sound perception as well and, most likely, the relative quality of the auditory sensation.

### 1.1.4 Sound Versus Noise

"Whenever society happens there is sound" (Maeder 2013, p. 424). Acoustic environments are full of sound sources that specifically shape the environment. In general, when we think about an acoustic environment, we refer to noise. Noise has many faces: it is any unwanted sound that is disturbing and can be violating when extreme. Achieving environmental quietness seems to be a frequently stated goal, but the question is whether a quiet environment is a good acoustic environment? What makes a good acoustic environment, and is there a common understanding of an acoustic environment that is good?

More and more people are living in highly dense cities, often packed in high rise buildings positioned between commercial activity and high amounts of all kinds of traffic. Schafer (1977) described this process of increasing urbanization in the 1970s as towns have grown into cities, and cities have expanded to cover much of what was formerly rural land. This development leads to many challenges, including social conflicts, environmental impacts, and the problems caused by noise. For years when people complained about noise, the reaction was to measure the noise physically in terms of basic level indicators, which did not provide insight into the actual details of noise that burdened the local population. A different approach is taken in soundscape research and data analyses.

Today's permanent technological changes at sources lead to the expectation that reduction of noise at the source will make the living environments become quieter. Yet, the opposite is the case: due to increasing traffic volume, increasing noise is the daily experience. Therefore, a different structure of urban areas may change the burden of noise. Eventually, adding more green areas, designating more pedestrian zones, and reducing the use of personal cars to use public transportation may lead to a new acoustics in urban environments.

For many years, transportation noise from diverse sources has been considered to be a type of environmental pollution that affects human health and well-being. Numerous studies determined limit values that, when exceeded, increased the risk for certain adverse health effects. These limit values were identified for each source separately as the relationship between response and SPL varies significantly from source to source (WHO 2018). As important as those health-related studies for effective health protection are, the rich and multidimensional experience of acoustic environments is not sufficiently covered by those simplified exposure-response perspectives that consider the impact of annoyance and sleep disturbance from only single sources. Acoustic environments are usually full of contrasting sound sources that shape the environment in site-specific ways.

Humans living in cities are not exposed separately to each unwanted source. They are exposed to an acoustic environment created by the complex superposition of several sound sources. Those acoustic environments can elicit numerous sensations and emotions that are not limited to annoyance. Moreover, certain sound sources cannot be described as more or less annoying and disturbing because they can be perceived as pleasant or promoting a feeling of restoration and relaxation. In addition, specific places (e.g., parks) and contexts (e.g., a Sunday afternoon) in which sounds are perceived can modify the response to an acoustic environment.

Studies of acoustic environments should be always related to quality of life that includes understanding human perception, experience, and expectations. If humans are perceiving and considering their acoustic environments, they are also cognizant of the overall "gestalt" of all sensory components of that environment. From Schafer (1977) we learn that acoustic environments must also be understood as resources for social life.

To understand the complex perception of acoustic environments more comprehensively, it is necessary to go beyond statistical considerations of dose (in terms of level) and response (in terms of the degree of annoyance) functions. The expertise of the population concerned is required to identify location-specific peculiarities. Obviously, popular places and parks in cities often aren't quiet. Nature settings appraised by most people for their recreational and restorative potential are not necessarily silent, for example, coastal sites with the sounds of ocean waves or sites with frequent bird calls.

A quiet and calm outdoor area implies a pleasant soundscape where people enjoy staying for a while (Salomons et al. 2013). Birdsong or water features improve the perception of an urban soundscape (Galbrun and Ali 2012; Zhao et al. 2020). Greenery and vegetation also can improve the perception of environmental noise and often outperform conventional noise mitigation measures. For example, based on quantitative estimates by van Renterghem (2019), the equivalent level reduction with (high quality) visible greenery from home could reach −10 dB(A), which is in addition to any physical SPL reduction one might obtain behind vegetation belts.

The COVID-19 pandemic of the 2020s revealed how the acoustic environment around the world could be affected by a reduction in human activities. Significant changes in soundscape happened at different scales (Aletta et al. 2020), and those changes affected a broad range of sites, including historic soundscapes and heritage sites that attract locals and tourist populations as documented by Jordan and Fiebig (2021). This sudden change in human activities showed that a broad approach is needed to study environmental noise and noise protection cannot be isolated from social context and changes in human behaviors.

### *1.1.5 The Soundscape Approach*

Some understand soundscape as a kind of "umbrella term" for a more comprehensive way of assessing noise that is more related to the respective context. This understanding comes along with varying definitions and notions that depend on the research discipline, which shapes how the term is applied. Although there was, and still is, a broad variety of meanings attributed to the term soundscape, the motivation to create the term seems clear. Noise needs to be studied for its complex effect on humans and animals. Following a negative approach and dealing with noise simply as pollution is too narrow (Schafer 1977).

Schafer (1977) wanted to treat the world as a macrocosmic musical composition. This simple idea triggered numerous applications and paradigm shifts in different scientific fields and disciplines. Schafer wondered about the dominance of visual culture and the loss of sonological competence within modern societies. One of his prominent techniques for understanding a sonic environment was the development of the so-called Isobel maps, as shown by his log notes of sound events during a full day in the countryside of British Columbia, Canada (Fig. 1.1). The field study included SPL measurements and the description of a wide range of sonic features (Truax 1978). According to Maeder (2013), the Isobel map holds differentiating information about the distribution of acoustic intensity and looks very similar to a geographic map produced with elevation contour lines. Decades later, noise mapping is a major issue regarding community noise and is part of the European Directive on Noise (END 2002).

Schafer's research strategies led to his classification of "hi-fi" and "lo-fi" soundscapes. A hi-fi environment is one in which sounds may be heard clearly without crowding or masking. Even sounds in the distance can be heard. In contrast, a lo-fi soundscape indicates that an environment is overcrowded with keynote sounds and signals that result in masking or lack of clarity for individual sounds. The listener cannot separate the different sound sources and cannot detect any sound events in the distance anymore (Schafer 1977). Such a qualitative understanding of an

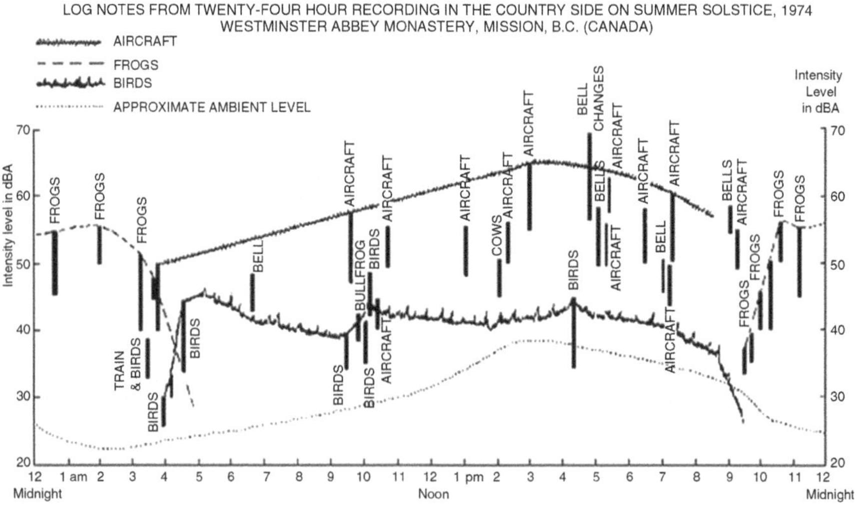


**Fig. 1.1** Schafer's short log notes of sound events taken during a 24-hour period in the countryside in British Columbia (Schafer 1977, p. 266)

acoustic environment may seem elitist, but such simple classifications may help to solve difficulties in assessing acoustic environments for enhancing quality of life.

In 1984, the Institute of Kanda was founded in Tokyo, Japan, by Keiko Torigoe and her colleagues to establish research in soundscapes (Hiramatsu 2006). They followed the understanding that fieldwork must be conducted under the subjective local view within the respective acoustic environment. According to Hiramatsu (2006), the concept of soundscape was also subject to contributions from musicologists, sociologists, philosophers of aesthetics, and environmental scientists reaching out for designing public gardens. Soundscape studies in Japan started with "fieldwork in urban areas than proceeded to soundscape design. [...] soundscape studies [...] are more or less related to the sonic environment with emphasis on the way it is perceived or understood by individual or by society" (Hiramatsu 2006, p. 863).

For example, the Yamahoko-cho area located in the city center of Kyoto is famous for the Gion Festival, which is one of the biggest and oldest festivals in Japan. Each year in July the ceremony and festival are undertaken for one month, and the soundscape in the city becomes dominated by a variety of sounds related to the festival. According to Hiramatsu (2000), the music is accepted as a characteristic of the city and, though the music is without any doubts the loudest sound ever heard in this area, there are no complaints.

Similar findings were reported from an Italian research study related to the Gigli di Nola folk ceremony in the little town Nola near Naples in Italy, which features a shoulder-borne procession celebrating the feast of Saint Paulinus. Alves et al. (2021) examined how the physical and spatial arrangement of Nola shaped the enactment of the festival's soundscape, atmosphere, and the behavior of its participants. They found that the procession soundscape dominated the atmosphere of the festival, and that the rhythmic qualities guided the parade for the participants at the Gigli festival. The soundscape analysis was an indicator of the value of the acoustic components in the festival.

Indeed, human perception is an important and firsthand measure for decisions about any assessments, initiatives for changes, or further development. Moreover, physical measurements play a different role compared to such measurements in noise research. In soundscape studies, physical measurements should be considered only as a follow-up to the analysis of perceptual evaluations.

Measuring individual perceptions with soundwalking and related procedures have become especially important methods (see Brambilla and Fiebig, Chap. 7). A perfect example for this procedure is the Nauener Platz project in Berlin. Through the systematic application of the soundscape process, a solution was found to change a public place suffering with acoustic and social problems into a place of social communication, relaxation, and safety that is well-accepted by the people living in the surrounding area (see Schulte-Fortkamp and Jordan, Chap. 3 and Fiebig and Schulte-Fortkamp, Chap. 11).

## 1.2 Measures and Measurements in Soundscape

Over time, there has been an increase in the demand for the soundscape approach for planning in urban areas. Based on the successes seen in soundscape research, the involvement and participation of people as local experts help to support the success of intended changes (de Coensel et al. 2010, 2017). The soundscape procedure gives the same level of relevance to people's thoughts and feelings as to physical measurements. These complementary methods have been codified in the standardization process.

The standardization process for soundscape (ISO 12913 series) helps researchers make the best methodological decisions for the soundscape measurement needed to support individual participation as a basic component in strategic planning for an acceptable acoustic living environment (see Schulte-Fortkamp and Jordan, Chap. 3). In addition to providing definitions and a conceptual framework, the soundscape standard offers appropriate measurements for soundscapes and analysis tools to be used for any activity considered in a soundscape.

The soundscape of a classroom (Brill et al. 2018) and restaurant (Roy and Siebein 2019) may use the participation of people concerned with the description and analysis of the space, but there is a further issue that counts: the context. Another example (Schulte-Fortkamp et al. 2007; Brooks et al. 2014) is the planning process for a public space where both residents and visitors will be involved as the understanding of any intervention will be different for these two groups who would participate in the process.

### *1.2.1 Context*

The importance of context in a soundscape is described in the ISO 12913-1 (2014). Context is defined there as the physical place where the acoustic environment exists. Contextual studies consider the interrelationships between person, activity, and place in space and time. Consequently, the context may influence soundscape due to what is heard, the interpretations of those auditory sensations, and the responses to that acoustic environment (Kang et al. 2016). Included in context is the meaning of the specific place to the individuals involved and its use by those individuals.

Other factors can influence the auditory sensation in addition to the acoustic environment: visual impressions, scents or odors, time of day, lighting, meteorological conditions that vary seasonally, and even individual hearing impairments and hearing aids (ISO 12913-1 2014). The interpretation of auditory sensation can be influenced by the specific sound sources, previous experiences with those sources, and individual expectations that include intended use of a space. Expectations can also vary with cultural background, personal activity preferences, and individual capacities to deal with variable situations (see ISO 12913-1 2014).

### 1.2.2 Acoustic Measurements

Classical noise control solely relies on the measurement of loudness determined in terms of SPL indicators like the energy equivalent sound pressure level ($L_{Aeq}$) or the day-evening-night level ($L_{DEN}$). The general understanding is that the higher the level, the more annoying the environmental noise should be. In some cases, the SPL alone is not sufficient to predict accurately the human response to an unwanted sound (noise) source. The concept of "rating level" $L_r$ was introduced to correct the physical level value by bonuses or penalties to predict noise annoyance more reliably. For example, sounds from unwanted sources with prominent tonal components are more annoying than sounds without prominent tones at the same SPL. Thus, simply a correction by a few dB, called the tone penalty, is applied to adjust the rating level to indicate the resulting noise annoyance properly.

In the context of soundscape, a simple link between SPL and annoyance is not assumed as sound is understood as a potential resource that can be beneficial. For example, the pleasantness of wanted sounds cannot be determined by loudness measures or the annoyance from particular music cannot be measured solely by the magnitude of its tonal components.

In contrast to the basic concept of noise control and noise abatement that is aiming to reduce the SPL, Schafer (1977) defined a different approach that focused on interventions that cannot be described simply by the magnitude of level reduction. The notion of the soundscape concept is that there is no assumption that the sounds that constitute the acoustic environment must be of low intensity (Brown et al. 2011). This fundamental difference has also had a significant effect on the way acoustic environments have to be measured, characterized, and analyzed. Schafer (1977) proposed the use of both acoustics and psychoacoustics to learn about the physical properties of sound and the way that sound is perceived and interpreted by the human brain. Therefore, acoustic measurements and analyses must strive for a more sophisticated characterization of the properties of the acoustic environment and their relationship to perception.

### 1.2.3 Measuring Human Perception

When considering the focus on perception in soundscape, any recording method must consider the way humans perceive the acoustic environment. In addition to established binaural measurement systems, which are the most used recording techniques for soundscape studies (Hong et al. 2017), other recording technologies (e.g., microphone arrays) are frequently used in soundscape investigations as well. Those measurement systems strive for a playback based on multi-loudspeaker arrays that should provide a good level of immersion (ISO 12913-2 2018). However, as those systems lack international standards, the comparability of acoustic analyses based on microphone arrays and ambisonics are limited (see Brambilla and Fiebig, Chap. 7).

Psychoacoustic and other perception-related parameters are measured and analyzed to describe the acoustic environment. The emotions and feelings elicited are measured by questionnaires that assign descriptive terms for the perception of the acoustic environment in all facets beyond the degree of annoyance (see Brambilla and Fiebig, Chap. 7). The approach to measure sound as perceived by humans led to increased attention to multiple parameters in psychoacoustic measurements within the realm of soundscape research since the early 2000s (Engel et al. 2021). Consequently, the ISO/TS 12913-3 (2019) called for consideration of binaural data analysis that include psychoacoustic indicators to enable the quantification of the acoustical impact on the listener and the exploration of relationships between physical properties of the environments and human response behavior. Other perception-related indicators considered range from eco-acoustic indices (Lawrence et al. 2022) to Mel Frequency Cepstral Coefficients (MFCC) based indicators (Lunden et al. 2016) to the second derivative of specific time and frequency parameters (Aumond et al. 2017).

Currently, searches for meaningful physical parameters are supported by machine learning and neural network approaches to predict human perception more reliably (e.g., Verma et al. 2019; Quinn et al. 2022). However, as the complexity of human perception seems almost infinite and is influenced by many intrinsic and extrinsic factors, the hunt for the most powerful (psycho)acoustic indicators will continue for a long time. For example, how humans make assessments based on bounded affective episodes is still not well understood; the nature of cognitive heuristic or normative operations converting patterns of experiences into overall assessments is surprisingly unclear (Fiebig 2019).

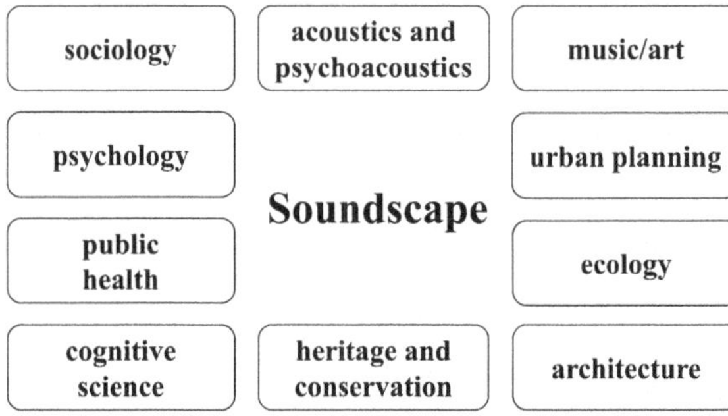


**Fig. 1.2** Overview of disciplines dealing with the concept of soundscape from different points of view

## 1.3 Disciplines Using Soundscape Methods

The concept of soundscape is applied in several contexts and disciplines as illustrated in Fig. 1.2. Applications range widely from underwater acoustics and bioacoustics that are used for environmental noise assessments and acoustic ecology to terrestrial designs for sound art and meditation, but soundscape methods can also be applied to other areas such as sociology, psychology, and public health. The scientific rigor and the fundamental theories underlying the use of soundscape analyses in those areas have basic differences. For example, in underwater acoustics the term soundscape is used as a "characterization of the ambient sound [...] in terms of its spatial, temporal and frequency attributes, and the types of sources contributing to the sound field" (ISO 18405 2017).

The first part of the soundscape standard ISO 12913 became available in 2014 and influenced research thereafter. However, in 2020, the use of soundscape standards was limited by the worldwide COVID-19 pandemic, which stopped many projects and initiatives in soundscape. However, in 2022, we have found new approaches in urban planning that are integrating the soundscape approach.

Di Loreto et al. (2022) found that the soundscape approach enabled a determination of the connections between the sensations of human beings and the environment during the planning phase of new attractions for an urban environment. This goes along with earlier findings by De Coensel et al. (2017) and Sun et al. (2018, 2019). When De Coensel (2017) was carrying out *The Urban Soundscapes of the World Project*, he stated that designing urban acoustic environments is a considerable challenge, especially regarding adequate measurements and data collection for architects who still work by example.

### *1.3.1 Eco-Acoustics*

In the context of soundscape ecology, sound is considered from an ecological perspective, investigating natural and anthropogenic sounds and their relationships with the environment over multiple scales of time and space (Farina and Gage 2017). The discipline of eco-acoustics comprises the study of populations, communities, ecosystems, landscapes, and biotic regions of the earth, including terrestrial, freshwater, and marine systems. Thus, according to Farina and Gage (2017), eco-acoustics extends the scope of acoustic investigations by including bioacoustics and soundscape ecology.

Clearly, when disciplines relate their research to soundscape, the approach is influenced by evaluation procedures that rely on perception (see Fiebig, Chap. 2). In sociology and psychology, the impact is given through qualitative research with narrative interviews (Hollstein 2011). According to Knoblauch (2013), the field of sound studies was largely ignored in qualitative research in sociology. Nevertheless, there are first steps for culture studies that show how the tunes of the world are analytically transformed into the sounds in and of society (Maeder 2013). However, soundscape research uses the soundwalk procedure (Sect. 1.3.2) as a qualitative research segment (ISO/TS 19123-2 2018).

### *1.3.2 Soundwalks*

In the context of the typical application, the soundwalk method is used to collect context-sensitive data, environmental noise assessment, and urban planning. The soundwalk method is defined as a method that implies a walk in an area with a focus on listening to the acoustic environment (ISO/TS 12913-2 2018). Moreover, the input of local experts is expected in those evaluation procedures. A local expert is a person who is familiar with the area under scrutiny through either living in the area or having daily routines related to the area (ISO/TS 12913-2 2018). Here, the implication of "expertise" is connected to daily experiences and collected knowledge about an acoustic environment. "Local experts are those people […] who provide their expertise to researchers, investigators, and project designers through such processes as soundwalks and open interviews […]. The experience considers all conscious and unconscious influences sound makes in people's minds as they judge the appropriateness of sounds, sound sources, places, or situations." (ISO/TS 12913-2 2018, p. 14).

This appreciation of local knowledge led to more work on participatory approaches for which interventions and design options have been developed in cooperation with local experts and other stakeholders (Maag and Munck-Petersen 2018). The identification of environmental acoustic issues that need to be considered calls for collaborations with citizens as co-specifiers of projects (Xiao et al. 2017). For example, locals can be involved in participatory noise monitoring, empowering them to actively participate in improving their living environment by creating smartphone-based participatory soundscape maps (Brambilla and Pedrielli 2020).

As use of the place, the context, and expectations of its users is important to how the sound environment is perceived, local expertise must be involved in deciding what measures are appropriate and which characteristics require priority (Schulte-Fortkamp and Jordan 2016). The increasing interest in participatory approaches supported development of "co-creation concepts" and consideration of how those can be used in the context of environmental noise (van Renterghem et al. 2020). Botteldooren et al. (2020) envisioned that co-creation could go one step further to allow users of the space to augment the space with their own designs, and they concluded that co-creation opens a wealth of opportunities to improve public spaces and increase their use.

In any case, the paradigm shift in soundscape studies has occurred because everyone's experience is important and directly related to the area under scrutiny. However, it is essential that people are open for communication and willing to share their knowledge about a certain area. The ISO/TS 12913-2 (2018) provides methods and procedures that guarantee the needed communication through questionnaires and guided interviews. Such data collection must capture the mood, restoration opportunities, individual appreciation of the space, individual preferences, and document overt behavior to create an accurate representation of a specific location. This type of evaluation, according to the ISO/TS 12913-2 (2018), shall respect the way people are experiencing their environment.

### *1.3.3 Architectural Applications*

Important soundscape work is being accomplished in architecture (see Siebein and Siebein, Chap. 5). According to Brown et al. (2016), soundscape approaches can be applied to different places, such as malls and markets, transport stations, sports stadia, museums, and the balconies or terraces of our own dwellings. Indoor spaces can also benefit from the soundscape approach, for example, hospitals, educational institutions, restaurants, and homes.

Fowler (2015) criticized the traditional consideration of the acoustic environment within architectural design that focused mainly on concert halls or recording studios. He stated that any new approaches to auditory design in architectural practice must integrate critical listening as an important component. "To readily accommodate the acoustic impact of design decisions, particularly within a parametric paradigm, requires an immediacy between hearing the connection that visual form making has on the impact of the design's ability to communicate an intended acoustic signature. In such a framework, architecture gains the potential to become more than what is immediately seen and moreover, the case of whether sounds inhabit the space or space is produced by the sounds is a question only relevant to how one hears the design." (Fowler 2015, p. 70).

As discussed by Schafer (1977), studies in the arts, particularly in music, will support the creation of ideal soundscapes, especially when imagination and psychic reflection lay the foundation for a new interdisciplinary approach: the acoustic design (Schafer 1977). Work by Schulze (2019) on sound studies provided insight on a part of soundscape that is strongly related to art and music. Wondering about sound design and its function in the future, he presented a detailed overview of the modern history of sound design. In some cases, the creation of sound art and audio installations can meet urban sound planning and management expectations. In their sound art study, Steele et al. (2019) concluded that sound installations can change soundscape evaluations compared to the previous baseline condition; specifically, the installation increased calmness, provided a capacity for respite, and reduced the perceived overall sound level in the proximity of the (non-music) sound installation.

### 1.3.4 Roles of Soundscape in Human Health

The good practice guide on quiet areas of the European Environment Agency concluded that one should not focus on the quantitative health effects to be achieved, but instead one should offer people the opportunity to find calm (European Environment Agency 2014). It is necessary to know what makes an acoustic environment calming and restorative: we know that silence tends to frighten most people and the absence of unwanted sound does not automatically result in a pleasant soundscape (Nilsson and Berglund 2006).

Four different components have been considered important for creating a restorative environment, which underlines that restoration cannot be related only to low SPLs (Payne and Bruce 2019). Herranz-Pascual et al. (2019) observed that the soundscape characteristics that contributed to greater emotional restoration and a reduction in perceived stress were pleasantness, calm, fun, and naturalness, which shows the range of soundscape properties to be considered. They concluded that the capacity for psychological restoration is not unique to natural settings outside cities: properly designed urban places can significantly decrease negative emotions and perceived stress and can even increase positive emotions. Restoration depends on (acoustic) comfort and not exclusively on acoustic environments of low intensity. Thus, approaches that go beyond loudness or level are needed to study the impact of acoustic environments on humans in specific contexts. These requirements spur applications of the soundscape approach, which aims to encompass the perception and appraisal of acoustic environments in their entirety.

Soundscape methodology has provided important input for health-related research regarding noise and noise effects (see Lercher and Dzhambov, Chap. 9). As it is true in other applications, the influence of an acoustic environment is based on its associated contexts. Moreover, in health-related research, soundscape is considered through the lens of noise exposure and there is the expectation that adverse health effects can be prevented through "healthy soundscapes" designed within environmental planning. Agreement on sustainable methodological procedures is required for consistent application of the soundscape approach.

The soundscape standard ISO 12913 series provides support in three areas: ISO 12913-1:2014: Acoustics-Soundscape-Part 1. *Definition and conceptual framework*; ISO/TS 12913-2:2018: Acoustics-Soundscape-Part 2. *Data collection and reporting requirements*; and ISO/TS 12913-3:2019: Acoustics-Soundscape-Part 3. *Data analysis*. The platform for targeted measures is supported by a holistic approach in soundscape studies (see Schulte-Fortkamp and Jordan, Chap. 3).

Sound is a critical component of the environment that can give people a sense of place and time, but when an acoustic environment is unfamiliar, it adds to the anxiety of those who receive the sounds (Talebzadeh and Botteldooren 2022). For example, hospital sound levels have been increasing for decades not only due to the addition of more medical devices and the device's auditory alarm but also due to structural components of the physical environment, such as the nature of the flooring, doors, doorknobs, walls, and windows. Very seldom are calmness and restfulness provided by the environment of the hospital room; instead, the hospital soundscape is loud with a cacophony of various activities. As a result, the hospital soundscape affects patients and staff negatively through a continuous burden of noise.

According to Busch-Vishniac et al. (2005) and Busch-Vishniac and Ryherd (2019), hospital soundscapes affect staff and patients, potentially increasing stress in the staff and anxiety in the patients. For some years now, various interventions have been discussed that might improve hospital soundscapes by including the implementation of quiet times, architectural designs that reduce reverberation, the addition of sound absorption, the use of earbuds or headphones, and the use of nature sounds to mask some less appreciated hospital sounds (see Busch-Vishniac and Ryherd, Chap. 10). As proposed in the soundscape standard ISO 12913 series, investigations are suggested to determine whether there is a direct link between patient medical outcomes and elements of the hospital soundscape that could confirm the success of specific interventions that can be scaled across a broad range of hospitals (Busch-Vishniac and Ryherd 2019).

In related work with patients suffering with dementia, Talebzadeh and Botteldooren (2022) explained how a personalized soundscape can support people by providing a pleasant acoustic environment. The development of that project has shown that a thoughtfully designed, familiar soundscape can reduce behavioral and psychological symptoms of dementia and also improve sleep quality.

## 1.4 Chapter Overview

Communication about noise management is required to guarantee that the specific components of soundscapes and human perceptions are equally relevant and seriously considered during the entire process of urban planning. The ISO standard 12931-1 (2014) on soundscape provides an important, and rigorous, distinction in the use of soundscape. Unfortunately, some individuals and groups, particularly planners, designers, laypersons, and even those primarily interested in management of indoor and outdoor environments through environmental noise control, use soundscape as a synonym for the physical acoustic environment. Thus, the chapters in this book are intended to help these people and other interested groups to better understand the full meaning of soundscape.

The need to accurately measure auditory perception and the challenges presented, especially outside of the laboratory, are discussed in Chap. 2 by André Fiebig: "Soundscape: A Construct of Human Perception." The author points out that further work is needed to develop sophisticated theoretical concepts that will allow researchers and practitioners to test the applicability of different methods to measure perception of a soundscape and to evaluate the validity of experimental outcomes.

In Chap. 3, "Soundscape: The Holistic Understanding of Acoustic Environment," Brigitte Schulte-Fortkamp and Pamela Jordan introduce soundscape as a construct of human perception that factors in the entirety of an acoustic environment and the individual's responses to it. This stands in contrast to the acoustic environment alone, which is simply the composition of sound stimuli in an environment. The chapter begins by tracing the broad trajectory of soundscape studies to contextualize a holistic approach and concludes by highlighting various holistic research projects that sought to enhance the quality of acoustic environments and living situations.

Continuing with the importance of soundscape for quality of life in Chap. 4, "Soundscape and Urban Planning," Bennett M. Brooks considers how soundscape techniques can be applied to planning for communities. The author discusses urban planning as a key component in the process of actualizing the soundscape theory and implementing holistic improvements in the acoustical environment on a large scale. This chapter presents the basic concepts and principles of urban planning and urban design, which is the bridge between urban planning and architecture, regulation, smart growth, and a handbook toolbox for action.

In Chap. 5, "Architectural Soundscapes," Gary W. Siebein and Keely W. Siebein focus on the transformative steps that can be taken to translate soundscape data and analyses into the physical form of a building for which sound is conceived as a generator of form and is not necessarily a result of form nor of a series of elements added to the form. The links between architectural theories and soundscape theories are used to illustrate the basis of the elements and levels of the architectural soundscape design theory.

The importance of psychoacoustic data for a comprehensive evaluation of acoustic environments is considered in Chap. 6, "Psychoacoustics in Soundscape Research," by Klaus Genuit, Brigitte Schulte-Fortkamp, and André Fiebig. Moreover, a key point is made that humans perceive acoustic environments binaurally, which must be included in valid analyses because perception cannot be described adequately by simple sound level measurements. The authors argue that there is a critical need for aurally accurate measurements and psychoacoustic analyses with the distinct purpose of archiving and providing the ability to reexperience different acoustic environments.

In Chap. 7, "Measurements and Techniques in Soundscape Research," Giovanni Brambilla and André Fiebig describe techniques that include input from people who experience the environment under consideration, quantify various aspects of the acoustic environment, and evaluate the context of human interactions with the environment. Included in this survey of methodology are soundwalks, questionnaires, interviews, and recordings of sound that mimic the binaural way in which humans perceive sound, and how those methods are applied. In addition, the authors consider how international standards and technical specifications have led to a harmonization of data collection in soundscape investigations.

Dick Botteldooren, Bert De Coensel, Francesco Aletta, and Jian Kang discuss additional methodology in Chap. 8, "Triangulation as a Tool in Soundscape Research." Triangulating information has become an essential practice in soundscape studies. Indeed, the application of this analysis tool has important implications for soundscape data collection and also for theory development. Triangulation provides a useful lens through which research trends and future lines of investigation can be identified. The authors reveal that very few scientific works in soundscape studies explicitly refer to triangulation as a reference framework, but the reality is that the concept underlies most soundscape research and practice.

In Chap. 9, "Soundscape and Health," Peter Lercher and Angel M. Dzhambov touch on the theory, practice, and assessment of the current state of research that relates the acoustic environment to quality of life and severe health effects. The authors describe an integrated approach to consider and characterize the acoustic environment and its associated physical, structural, social, and cultural contexts. They summarize the current status of health-related soundscape research and suggest further research needs. The authors clearly show that soundscape approaches have provided useful input for small scale environmental assessment and planning.

More specific applications of health-related soundscape studies follow in Chap. 10, "Soundscape in Hospitals," by Ilene Busch-Vishniac and Erica Ryherd. They point out that hospital soundscapes are challenging because there are many noise sources that contribute to the soundscape at all hours, and that this can negatively affect a vulnerable population. They also consider the specific sounds of the hospital soundscape and the physiological and emotional responses experienced by the people exposed to them. The practical uses of holistic tools and triangulation are revealed, building on topics discussed in Chaps. 3 and 8.

In the final chapter, André Fiebig and Brigitte Schulte-Fortkamp discuss "How to Put Soundscape into Practice." Soundscape is frequently regarded as an academic area of research, with studies of indicators and descriptors, old and new technologies, and new conceptual frameworks. On the one hand, there are still challenges in transferring the soundscape concept with its inherent holistic demand and its interdisciplinary foundation to real-world application. On the other hand, there are reservations from some noise consultants about applying new, evolving methods and approaches to deal with environmental noise. Therefore, the authors provide a guideline for practitioners on how to assess soundscape data, how to determine the need for interventions to preserve and/or improve a soundscape, and how to implement a soundscape design and/or intervention.

## 1.5 The Future of Soundscape Research

As the vital role of soundscape for the quality of life, well-being, and health has been recognized widely, researchers and practitioners have continued to work on guidelines to improve soundscapes effectively. An urban soundscape can promote the psychological restoration of its users; therefore, urban planning and architectural design need to focus on improving the perception of urban places, as summarized by Herranz-Pascual et al. (2019). Although this demand is almost self-evident, questions remain as to how the perception of acoustic environments can be improved: how do we evaluate the relationship between an acoustic environment and a specific context, and what are the mechanisms for improvement? Therefore, the collaborations with local experts, persons familiar with the soundscape due to their daily routines related to the area, are needed to understand the site-specific perceptions of the acoustic environments, which are required to develop soundscape designs that reflect the needs of those local persons. Thus, urban development and planning need participatory processes and co-creation to be successful.

These developments to make the soundscape concept more popular are further promoted by the observations of the World Health Organization (WHO). In a review prepared in the framework of the WHO guidelines for environmental noise (Brown and Kamp 2017), different types of interventions were determined. In addition to the classical types, e.g., mitigation measures at the source or at the route/infrastructure level, the value of design and necessary communication between stakeholders were also considered. Accordingly, the authors of the review concluded that there is wide and increasing demand for innovative approaches that will decrease the negative impact of noise by using all types of interventions, including soundscape design. The utilization of the soundscape approach in the context of urban sound is still in its infancy and far more applications of the soundscape approach will lead to far more successful designs and interventions.

**Acknowledgments** The editors would like to thank Dr. Peggy Walton for working with us in editing this and many other chapters in the book. Dr. Walton did an exceptional job not only fixing grammar and formatting but also providing scholarly insight into ensuring that ideas were expressed clearly and organized in the most effective way.

**Compliance with Ethics Requirements** Brigitte Schulte-Fortkamp declares that she has no conflict of interest.

André Fiebig declares that he has no conflict of interest.

## References

Aletta F, Oberman T, Mitchell A, Tong H, Kang J (2020) Assessing the changing urban sound environment during the COVID-19 lockdown period using short-term acoustic measurements. Noise Mapp 7(1):123–134. https://doi.org/10.1515/noise-2020-0011

Alves S, Gabriele M, Carillo S, Masullo M, Maffei L (2021) Exploring the soundscape and the atmosphere of the Gigli di Nola cultural festival in Italy. Emot Space Soc 41:100848. https://doi.org/10.1016/j.emospa.2021.100848

Aumond P, Can A, De Coensel B, Botteldooren D, Ribeiro C, Lavandier C (2017) Modeling soundscape pleasantness using perceptual assessments and acoustic measurements along paths in urban context. Acta Acust united Ac 103(3):430–443. https://doi.org/10.3813/AAA.919073

Botteldooren D, De Pessemier T, Filipan K, Sun K, De Coensel B, Van Renterghem T (2020) Modifying and co-creating the urban soundscape through digital technologies. In: Smaniotto Costa C, Mačiulienė M, Menezes M, Goličnik Marušić B (eds) Co-creation of public open places. Practice – reflection – learning. C3Place, pp 185–200. https://doi.org/10.24140/2020-sct-vol.4-2.2, www.c3places.eu | https://jpi-urbaneurope.eu/project/c3places, last assessed at Sept. 08. 2022

Brambilla G, Pedrielli F (2020) Smartphone-based participatory soundscape mapping for a more sustainable acoustic environment. Sustainability 12(19):7899., 1-20. https://doi.org/10.3390/su12197899

Brill LC, Smith K, Wang L (2018) Building a sound future for students – considering the acoustics in occupied active classrooms. Acoust Today 14(3):14–21
Brooks B, Schulte-Fortkamp B, Voigt KS, Case AU (2014) Exploring our sonic environment through soundscape research and theory. Acoust Today 10(1):30–40
Brown AL, van Kamp I (2017) WHO environmental noise guidelines for the European region: a systematic review of transport noise interventions and their impacts on health. Int J Environ Res Public Health 14(8):873. https://doi.org/10.3390/ijerph14080873
Brown AL, Kang J, Gjestland T (2011) Towards standardization in soundscape preference assessment. Appl Acoust 72(6):387–392. https://doi.org/10.1016/j.apacoust.2011.01.001
Brown AL, Gjestland T, Dubois D (2016) Acoustic environments and soundscapes. In: Kang J, Schulte-Fortkamp B (eds) Soundscape and the built environment. CRC Press/Taylor & Francis Group, Boca Raton, pp 1–16, 978-1-4822-2632-4
Busch-Vishniac I, Ryherd E (2019) Hospital soundscapes: Characterization, Impacts, and Interventions. Acoust Today 15(3). https://doi.org/101121/AT.2019.15.3.11
Busch-Vishniac I, West J, Barnhill C, Hunter T, Douglas O, Chivukula R (2005) Noise levels in Johns Hopkins Hospital. J Acoust Soc Am 118(6):3629–3645
De Coensel B, Bockstael A, Dekoninck L, Botteldooren D, Schulte-Fortkamp B, Kang J, Nilson ME (2010) The soundscape approach for early stage urban planning: a case study. In: Internoise 2010, proceedings, Lisbon, Portugal, pp 2333–2342
De Coensel B, Sun K, Botteldooren K (2017) Urban soundscapes of the world: selection and reproduction of urban acoustic environments with soundscape in mind. In: Internoise 2017, proceedings, Hong Kong, China
Derryberry EP, Phillips JN, Derryberry GE, Blum MJ, Luther D (2020) Singing in a silent spring: birds respond to a half-century soundscape reversion during the COVID-19 shutdown. Science 370(6516):575–579. https://doi.org/10.1126/science.abd5777
Di Loreto S, Serpilli F, Lori V (2022) Soundscape approach in the seaport of Ancona: a case study. Acoustics 4(2):492–516. https://doi.org/10.3390/acoustics4020031
END (2002) European Directive 2002/49/EC 2002 of the European Parliament and of the Council of 25 June 2002 (2002) relating to the assessment and management of environmental noise. Official Journal of the European Communities L 189/12, Brussels
Engel MS, Fiebig A, Pfaffenbach C, Fels J (2021) A review of the use of psychoacoustic indicators on soundscape studies. Curr Pollut Rep 2021(7):359–378. https://doi.org/10.1007/s40726-021-00197-1
European Environment Agency (2014) Good practice guide on quiet areas, EEA Technical Report Nr. 4, Copenhagen, Denmark
Farina A, Gage SH (2017) Ecoacoustics: a new science. In: Farina A, Gage SH (eds) Ecoacoustics. The ecological role of sounds. John Wiley and Sons, Oxford, UK
Fiebig A (2019) The perception of acoustic environments and how humans form overall noise assessments. In: Internoise 2019, proceedings, Madrid, Spain
Fowler M (2015) Sounds in space or space in sounds? Architecture as an auditory construct. Archit Res Q 19(1):61–72. https://doi.org/10.1017/S1359135515000226
Galbrun L, Ali TT (2012) Perceptual assessment of water sounds for road traffic noise masking. In: Acoustics 2012, proceedings, Nantes, France, pp 2147–2152
Herranz-Pascual K, Aspuru I, Iraurgi I, Santander A, Eguiguren JL, García I (2019) Going beyond quietness: determining the emotionally restorative effect of acoustic environments in urban open public spaces. Int J Environ Res Public Health 2019(16):1284. https://doi.org/10.3390/ijerph16071284
Hiramatsu K (2000) Soundscapegraphy of old town of Kyoto with the Gion festival. J Acoust Soc Am 108:2497. https://doi.org/10.1121/1.4743217
Hiramatsu K (2006) A review of soundscape studies in Japan. Acta Acust united Ac 92(6):857–864
Hollstein B (2011) Qualitative approaches to social reality: the search for meaning. In: Scott J, Carrington PJ (eds) The sage handbook of social network analysis. Sage Publications, London, pp 404–416

Hong JY, He J, Lam B, Gupta R, Gan WS (2017) Spatial audio for soundscape design: recording and reproduction. Appl Sci 7:627. https://doi.org/10.3390/app7060627
ISO 12913-1 (2014) Acoustics-soundscape-part 1: definition and conceptual framework. Geneva, Switzerland, https://www.iso.org/standard/52161.html
ISO 18405 (2017) Underwater acoustics – Terminology. International Organization for Standardization. Geneva, Switzerland, https://www.iso.org/standard/62406.html
ISO/TS 12913-2 (2018) Acoustics-soundscape-part 2: data collection and reporting requirements. Geneva, Switzerland, https://www.iso.org/standard/75267.html
ISO/TS 12913-3 (2019) Acoustics-soundscape part 3: data analysis. Geneva, Switzerland, https://www.iso.org/standard/69864.html
Jordan P, Fiebig A (2021) COVID-19 impacts on historic soundscape perception and site usage. Acoustics 2021(3):594–610. https://doi.org/10.3390/acoustics3030038
Kang J, Aletta F, Gjestland TT, Brown LA et al (2016) Ten questions on the soundscapes of the built environment. Build Environ 108:284–294. https://doi.org/10.1016/j.buildenv.2016.08.011
Knoblauch H (2013) Qualitative Methoden am Scheideweg: jüngere Entwicklungen der interpretativen Sozialforschung. Historical Social Research, 38(4):257–270. https://doi.org/10.12759/hsr.38.2013.4.257-270
Lawrence BT, Hornberg J, Haselhoff T, Sutcliffe R, Ahmed S, Moebus S, Gruehn D (2022) A widened array of metrics (WAM) approach to characterize the urban acoustic environment; a case comparison of urban mixed-use and forest. Appl Acoust 185(108387):1–20. https://doi.org/10.1016/j.apacoust.2021.108387
Lunden P, Axelsson Ö, Hurtig M (2016) On urban soundscape mapping: a computer can predict the outcome of soundscape assessments. In: Internoise 2016, proceedings, Hamburg, Germany, pp 4725–4732
Maag T, Munck Petersen R (2018) Can participatory experience performances co-create qualification and design of audible public realm? In: Internoise 2018, proceedings, Chicago, USA, pp 1999–2009
Maeder C (2013) Analyzing sound. In: Flick U (ed) The Sage handbook of qualitative research. Sage, London-Singapore, pp 424–434
Nilsson ME, Berglund B (2006) Soundscape quality in suburban green areas and city parks. Acta Acust united Ac 92(6):903–911
Payne SR, Bruce N (2019) Exploring the relationship between urban quiet areas and perceived restorative benefits. Int J Environ Res Public Health 16(9):1611. https://doi.org/10.3390/ijerph16091611
Quinn CA, Burns P, Gill G, Baligar S, Snyder RL, Salas L, Goetz SJ, Clark ML (2022) Soundscape classification with convolutional neural networks reveals temporal and geographic patterns in ecoacoustic data. Ecol Indic 138:108831. https://doi.org/10.1016/j.ecolind.2022.108831
Roy KP, Siebein K (2019) Satisfying hunger, thirst, and acoustic comfort in restaurants, diners, and bars… is this an oxymoron? Acoust Today 15(2):20–28
Salomons E, Ögren M, van den Berg F, Schoonebeek C, Hillebregt M, Knape M, Gidlöf-Gunnarsson A (2013) QSIDE: quiet places in cities. Quiet façades and quiet areas in urban noise policy. Recommendations and examples, http://www.qside.se/proj/pub/QSIDE_Action5_Quiet_places_website.pdf
Schafer RM (1977) The tuning of the world. Alfred A. Knopf, Inc, New York. ISBN 978-0394409665
Schulte-Fortkamp B, Jordan P (2016) When Soundscape meets architecture. Noise Mapp 3:216–231. De Gruyter Open, https://doi.org/10.1515/noise-2016-0015
Schulte-Fortkamp B, Brooks B, Bray W (2007) Soundscape – an approach to rely on human perception and expertise in the post-modern community noise era. Acoust Today 3(1):7–15
Schulze H (2019) Sound works. A cultural theory of sound design. Bloomsbury Press, New York. 2019, ISBN 978-1-5013-3022-3
Slabbekoorn H (2018) Soundscape ecology of the Anthropocene. Acoust Today 14(1):42–49
Schultz TJ (1978) Synthesis of social surveys on noise annoyance. J Acoust Soc Am 852–861(64):377–405

Steele D, Legast E, Fraisee V, Guastavino C (2019) Sounds in the City: improving the soundscape of a public square through sound art. In: ICSV, proceedings, Montreal, Canada, pp 1–8

Sun K, Botteldooren D, De Coensel B (2018) Realism and immersion in the reproduction of audio-visual recordings for urban soundscape evaluation. In: Internoise 2018, proceedings, Chicago, USA, pp 3432–3441

Sun K, De Coensel B, Filipan K, Aletta F, Van Renterghem T, De Pessemier T, Joseph W, Botteldooren D (2019) Classification of soundscapes of urban public open spaces. Landsc Urban Plan 189:139–155. https://doi.org/10.1016/j.landurbplan.2019.04.016

Talebzadeh A, Botteldooren D (2022) Designing personalized soundscape for care facilities. J Acoust Soc Am 151:A72. https://doi.org/10.1121/10.0010695

To WM, Chung A, Vong I, Ip A (2018) Opportunities for soundscape appraisal in Asia. In: Euronoise 2018, proceedings, Crete, Greece

Truax B (1978) The world soundscape project's handbook for acoustic ecology. A.R.C. Publications, Vancouver. ISBN 978-0889850118

Van Renterghem T (2019) Towards explaining the positive effect of vegetation on the perception of environmental noise. Urban For Urban Green 40:133–144. https://doi.org/10.1016/j.ufug.2018.03.007

Van Renterghem T, Dekoninck L, Botteldooren D (2020) Multi-stage sound planning methodology for urban redevelopment. Sustain Cities Soc 62:102362. https://doi.org/10.1016/j.scs.2020.102362

Verma D, Jana A, Ramamritham K (2019) Classification and mapping of sound sources in local urban streets through AudioSet data and Bayesian optimized neural networks. Noise Mapp 6:52–71. https://doi.org/10.1515/noise-2019-0005

WHO (2018) Environmental noise guidelines for the European region. WHO, Copenhagen

Xiao J, Lavia L, Kang J (2017) Towards an agile participatory urban soundscape planning framework. J Environ Plan Manag 61(4):677–698. https://doi.org/10.1080/09640568.2017.1331843

Zhao W, Li H, Zhu X, Ge T (2020) Effect of birdsong soundscape on perceived restorativeness in an urban park. Int J Environ Res Public Health 17(16):5659. https://doi.org/10.3390/ijerph17165659

# Chapter 2
# Soundscape: A Construct of Human Perception

**André Fiebig**

**Abstract** A soundscape is a perceptual construct of an acoustic environment and, therefore, must be distinguished from the actual physical environment. Because the definition of soundscape is based on the perception or experience by a person or people in context, the study of human perception has become a major objective of soundscape research. The soundscape approach has roots in environmental psychology and goes beyond the simplistic notion of conventional environmental noise assessment. A listener is not only a passive receiver of environmental noise; instead, a listener becomes part of a dynamic system of information exchange and is involved in its creation. The listener responds not only to sound in terms of wasteful annoyance, but within the soundscape paradigm, environmental sound can be interpreted as a resource composition that can elicit diverse affective reactions. These affective reactions are believed to reflect evolutionarily shaped responses that prepare humans for action and are accompanied by physiological responses and behavioral changes. This perspective has stimulated multiple investigations regarding the main affective descriptors and the underlying indicators of soundscape appraisal, with some affective factors increasingly being acknowledged as the driving factors of emotional apprehension. Soundscape research surpasses the simplistic realm of environmental noise assessment solely in terms of sound pressure level indicators and annoyance and promotes the idea that multidimensional responses to sound cannot fully be understood without contemplating the context.

**Keywords** Affect · Affective quality · Emotion · Sound perception · Saliency

---

A. Fiebig (✉)
Department of Engineering Acoustics, Technische Universität Berlin, Berlin, Germany
e-mail: andre.fiebig@tu-berlin.de

 
B. Schulte-Fortkamp et al. (eds.), *Soundscapes: Humans and Their Acoustic Environment*, Springer Handbook of Auditory Research 76,
https://doi.org/10.1007/978-3-031-22779-0_2

## 2.1 Introduction: The Measurability of Sensations

The nature of human perception has been the subject of scholarly and scientific inquiry over centuries and the enthusiasm of researchers has not waned. Ancient philosophers, as early as Plato, speculated about the relationship between the conscious mind and the physical body, which is now considered as the mind–body issue and has caused intense philosophical debates (Hassett 1978).

In the late nineteenth century, the emerging discipline of psychophysics became increasingly popular and with it came the view that sensations could be measured quantitatively and mapped. Ernst Heinrich Weber postulated that the amount by which the intensity of a stimulus feature must change for a difference to be perceived is a fixed proportion of its magnitude (Weber's Law). This formed the basis for Gustav Theodor Fechner to propose that psychological magnitudes can be measured based on "just noticeable difference." Consequently, the belief grew that empirical psychology (studying the phenomena of mental life) can be an "exact science" in the fullest sense of the term (Titchener 1922), measuring psychological events with the aid of comparative scales (e.g., Lockhead 2004).

That view attracted severe criticism as it contrasted strongly with the traditionalist view that human perception is inherently qualitative and cannot be expressed in terms of magnitudes, especially about feelings, emotions, and affects. For example, James (1884), in a precursor of behaviorism, described empirical psychology as limited to cognitive and volitional aspects of the brain, thereby ignoring pleasures and pains, and its emotions.

Sellars (1907) later remarked that different individuals cannot have identical experiences in a numerical sense; they are not able to provide reliably exact numbers to compare individual experiences. Thus, in the area of sensation and perception, introspective reports were often limited to simple judgments of size, intensity, and duration of physical stimuli in the context of experimental psychology (Danziger 1980). Even today, the *quantity objection argument* claims that experiments to measure sensation only determine the estimation of physical differences instead of quantifying the sensation itself.

Another argument against empirical psychology was expressed by Cattell (1893), who asked whether we do in fact judge differences in the intensity of sensations or whether we merely judge differences in the stimuli determined by association with their known objective relations. This criticism is frequently called the *stimulus-error problem* and is a serious methodological pitfall in studying human perception (Chirimuuta 2016). According to Boring (1921), psychological reports must be based on the mental material itself to study sensation rather than on objects that judge the stimulus magnitude. Although the term "stimulus-error" is rarely in use today, the significance of this potential error has not disappeared (Chirimuuta 2016). If participants use their knowledge of the stimulus for judgements, this processing is labeled "cognitive," referring unwittingly to the stimulus-error.

Even today, sensations are regarded as difficult to measure, not because they are mental, subjective, or inaccessible but simply for want of an adequate psychophysical theory (Marks and Algom 2001). To make things worse, the absolute nature of perception is frequently questioned: perception in terms of responding to stimuli is frequently understood as merely based on relationships between the stimulus and its context rather than on the intensity of an attribute of the stimulus (Lockhead 2004). In other words, if a stimulus has a certain magnitude, the response cannot be predicted without considering its context. Consequently, Lockhead (2004) conjectured that people do not identify a stimulus and its magnitude as presented but instead search for its relation to memories of other, preceding events, thereby broadening the scope of psychophysics.

Overcoming the barriers imposed by previous beliefs, Stevens (1975) forged a fresh paradigm that was unimaginable 100 years ago and laid the groundwork for modern psychophysics. He claimed that nearly every sensory continuum can be described by means of a very basic principle: power functions with varying exponents. Nonetheless, discussions about the proper quantification of human perception are still ongoing. Researchers still struggle to draw the right conclusions regarding the relationship between stimulus strength and perceptual magnitude. They still seek clearer distinctions between body and mind and between the physiological and psychological processes inherent to humans; however, there is increasing acceptance that, in principle, genuine sensations and genuine perceptions are measurable.

### *2.1.1 History of Soundscape*

The origin of the soundscape concept dates back to Southworth (1967), who analyzed the perceptual form of the soundscape with the purpose to study the possibilities and relevance of sonic design for cities and to establish criteria for design. By stressing the need for a holistic concept and concluding that (optimized) soundscape design may be a way of improving the aesthetics and acceptability of cities for their inhabitants, he also pioneered the notion that it is no longer sufficient to design environments which only satisfy the eyes (Southworth 1969).

Schafer (1977), a Canadian musician, popularized this idea a few years later with his book on the sonic environment and tuning of the world. At the time of Southworth's publication, Schafer had begun working in the newly established, interdisciplinary department Centre for Studies in Communication and the Arts at the Simon Fraser University in Vancouver with a simple but novel inspiration: To study all sounds, not merely those that were unpleasant or dangerous (Schafer 2012).

According to Schafer (2012), "soundscape" was not recognized in the early 1970s, or at least the definition of the term was unknown to almost everyone, allowing it to become an umbrella term for the study of diverse sounds: past and present, useful and useless, beautiful and ugly, exciting and boring. Until then, acousticians had not understood the merit of the word *soundscape*, because in their view, sound could be adequately described by phon, decibel, and other technical terms. However, the new, invented term united the practical and aesthetic aspects of sound, allowing researchers to study and describe acoustic environments on a more pragmatic level that was closer to daily experiences (Schafer 2012).

Schafer argued that up until that point, researchers engaged with the same question used different approaches while attempting to answer it: what is the relationship between humans and the sounds of the environment? Soundscape studies would aim to unify these various themes, drawing conclusions from each other (Schafer 1977). Because the concept of soundscape was originally rooted in music and acoustic ecology (Kang et al. 2016), understanding health-related issues (like well-being and restoration) required an enhanced approach to discover new relevant relationships that constitute a supportive environment (Schulte-Fortkamp 2002). Therefore, soundscape research can be seen as a countermovement to the conventional *noise control perspective*. Up until that point the preoccupation with noise as a disease that could somehow be cured had overshadowed the understanding of how healthy soundscapes function (Truax 1984).

Schafer (1977) observed that the efforts to reduce noise pollution were primarily focused on noise abatement with every additional sound source being regarded as a negative addition. Therefore, he called for a more positive study of environmental acoustics. This simple change of perspective may constitute the reason for the success of the soundscape concept. Acceptance of this paradigm shift may have been supported due to the inefficacy of the conventional sound level reduction measures, which did not lead to a noticeable improvement for quality of life in urban and rural areas (Kang et al. 2016). The soundscape approach involved not only physical measures but also actively sought the contribution of humanities and social science to account for the diversity of soundscapes across countries and cultures. In doing so, environmental sound was understood as a resource rather than as an environmental burden only. The soundscape approach emphasizes an analysis of how an environment is understood by those creating it and those living within it. The listener is no longer merely a passive receiver of sound; instead, the listener becomes part of a dynamic system of information exchange (Truax 1984).

The increasing interest in this alternative approach is illustrated by numerous publications in several special issues of peer-reviewed journals on acoustics. In addition, since the late 1990s, special sessions related to the soundscape topic have become an integral part of conferences and congresses on acoustics. Moreover, a European Cooperation in Science and Technology (COST) action *Soundscape of European Cities and Landscapes*, an intergovernmental network for cooperation in research, enabled researchers to connect with each other across Europe and beyond, with the aim of providing a scientific underpinning and practical guidance for intensified international research activities in this field (Kang et al. 2013).

Today, studies referring to the soundscape concept are more and more popular. As the systematic literature of To et al. (2018) has shown, the number of soundscape publications regarding non-open-access and open-access documents grew rapidly from 1985 to 2018. Due to all of these developments, the soundscape approach has received significant attention in particular since the early 2000s until today in the field of community noise and environmental acoustics by researchers, policy makers, and practitioners (Kang et al. 2016).

### 2.1.2 The Definition of Soundscape

The international standard on soundscape (ISO 12913-1) defines soundscape as a perceptual construct that must be distinguished from the physical phenomenon: "*Soundscape is an acoustic environment as perceived or experienced by a person or people in context.*" (ISO 12913-1 2014, p. 1). This definition resembles the one from Truax (1984), a former colleague of Schafer, who wrote in 1984 that the term soundscape emphasizes how an environment is understood by those living within it and therefore must be distinguished from the physical sonic environment. The term soundscape is also used in other contexts, for example, as in underwater acoustics to characterize ambient sound (ISO 18405 2017) and thus differs from the definition that refers to sound perceived by humans in context (ISO 12913-1 2014).

For several decades there was no universal agreement about the meaning of the term soundscape. Even working group 54 for ISO/TC 43/SC 1 struggled with diverse views, concepts, and levels of understanding until a consensus was reached (Brown et al. 2011). According to the working group, standardization was initiated after the realization that progress in soundscape research was impeded by the lack of a clear, shared understanding of what was meant by the term (Schomer et al. 2010). Noise annoyance research with its different instructions, different annoyance scales, and unreported contextual factors had shown that a lack of standards significantly impedes meta-analyses (Brown et al. 2011).

Today, the term soundscape is acknowledged widely as a reference to the construction of an acoustic environment based on human perception. This was supported by the standardization efforts and many researchers cite the international standard (Fiebig 2018), although deviations from this established notion are still found frequently. Even with a rigorous ISO definition, the term soundscape is sometimes used as a synonym for the physical acoustic environment.

According to Kang et al. (2016), this may be admissible, if such equivocal usage of the term soundscape does not introduce confusion in communication. In general, the working group expected that by the introduction of standardized definitions and the description of soundscape investigation methods to be utilized by researchers, the outcomes from various studies dealing with relationships between perceived soundscape quality and acoustic, physical, and visual properties of areas would be more compatible, thereby ensuring comparability (Brown et al. 2011). This is true to a certain extent. There are three key components, which are understood to constitute a *soundscape*: people, acoustic environment, and context (see Fig. 2.1). According to the ISO/TS 12913-2 (2018), data from all key components must be collected for a study to be acknowledged as a full-featured soundscape study (see Schulte-Fortkamp and Jordan, Chap. 3). Some researchers deviate from this key understanding by using slightly different terms, such as Kogan et al. (2017), who called the key components of soundscape *experienced environment*, *acoustic environment*, and *extra-acoustic environment*. However, those alternative components are usually like the components proposed in the ISO 12913-1 (2014).

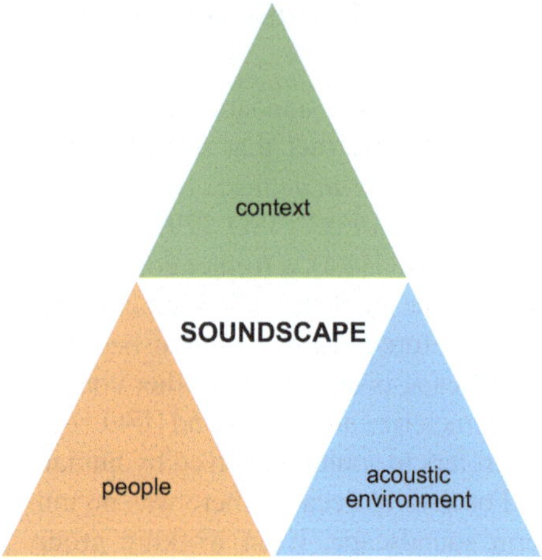


**Fig. 2.1** Key components of soundscape studies

Due to the emphasis on perception and cognitive construction processes by the established soundscape concept, the study and understanding of human perception becomes the main objective of all research dedicated to soundscapes. Since this concept of soundscape emerged, researchers have investigated how acoustic environments affect the perceived (sonic) quality of cities and how sounds can be effectively used in urban planning and design (Aletta et al. 2016a). Soundscape research aims for an understanding of the relationship between people and their acoustic environments by examining the sounds that people value or disapprove and their reactions to them in specific contexts of place and activity (Kamp et al. 2016).

## 2.2 Perceptions of Environments

Environments affect humans and humans affect environments. This simple statement seems trivial, but it is not. The impact the physical environment has on human beings is classically understood to be a bottom-up process. Noise control engineering still uses this simplified concept and still strives to reduce annoyance strictly by minimizing the sound pressure level of unwanted noise. Any interactivity of this process is neglected, and any multidimensionality is discarded. In the area of environmental psychology, however, the notion that the perception of an environment involves an interaction between the individual and said environment has long been established (Fisher et al. 1984), as Fig. 2.2 illustrates. This simple diagram highlights that an individual is not simply a reacting and adapting organism trying to make sense of the fast-changing environment. An individual is not only struggling to cope successfully with the environment in order to survive, but the person is also an agent based on individuality and actively influences the environment (Barandiaran et al. 2009).

**Fig. 2.2** Basic concept of interaction between individual and environment in environmental psychology

Some of the explanations for perceptual experience lie within the environment, some rest within the individual, and some are the outcome of their interaction. A simple example can illustrate the need for widening the theory of perception based on solely bottom-up mechanisms: There are numerous examples of auditory or visual illusions that demonstrate elementary sensations are not sufficient to provide an explanation for these perceptions, such as the influence of the color of a train on judgements of loudness (Fastl 2004).

Humans do not simply react to physical stimuli exciting the senses; they partially create their surrounding environment. They use heuristics and knowledge to extract and manage useful information out of a stream of continuous input. According to Brosch et al. (2010), a certain category in the mind is activated, which supports the process to give meaning to the world.

The person cognitively constructs the environment to a certain extent through knowledge-driven information processing. In contrast to the simplistic bottom-up concept, top-down controlled perception causes human perception and responses are not determined only by external physical stimuli. Thus, the knowledge-driven component within the perception process (top-down processing) should not be underestimated. Consequently, Schafer claimed that humans are anti-entropic creatures; humans are a random-to-orderly arranger and they perceive patterns in all things (Schafer 1977). Sensory perception could even be regarded as a factor that corrects and fine-tunes mental representations. Humans influence their perceptions of the surrounding world by top-down processes in which they pay attention to particular aspects of the environment. This is sometimes described as the *attended stimulus* (Goldstein 2002). Thus, any theory of perception must consider the adjustive process of the organism that contributes to adaptation to its environment (Helson 1967). Disentangling bottom-up from top-down processes is difficult, as is understanding how physical stimuli give rise to mental representations, but both processes must be addressed.

Attention guides how humans perceive their environments. Any information that becomes part of working memory (due to mechanisms of attention) is evaluated and analyzed, allowing decisions about that information to be made and plans for action to be elaborated (Knudsen 2007). Attention-related factors (i.e., sustained attention to an entire auditory scene, selective attention to particular objects or streams within a scene, attention switching, or attention limitations) can have dramatic influences on the perceptual organization of scenes and the ability to detect important events in the environment. At the same time, there is evidence that other high-level mental processes, such as intention and previous knowledge, also greatly impact auditory perception (Snyder et al. 2012). Moreover, in this context the process or act of *recognition* needs special attention because, through recognition, perceived objects are placed in a category and are given meaning (Goldstein 2002).

Human perceptions of physical environments lead to a few basic emotional dimensions, which are indicative of human feelings and are thought to be independent of each other. According to Mehrabian and Russell (1974), these dimensions are *pleasure*, *arousal*, and *dominance*. Theses emotional dimensions are still used by numerous researchers in the field of environmental psychology (Bakker et al. 2014).

*Affect* is understood as the semantic representation of emotion and can be distinguished from the perceptual or cognitive processing of the environment (Russell et al. 1981). In contrast to sensations, which are induced by the excitation of sensory cells, emotions can be considered as an affective response of the body to an external stimulus. Traditionally, the study of perception has been quite distinct from the study of emotion, but these basic processes of *perception*, *cognition*, and *emotion* (*affect*) seem to be highly interrelated and must be studied beyond isolation (Zadra and Clore 2011). An affective response depends upon the way in which it is first perceived and recognized (Russell et al. 1981).

Are pleasure and arousal conceived as indicators of affect? Is dominance as a feeling that the environment is dominant or fully in control more or less a cognitive product? Simplifications such as these are highly questionable. Bakker et al. (2014) assumed that pleasure is an affect, and arousal has a cognitive nature, whereas dominance underlies a conative concept, which is connected with the wish, intention, or effort to achieve something. The term dominance representing a conative dimension was not used by Mehrabian and Russell and should emphasize effects on behavior beyond affect and cognition according to Bakker et al. (2014).

Positiveness or negativeness of affect are assumed to refer to satisfaction and well-being. Thus, affective responses play a major role in the perception and appraisal of environments. There seems to be an almost unlimited array of affective descriptors, but environmental psychologists focused their attention on only a few (such as degree of comfort, annoyance, pleasantness, or psychological stress). In the context of the perception of acoustic environments, a relatively simple conceptualization that encompasses diverse affective concepts has been increasingly preferred (Aletta and Kang 2018) due to the observation that several researchers repeatedly explored the same dimensions in the context of soundscape (see Fiebig et al. 2020). On the other hand, human sensations, responses, and outcomes cannot easily be reduced to singular values of physical units because responses to sounds can depend on the listener's mental, social, and geographical relationships with the sound source (ISO/TS 12913-2 2018).

## 2.3 Perceptions of Acoustic Environments

### *2.3.1 How Sound Shapes Human Life*

As early as 1914, Hollister (1914) observed the effect of attitude in dealing with controllable and unavoidable noises in the context of nursing. She concluded that the more passive our stance towards said noise, the better we can cope with it, such as for sleeping. If we can go as far as to convince ourselves that we like the noise, the noise may even prove to be a source of help. Although this statement is partly questionable from the viewpoint of noise effect research, it underlines the fact that human perception of sound is based on an interaction between the perceiving individual and the acoustic environment. Through this interaction, human behavior is enriched by aspects of social characteristics and the environment. Thus, acoustic environments can affect human perception, and human perceptions can in turn influence environments and other humans in both indoor and outdoor spaces (Meng et al. 2018).

It is beyond question that the appraisal of sound is more complex than loudness of the sound. The assessment of a sound's desirability has no obvious relationship to this simple unit (van den Bosch et al. 2018). If, for example, noise annoyance is considered, only one-third of variance in annoyance data can be explained by acoustical properties of the sound, usually determined in terms of overall, time-averaged, sound pressure levels (Guski 1999). The amount of variance in data on unpleasantness that could be explained by loudness-related metrics (e.g., level, psychoacoustic loudness) was also evaluated by repeated soundwalks, which are on-site evaluations at different points by local experts, where a similar amount of variance was explained (Fiebig 2018).

Understanding the relationship between people and their soundscapes in an urban context of diverse sensory stimulations is a difficult endeavor (Bild et al. 2018); thus, research on noise annoyance has slowly broadened to include an increasing number of physiological and psychological aspects (e.g., Taghipour and Pelizzari 2019). In the context of soundscape, auditory attention, which allows humans to focus mental resources on a particular auditory stream of interest while ignoring other acoustic elements, is of particular interest (de Coensel and Botteldooren 2010). Due to the crucial role for the perception of acoustic environments, several researchers have investigated the *saliency* of sounds (Botteldooren and de Coensel 2009; Filipan et al. 2019).

Sound events that are salient and stand out in the sonic environment capture our attention and contribute highly to the perception and the appraisal of the soundscape (Filipan et al. 2019). In general, auditory saliency can be distinguished into two non-exclusive dimensions: *sensory saliency*, referring to specific sound features meeting the enhanced sensitivity of human hearing; and *semantic saliency*, which is based on recognition of the sound and its potential incongruency within the environment (Filipan et al. 2019). In this context, the mechanisms between *signal-driven attention*, which is due to the prominent acoustic features of the sound (Genuit and Fiebig 2006; Oldoni et al. 2013), and *meaning-related attention*, which is responsible for drawing or losing interest in auditory events, are still not fully understood (Botteldooren et al. 2012). Clearly, attention and the cognitive load of listeners affect noise ratings for complex sound scenarios (Steffens et al. 2019).

Moreover, the perception and assessment of environmental noise depends on the social and cultural background of the individual, indicating another limitation of simplified bottom-up concepts (Schulte-Fortkamp and Fiebig 2006). Bruce and Davies observed that assessment of a soundscape is affected by expectations in several different ways, including the aspects of *behavior* and *control*, which are partly based on an acquired set of social rules or norms (Bruce and Davies 2014). Sun et al. (2019) noticed that soundscape perception also depended on potential interference with other possible activities at the respective site.

There are numerous studies revealing the crucial role of environmental noise on human life in all facets:

- Noise affects behavior. Individuals were less affiliative and less helpful with increasing noise (Moser and Uzzell 2003).
- Noise affects environmental awareness. Attentiveness dropped with increasing noise (Korte and Grant 1980).
- Noise affected the length of stay at public places (Aletta et al. 2016b).
- Noise (music and speech) affected human serial recall performance (Schlittmeier et al. 2008).
- Environmental and classroom noise had a detrimental effect on learning and performance by children (Shield and Dockrell 2008). Relationships with their peers and teachers and their motivation for achievement were also negatively affected (Klatte et al. 2010).
- Listening to bird songs decreases walking speed compared to walking with city noise conditions (Franek et al. 2019).
- Moderate levels of ambient noise positively affected creative cognition in contrast to low or high ambient noise levels (Mehta et al. 2012).
- Sound can mitigate antisocial behavior, can lead to pro-social effects and increase the feeling of safety (Lavia et al. 2016) or vice versa.
- Sound can reduce agitated behavior in older adults with dementia (Lin et al. 2018).

Worth noting is that these soundscape influences on humans and their behavior in response to the soundscape often take place without being noticed. In many cases, humans are unaware of this special role of sound and they underestimate its impact on behavior and well-being. Thus, a soundscape has the potential to evoke responses in the individual and to induce certain outcomes within a particular context (Brown et al. 2016).

### 2.3.2 *Affective Qualities and Emotions Attributed to Acoustic Environments*

A core doctrine of the soundscape concept is that acoustic environments can elicit different affective responses and emotions. This differentiates soundscapes from classical environmental noise assessment, which only deals with levels of annoyance. There is a long-running debate about human emotions, their causes and effects, which must be addressed if psychology is ever to explain them (Hasset 1978). Unfortunately, the definition of the construct *emotion* is in a state of conceptual and definitional chaos and remains a heavily freighted term full of imprecision (e.g., Gross 2010).

According to the Merriam-Webster dictionary (2022), emotion is defined as "*a conscious mental reaction (such as anger or fear) subjectively experienced as strong feeling usually directed toward a specific object and typically accompanied by physiological and behavioral changes in the body*." This definition illustrates the building blocks of emotion: a triggering stimulus (object), physiological and psychological responses, and behavioral changes.

Lang and Bradley (2000) observed the close relationships between the building blocks of emotion and noticed that the affective quality of sounds that elicit physiological reactions does not correlate with the intensity of those sounds. Accordingly, van den Bosch (2015) argued that the understanding of the acoustical properties of a certain place is far less important than understanding how that place influences a person emotionally. The question remains: Does an environment elicit a pattern of bodily changes leading to emotions or do humans actively develop emotions to prepare the organism to deal successfully with an environment? How are cognitive processes and emotional regulation involved in this bottom-up related notion of emotion?

Emotions are short-lived affective processes (in contrast to attitude and mood which are understood to be more stable, less affected by the moment, and long-lasting). Emotions are responses to situations that are perceived as relevant to an individual's current goals and consist of appraisals that give rise to changes in feelings, behavior, and physiological processes (Gross 2010). Most researchers dealing with emotion theories agree that emotional stimuli and emotional responses represent a special type of stimulus–response as they possess high relevance for survival and well-being, potentially preparing the individual for action (Brosch et al. 2010). It is assumed that by means of elicited emotions, humans rapidly recognize and quickly adapt with the necessary behavioral output.

This complexity shows that it is necessary to study not only the objective environment but also the internal representation of that environment: the meaning people attribute to it (Russell et al. 1981) and the psychophysiological concomitant effects. A milestone of research in the context of affective qualities attributed to physical environments is the work of Mehrabian and Russell (1974). The authors proposed a conceptual framework that bears some resemblance to the conceptual framework proposed in the ISO 12913-1 (see Fig. 2.3 and Sect. 2.3.3). Fiebig et al.

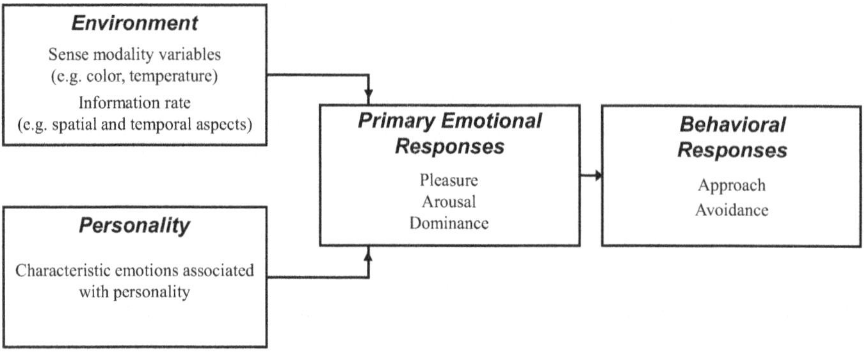


**Fig. 2.3** Outline of the conceptual framework of environmental psychology. (Adapted from Mehrabian and Russell 1974)

(2020) combined the insights of emotion research with the conceptual framework proposed in ISO 12913-1 in order to highlight the important role of emotion in the concept of soundscape. They explained that soundscapes frequently elicit unconsciously emotions, which exert influence on individuals' behavior, well-being, and health.

Mehrabian and Russell (1974) believed in a common core of responses that are the immediate result to stimulation of all types, regardless of the modality excited. They concluded that there is a limited set of emotional (connotative, affective, feeling) responses to all stimulus situations: responses of evaluation and activity correspond to emotional responses of *pleasure* and *arousal* and the response of potency corresponds to *dominance* (vs. submissiveness) (Mehrabian and Russell 1974). Although meaning attributed to environments contains both affective and perceptual-cognitive components, with the two highly interrelated, the basic dimensions described by Mehrabian and Russell (1974) focus specifically on emotions (Russell and Pratt 1980).

Russell (1980) believed that affective states elicited by environments are best represented as a circle in a two-dimensional bipolar space based on the dimensions of *pleasure–displeasu*re and *degree of arousal*. This representation leads to a circumplex model of affect: pleasure, excitement, arousal, distress, displeasure, depression, sleepiness, and relaxation. Usually, these dimensions are obtained by the results of factor analyses based on a set of data consisting of a heterogenous sample of adjectives and a set of rated stimuli. The deduced factors denote some of the most fundamental affective or perceptual components.

For example, Russell et al. (1981) used a list of 105 adjectives, analyzed the ratings of 323 environments by means of a common factor analysis, and detected three factors (*pleasure*, *arousal*, *potency*) accounting for 47% of the total variance in the data set. These fundamental affective dimensions attributed to environments could be ecologically interpreted in their combinations: "[…] *exciting places are both pleasant and arousing. Peaceful and comfortable places are also pleasant but unarousing. Frightening and harsh places are unpleasant and high in arousal quality. Depressing places are unpleasant and unarousing.*" (Russell et al. 1981, p. 280).

The affective concepts attributed to environments have been subject to intensive research in the context of soundscape. For example, Axelsson et al. (2010) proposed a few basic dimensions of affective qualities that resemble the widespread circumplex model of Russell et al. (1981) as shown in Fig. 2.4. Interestingly, the underlying dimensions of affective appraisal for acoustic stimuli are like those determined for image processing (Bradley and Lang 2000; Axelsson 2011).

In the context of soundscape, Axelsson et al. (2010) found three basic dimensions: *pleasantness*, *eventfulness*, and *familiarity*. Because of the low variance explained by *familiarity*, this component is often disregarded. Botteldooren et al. (2016) noted that one could argue that these dimensions are related to the individual person rather than to the sonic environment, but with soundscape interpreted as an object in the mind, this does not pose any problem.

According to van den Bosch et al. (2018), the two frequently observed independent dimensions *pleasantness* and *eventfulness* reflect characteristics with evolutionary significance that would promote survival by causing a preference for certain environments and avoidance of others. The descriptors of *emotional* responses to the environment (valence, arousal) substantially differ from those used to express its *affective* quality (pleasantness, eventfulness). However, they can be related to the appetitive and defensive motivational systems that underlie affective judgments: valence indicates which system is active; arousal indicates the intensity of activations of these systems (Bradley and Lang 2000; van den Bosch et al. 2018).

Tarlao et al. (2019) used slightly different terms for the two basic dimensions, naming them *appreciation* and *dynamism*. Despite apparent similarities, they could not fully confirm Axelsson's circumplex model because they found that *monotony* was an independent factor. Davies et al. (2013; also Cain et al. 2013) observed two principal dimensions of emotional responses to soundscapes: *calmness* and *vibrancy*

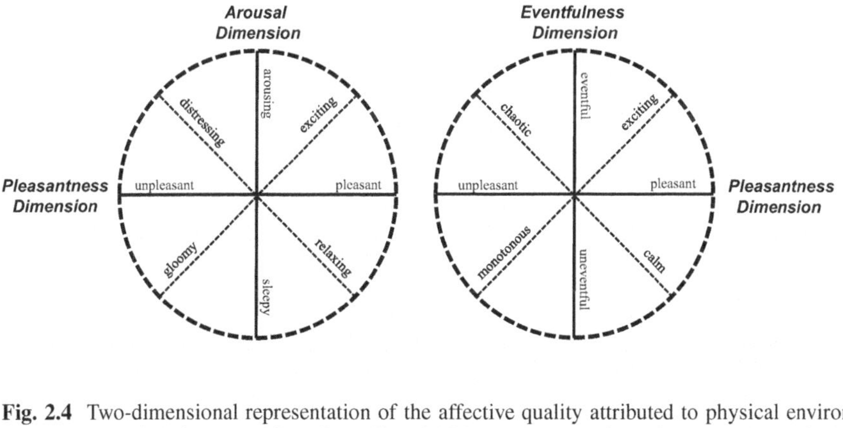


**Fig. 2.4** Two-dimensional representation of the affective quality attributed to physical environments in general. (left, adapted from Russell et al. 1981) and to acoustic environments in particular. (right, adapted from Axelsson et al. 2010) (cf. Fiebig et al. 2020)

which are close to the pleasantness–eventfulness model, if one were to rotate the axes in the components analysis result of the circumplex model by 45° as shown in Fig. 2.4.

Andringa and van den Bosch (2013) referred to the main dimensions of *pleasure* and *activation*, which according to the authors belong to the *core affect*. The core affect characterizes a relationship to the world as a whole and not to something specific within that world (van den Bosch et al. 2018). Welch et al. (2019) observed five affective dimensions for soundscapes (*calming*, *protecting*, *hectic*, *belonging*, and *stability*) by applying a factor analysis on questionnaire data.

Yu et al. (2016) extracted five major perceptual factors of soundscape perception in urban shopping streets, using these factors: *preference*, *loudness*, *communication*, *playfulness*, and *richness*. In very specific locations, further affective dimensions are conceivable. For example, Sudarsono et al. (2019) identified the dimensions *privacy*, *disturbance*, *dynamic*, *fear*, and *satisfaction* in crowded third-class hospital wards.

According to Aletta et al. (2016a), the *appropriateness* of a soundscape could be a potential third dimension. Since an encountered situation is usually matched against existing cognitive schemes, appropriateness viewed as the level of congruency between a scheme and a real-world situation will influence positive affective responses. Inappropriate matches consequently lead to negative affective responses (van den Bosch et al. 2018).

Referring to Jeon et al. (2018), the components *pleasantness* and *eventfulness* have been commonly identified in several studies across different countries, demonstrating their robustness across languages, cultures, and environments (see Table 2.1). Because of their universality, these two components have gained recognition by several researchers and have recently been included in a questionnaire defined in the ISO/TS 12913-2 that consists of response scales related to eight affective attributes: pleasant, chaotic, vibrant, uneventful, calm, annoying, eventful, monotonous. Although the developed dimensions of affective qualities have been

**Table 2.1** Soundscape descriptors as emotion dimensions. Dimensions proposed by selected publications are related to the Mehrabian and Russell's pleasantness and arousal dimensions

| Publication | Main dimensions related to circumplex model from Mehrabian and Russell (1974) | | Further dimensions |
|---|---|---|---|
| Mehrabian and Russell (1974) | Pleasantness | Arousal | |
| Truax (1984) | | Variety | Coherence |
| Axelsson et al. (2010) | Pleasantness | Eventfulness | Familiarity |
| Cain et al. (2013) | Calmness | Vibrancy | |
| Andringa and van den Bosch (2013) | Pleasure | Activation | |
| ISO/TS 12913-3 (2019) | Pleasantness | Eventfulness | |
| Tarlao et al. (2019) | Appreciation | Dynamism | Monotony |

applied by numerous researchers, a debate about the interpretation of the dimensions and their underlying mechanisms continues (Bakker et al. 2014).

Van den Bosch et al. (2018) assumed that the affective quality dimensions that are typically observed reflect motives with evolutionary significance, such as survival (*coping mode*) and flourishing (*co-creation mode*). The authors relate all affective qualities to the indicators of *affordance* and *complexity*, allowing for the establishment of *audible safety* to various degrees. *Affordance* can be understood as cues from the environment that immediately allow us to detect a function. These cues furnish behavior (Gibson 1979). According to van den Bosch et al. (2018), the evolutionary perspective of *audible safety* is an important component of auditory environments, warning humans of potential danger. In an acoustic environment lacking a high level of audible safety, people become vigilant and are alerted more easily, which results in stress and appraised unpleasantness. Simply said, people appraise their soundscapes based on the level of safety they attribute to them (van den Bosch 2015). This means that next to an emotional appraisal, a semantic one (e.g., in terms of the attribution of audible safety to an environment) is also involved.

Like the complexity indicator postulated by Van den Bosch et al. (2018), Axelsson (2011) referred to the amount of *information load* that drives affective responses to stimuli. The components of *pleasantness* and *activation* are then a direct result of information load. Bakker et al. (2014) explained pleasure and arousal as related to the degree of *order* and *variation*.

Truax (1984) proposed *variety* and *coherence* as soundscape-related indicators that are important to consider in acoustic design; in contrast, Llorca (2018) refers to *congruence* in the context of multisensory attention, which is closely related to soundscape appraisal. According to Llorca, congruence moderates the level of valence of a soundscape.

There is a lot of evidence that multisensory interactions can play a dominant role with respect to annoyance, pleasantness, or perceived quality of sound in daily environments (e.g., Fastl and Florentine 2011). Therefore, the dimension of congruence or coherence related to the multisensory experience might be relevant in the context of soundscape as well. Eventfulness would then be interpreted as a semantic dimension of (auditory) *order* and *variation*. Doubtless, future research must also further explore the fundamental affective dimensions involved in soundscape dealing with different mechanisms of affect such as *incidental affect* (affect unrelated to a judgment or decision such as a mood) versus *integral affect* (affect as part of the individual's internal representation) (see Västfjall et al. 2016). In particular, the generalizability of the latent dimensions observed and the association of those dimensions with (physical) indicators seems of utmost importance (Aletta and Kang 2018). Research on emotion and its source is of vital interest because as Brosch et al. (2010) concluded: "*Emotional stimuli are prioritized in perception, are detected more rapidly and gain access to conscious awareness more easily than non-emotional stimuli.*" (Brosch et al. 2010, p. 385).

Having a broader view of soundscape appears necessary. The dimensionality involved may gradually have to be increased or different dimensionalities may have to be applied to a soundscape. Currently, it is still challenging to disentangle all the intertwined emotion-related approaches, theories, and concepts, where contrary opinions of cause and effect can be found. Researchers will be kept busy with the further elaboration of emotion theory in the context of soundscape. However, the consideration of different emotion-related dimensions—in contrast to the simplistic, annoyance-focused community noise approach—is undoubtedly one of the core principles in soundscape research. The general aim of understanding perception requires us to examine its building blocks (emotion, affect, behavior, physiological responses) in more detail. By currently accepting *pleasantness* and *eventfulness* as main affective descriptors of soundscape appraisal, the hunt for the underlying indicators has begun (van den Bosch et al. 2018).

### *2.3.3 Human Perception of Acoustic Environments: ISO 12913-1*

As described in Sect. 2.1.1, a soundscape is defined as an acoustic environment that is perceived or experienced and/or understood by a person or people in a context. ISO 12913 applies this definition as a perceptual construct (ISO 12913-1 2014). Figure 2.5 shows this perceptual process and highlights the important elements involved. An acoustic environment triggers auditory *sensations*. The interpretation of these sensations creates useful information (called auditory *perception*) that results in responses and an outcome. A *response* is considered to be related to short-term reactions, emotions, and behaviors that may change the context. An *outcome*, on the other hand, is understood as an overall, long-term consequence of attitudes, beliefs, judgments, or habits that are facilitated or enabled by the acoustic environment.

The *context* refers to the interrelationships between person, activity, and place in space and time, according to the ISO 12913-1 standard. Context simply influences perception at all perceptual and cognitive stages. Clearly, the term *context* covers a

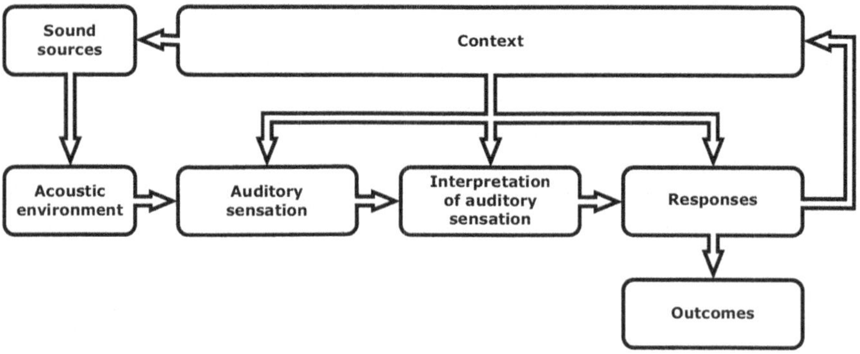


**Fig. 2.5** Elements in the perceptual construct of soundscape according to ISO 12913-1 (2014), p. 2

wide range of factors that potentially affect soundscape perception (Bruce and Davies 2014).

A soundscape is formed within a context (Botteldooren et al. 2016) and soundscape preferences critically depend on context (Brown et al. 2011). Consequently, the perception of an acoustic environment also depends on the context in which the perception process is embedded (ISO 12913-1 2014). The apparent difficulty of this consideration quickly becomes obvious: in a sense, a full and viable theory of context could be a theory of nearly everything (Marks and Algom 1998). Context affects processes occurring at every stage, from early sensory transduction, perceptual encoding to cognitive recoding, and decision-making (Marks and Algom 1998). Although a comprehensive theory of context is lacking, at least the ISO 12913-1 standard provides a starting point to address the context factor sufficiently.

No fixed relationship exists between the physical stimulus and the human reaction. An act of interpretation takes place, which depends on the way people accept certain sound sources and trust the authorities who are responsible to protect them against harmful noise (Schulte-Fortkamp and Fiebig 2006). A person's response to an environment cannot be completely understood within a strict stimulus–response framework because responses depend on moods or intentions often formed in whole or in part before encountering a specific stimulus and environment (Snodgras et al. 1995). Humans construct their perceptions by interpreting sensations and apparently rely on normative principles and on heuristic principles (Fiebig 2019). They usually do this by considering components of acoustic environments as carriers of meaning (Fiebig 2015). High-level cognitive effects using heuristic principles are at least as significant as low-level percepts or physical attributes of the signal; humans extract meaning from a soundscape through information conveyed by the soundscape and in terms of human behavior (Davies et al. 2013). In fact, any acoustic environment can be viewed as a kind of composition where sounds play an informative role (Schulte-Fortkamp and Fiebig 2016).

## 2.4 Appraisal of Soundscapes: Processing Streams of Experiences

A particular aspect of human perception concerns the construction of overall assessments of complex, prolonged experiences. Frequently, the perception of acoustic environments is considered in terms of the instantaneous response to acoustic stimuli; but in everyday life it is a viable need to cope with complex environments that continuously excites human receptor cells. Indeed, a soundscape is considered to be a dynamic system that is characterized by the time-dependent occurrence of particular sound events embedded in specific environments (Schulte-Fortkamp and Fiebig 2016).

With regard to data collection, ISO/TS 12913-2 recommends that participants listen to a given sound in silence for a defined period of time (e.g., 3 min) and that they then assess their experiences on different rating scales. This task involves a recollection of the past period with perceived intensities that probably varied over time. Retrospective summarized evaluations seem very natural because they form the basis on which decisions are made to repeat or avoid past experiences that have direct hedonic consequences (Ariely and Carmon 2000).

Most investigations in the context of perception and assessment of environmental noise request ratings, evaluations, or descriptions of how a certain acoustic environment was perceived over a certain period of time in total, whether in the context of noise annoyance or a soundscape. The requested *summary assessment* of a past episode requires a significant reduction of complex streams of varying momentary sensations into a simple category (like a judgment on a rating scale). This reduction of complexity is needed to avoid cognitive overload. Humans would experience stress if they had to recount every moment of a longer episode to conclude whether they want to repeat or avoid it in future (Fiebig 2019).

Sensations endure and inform cognition well beyond the physical presence of the triggering stimulus (Algom 2001). In cognitive processes, memories and experiences are coupled with past and present judgments to help with recognition and organization of the layout of an environment (Bell et al. 2001). Moreover, there was a strong connection between what people felt and how they appraised the acoustic environment surrounding them (van den Bosch 2015).

On the other hand, a dissociation of retrospective evaluations from immediate experience is frequently observed in different sensory domains (Kahneman et al. 1993). Indeed, there is substantial evidence that people tend to use selected moments of extended experiences to form overall assessments, which is described by Kahnemann (2000) as *judgment by prototype*. This could be explained by an underlying economy principle for processing complex perceptual data; however, some questions remain. What are the possible determinants of retrospective evaluations of time-variant (noise) sequences? How are complex feelings transformed into overall appraisals? Is the construction of an overall soundscape appraisal based on normative or heuristic principles? The construction of soundscape appraisals does not occur only when overall assessments are requested. Appraisals are the product of an ongoing, unconscious process that is based on the omnipresent need of humans to constantly reevaluate past experiences.

Despite their naturalness, retrospective reports harbor serious methodological problems. These problems include possible distortions of memory and the evidence that human memory can be full of gaps (Danziger 1980). For example, with respect to noise annoyance surveys, self-reported long-term noise annoyance judgments were significantly affected by the very moment of questioning (Brink et al. 2016). Studies have shown that the season where a noise annoyance survey is performed has an impact on the annoyance ratings, although the respondents are always requested to consider the last 12 months (Brink et al. 2016). For example, the environmental noise of the last 12 months at home is rated in average as less annoying and disturbing in spring than in autumn, which illustrates a bias in retrospective reports.

Although a satisfying solution to this problem has yet to be found, there is strong evidence that the construction of an overall assessment goes beyond pure "cognitive averaging" when evaluating a soundscape as a whole (Steffens and Guastavino 2015). Humans must rapidly make sense of their environment to successfully move in the world and respond to its challenges, which might be achieved by creating an efficient mental representation of sensory stimuli by grouping certain objects as equivalent and reducing the complexity of information in the external world. Information about a particular stimulus is then inferred due to its association with a category (Brosch et al. 2010).

In laboratory experiments, participants often rely on normative principles and avoid ignoring larger parts of experienced episodes and their respective intensities (Fiebig and Sottek 2015). However, it is very likely that in everyday situations, humans tend to use selected moments of extended experiences to form overall assessments (Kahneman et al. 1993). People compare each stimulus event with known possibilities and judge it in comparison to perceived relationships and remembered alternatives (Lockhead 2001). Although it is conceivable that humans reduce complex episodes to a few properties, preserving only the information needed to navigate the real world and to form a stable global percept (Ariely 2001), the idea of mental representation with statistical properties as a kind of hedonic calculator has its limits. It appears likely that experienced episodes and their components are summarized into a global percept.

According to Schulte-Fortkamp (2014), the assigned meaning to sounds significantly affects the evaluation of sounds. Clearly, this notion must push the body of empirical work beyond the study of only one type of affect into studies of experiences characterized by multiple or mixed affective states, including cognitive processes and meaning attribution (Fredrickson 2000). At the same time, gaining a comprehensive understanding of human cognitive processing streams that are involved in the perception of acoustic environments might be a difficult endeavor, especially if we accept that humans possess a variety of cognitive schemes, each of which can be evoked or suppressed by subtle contextual features (Frederick and Loewenstein 2008). Without doubt, in the context of soundscape, studying the way humans summarize long-term experiences of acoustic environments in ecological settings is essential because that is the way humans report on their experienced acoustic environments. Unfortunately, there still seems to be a significant gap in understanding of which elements of perception contribute to retrospective assessments of time-variant experiences of acoustic environments.

## 2.5 Summary

The soundscape approach moves beyond current noise control engineering and retrofitting of the acoustic environment (Aletta et al. 2016a). Schafer laid the basis for this approach by interpreting environmental noise as a musical composition that could sound pleasant or terrible. Schafer wanted to establish a novel way of thinking, and he believed that it is up to the composers to tune the world (Schafer 1977).

This creative paradigm shift when dealing with acoustic environments lead to a focus on perception instead of treating the physical aspects of unwanted noise sources as linked to noise annoyance. It became evident that a listener within a soundscape is not simply engaged in a passive type of energy reception, but rather is part of a dynamic system of information exchange (Truax 1984). Accordingly, soundscape research embraced environmental psychology's understanding that (environmental) perception is constituted by the individual, the environment, and the interaction between the two.

Indeed, humans and soundscapes have a dynamic bidirectional relationship: humans affect soundscapes with their behavior and humans are in turn influenced by their soundscapes (Erfanian et al. 2019). These interactions between individuals and acoustic environments are broadly acknowledged and incorporated in the available soundscape standards and technical specifications (e.g., ISO 12913-1).

As humans presumably like places that allow them to carry out their plans and dislike places that violate their expectations (Snodgras et al. 1995), the concept of soundscape must deal with the function of sound in context and must include examination of the expectations and attitudes of the individuals experiencing acoustic environments. Therefore, it seems advisable to focus more strongly on the individual with his or her inherent traits, beliefs, moods, and desires instead of only considering the average person (Botteldooren et al. 2016). In this context, emotions elicited by the (acoustic) environment play a major role in well-being. Instead of debating the nature of "true" emotions and whether they do or do not require cognition, research must be focused on the details of cognition–emotion interaction and the function of these human processes in everyday life (Gross 2010).

In the past, adverse health effects of environmental noise on people and communities have been thoroughly investigated, primarily addressing unwanted sound. In contrast, the aspects of environmental noise that could induce potentially positive moods have been disregarded entirely (Aletta et al. 2018). As a matter of fact, preventative health research involving positive health outcomes from exposure to urban sounds is still limited (Payne and Bruce 2019). In particular, soundscape research requires more scientific evidence on the potential to use cognitive restoration to promote healthy urban environments (Kang et al. 2016). Some studies suggest that recovery from psychological stress and physiological recovery from sympathetic activation is faster during exposure to pleasant than to unpleasant sounds (Alvarsson et al. 2010).

The identification and preservation of "quiet areas" is enforced by Directive 2002/49/EC of the European Parliament and of the Council relating to the assessment and management of environmental noise (END 2002), and local authorities are required to seriously address this issue. Quiet areas in urban context are assumed to be beneficial and to have a health-promoting function by acting as buffers against the adverse health effects of traffic-noise exposure (Gidlof-Gunnarsson and Ohrstrom 2007). However, it remains astonishingly unclear how restoration and health benefits are supported by (acoustic) elements of a quiet area. Natural sounds, for example, have the ability to evoke pleasant feelings (Gidlof-Gunnarsson and Ohrstrom 2007), but classical indicators related to sound-pressure-level threshold values cannot accurately explain this induced feeling. Only limited correlations have been found between sound pressure levels and feelings of restoration in urban parks (Brambilla et al. 2013).

In this context, it is likely that the soundscape approach, with its implicit orientation to positive aspects of sounds, will gain in significance to support investigations of quiet areas and explorations of the (acoustical and perceptual) elements that are fundamental for restoration and health. Accordingly, Aletta et al. (2018) concluded in their literature review that positive perceptions of acoustic environments (soundscapes) can be associated with positive health effects. Thus, it is the long-term objective of soundscape research to design acoustic environments that specifically evoke emotional responses composed of positive affective qualities and to reduce adverse noise effects. However, the current understanding of the entanglement between psychology and soundscape is mostly limited to the link between acoustic characteristics and overt appraisal of soundscapes. This view lacks clarification of the corresponding responses in the physiological domain, which is probably relevant for genuine restoration and recovery (Erfanian et al. 2019). The literature on beneficial effects of sound is still scarce (Kamp et al. 2016). Most soundscape research has provided little insight into the promotion of physical and mental health, but mostly dealt with adverse effects (see Lercher and Dzhambov, Chap. 9).

The major challenge of soundscape research will continue to lie in determining ways to measure perception, especially outside of laboratories, which is the basic tenet of soundscape. There is still a significant lack of clarity and consensus regarding the use of terms such as *sensation*, *perception* (including *emotion*), and *cognition*, which impedes scientific progress. In addition, future research efforts must focus on improving the ecological validity of the experimental settings (Steffens and Guastavino 2015). Narrowly conceived studies that are based on empiricism and that ignore relevant conceptual and philosophical issues are not informative (Michell 1997). Because experimental design and data interpretation are fundamentally shaped by the theoretical commitments of the researchers (e.g., Chirimuuta 2016), further work is needed to develop sophisticated theoretical concepts that will allow researchers and practitioners to test the applicability of different methods to measure perception of a soundscape and to evaluate the validity of experimental outcomes.

**Compliance with Ethics Requirements** André Fiebig declares that he has no conflict of interest.

# References

Aletta F, Kang J (2018) Towards an urban vibrancy model: a soundscape approach. Int J Environ Res Public Health 15:1712

Aletta F, Kang J, Axelsson Ö (2016a) Soundscape descriptors and a conceptual framework for developing predictive soundscape models. Landsc Urban Plan 149:65–74

Aletta F, Lepore F, Kostara-Konstantinou E, Kang J et al (2016b) An experimental study on the influence of soundscapes on people's behavior in an open public space. Appl Sci 6:276
Aletta F, Oberman T, Kang J (2018) Associations between positive health-related effects and soundscapes perceptual constructs: a systematic review. Int J Environ Res Public Health 15:2392
Algom D (2001) Memory psychophysics. In: International Encyclopaedia for the social and the behavioral sciences, Elsevier, London, UK, pp 9610–9613
Alvarsson JJ, Wiens S, Nilsson ME (2010) Stress recovery during exposure to nature sound and environmental noise. Int J Environ Res Public Health 7:1036–1046
Andringa TC, van den Bosch KAM (2013) Core affect and soundscape assessment: Fore- and background soundscape design for quality of life. In: Proceedings of InterNoise 2013, Innsbruck, Austria
Ariely D (2001) Seeing sets: representation by statistical properties. Psychol Sci 12(2):157–162
Ariely D, Carmon Z (2000) Gestalt characteristics of experiences: the defining features of summarized events. J Behav Decis Making 13:191–201
Axelsson Ö (2011) Aesthetic appreciation explicated. Doctoral thesis, Stockholm, Sweden
Axelsson Ö, Nilsson ME, Berglund B (2010) A principal components model of soundscape perception. J Acoust Soc Am 128(5):2836–2846
Bakker I, van der Voordt T, Vink P, de Boon J (2014) Pleasure, arousal, dominance: Mehrabian and Russell revisited. Curr Psychol 33:405
Barandiaran X, Di Paolo EA, Rohde M (2009) Defining agency: individuality, normativity, asymmetry, and spatio-temporality in action. Adapt Behav 17:367–386
Bell PA, Greene TC, Fisher JD, Baum AS (2001) Environmental psychology. Harcourt College Publishers, New York
Bild E, Pfeffer K, Coler M, Rubin O et al (2018) Public space users' soundscape evaluations in relation to their activities. An Amsterdam-based study. Front Psychol 9:1593. https://doi.org/10.3389/fpsyg.2018.01593
Boring EG (1921) The stimulus-error. Am J Psychol 32(4):449–471
Botteldooren D, de Coensel B (2009) The role of saliency, attention and source identification in soundscape research. In: Proceedings of the InterNoise 2009, Ottawa, Canada
Botteldooren D, Boes M, Oldoni D, de Coensel B (2012) The role of paying attention to sounds in soundscape perception. In: Proceedings of the acoustics 2012, Hong Kong, China
Botteldooren D, Andringa T, Aspuru I, Brown AL et al (2016) From sonic environment to soundscape. In: Kang J, Schulte-Fortkamp B (eds) Soundscape and the built environment. CRC Press/Taylor & Francis Group
Bradley MM, Lang PJ (2000) Affective reactions to acoustic stimuli. Psychophysiology 37:204–215
Brambilla G, Gallo V, Asdrubali F, D'Alessandro F (2013) The perceived quality of soundscape in three urban parks in Rome. J Acoust Soc Am 134(1):832–839
Brink M, Schreckenberg D, Vienneau D, Cajochen D et al (2016) Effects of scale, question location, order of response alternatives, and season on self-reported noise annoyance using ICBEN scales: a field experiment. Int J Environ Res Public Health 13:1163
Brosch T, Pourtois G, Sander D (2010) The perception and categorisation of emotional stimuli: a review. Cognit Emot 24(3):377–400
Brown AL, Kang J, Gjestland T (2011) Towards standardization in soundscape preference assessment. Appl Acoust 72:387–392
Brown AL, Gjestland T, Dubois D (2016) Acoustic environments and soundscapes. In: Kang J, Schulte-Fortkamp B (eds) Soundscape and the built environment. CRC Press/Taylor & Francis Group
Bruce NS, Davies WJ (2014) The effects of expectation on the perception of soundscapes. Appl Acoust 85:1–11
Cain R, Jenning P, Poxon J (2013) The development and application of the emotional dimensions of a soundscape. Appl Acoust 74(2):232–239
Cattell JM (1893) On errors of observation. Am J Psychol 5:285–293
Chirimuuta M (2016) Why the "stimulus-error" did not go away. Stud Hist Philos Sci 56:33–42

Danziger K (1980) The history of introspection reconsidered. J Hist Behav Sci 16:241–262
Davies WJ, Adams MD, Bruce NS, Cain R et al (2013) Perception of soundscapes: an interdisciplinary approach. Appl Acoust 74:224–231
De Coensel B, Botteldooren D (2010) A model of saliency-based auditory attention to environmental sound. In: Proceedings of the ICA 2010, Sydney, Australia
END (2002) European Directive 2002/49/EC 2002 of the European Parliament and of the Council of 25 June 2002 (2002) relating to the assessment and management of environmental noise. Official Journal of the European Communities L 189/12, Brussels
Erfanian M, Mitchell AJ, Kang J, Aletta F (2019) The psychophysiological implications of soundscape: a systematic review of empirical literature and a research agenda. Int J Environ Res Public Health 16:3533
Fastl H (2004) Audio-visual interactions in loudness evaluation. In: Proceedings of the ICA 2004, Kyoto, Japan
Fastl H, Florentine M (2011) Loudness in daily environments. In: Florentine M, Popper AN, Fay RR (eds) Springer handbook of auditory research book series: loudness, vol 37. Springer, New York, pp 199–222
Fiebig A (2015) Cognitive stimulus integration in the context of auditory sensations and sound perceptions. epubli, Berlin
Fiebig A (2018) Recent developments in the standardization of soundscape. In: Proceedings of InterNoise 2018, Chicago, USA
Fiebig A (2019) The perception of acoustic environments and how humans form overall noise assessments. In: Proceedings of InterNoise 2019, Madrid, Spain
Fiebig A, Sottek R (2015) Contribution of peak events to overall loudness. Acta Acust united Ac 101(6):1116–1129
Fiebig A, Jordan P, Moshona CC (2020) Assessments of acoustic environments by emotions - the application of emotion theory in soundscape. Front Psychol 11:573041. https://doi.org/10.3389/fpsyg.2020.573041
Filipan K, De Coensel B, Aumond P, Arnaud C et al (2019) Auditory sensory saliency as a better predictor of change than sound amplitude in pleasantness assessment of reproduced urban soundscapes. Build Environ 148:730–741
Fisher JD, Bell PA, Baum A (1984) Environmental psychology. CBS College Publishing, New York
Franek M, Rezny L, Sefara D, Cabal J (2019) Effect of birdsongs and traffic noise on pedestrian walking speed during different seasons. Peer J 7:e7711. https://doi.org/10.7717/peerj.7711
Frederick S, Loewenstein G (2008) Conflicting motives in evaluations of sequences. J Risk Uncertain 37:221. https://doi.org/10.10070/s11166-008-9051-z
Fredrickson BL (2000) Extracting meaning from past experiences: the importance of peaks, ends, and specific emotions. Cognit Emot 14(4):577–606
Genuit K, Fiebig A (2006) Psychoacoustics and its benefit for the soundscape approach. Acta Acust united Ac 92(6):952–958
Gibson JJ (1979) The ecological approach to visual perception. Houghton Mifflin, Boston
Gidlof-Gunnarsson A, Ohrstrom E (2007) Noise and well-being in urban residential environments: the potential role of perceived availability to nearby green spaces. Landsc Urban Plan 83:115–126
Goldstein EB (2002) Sensation and perception, 6th edn. Wadsworth, Pacific Grove
Gross JJ (2010) The future's so bright, I gotta wear shades. Emot Rev 2(3):212–216
Guski R (1999) Personal and social variables as co-determinants of noise annoyance. Noise Health 3:45–56
Hasset J (1978) A primer of psychophysiology. W.H. Freeman and company, San Francisco
Helson H (1967) Perception. In: Helson H, Bevan W (eds) Contemporary approaches to psychology. van Nostrand, Princeton/New Jersey/Toronto/London
Hollister CM (1914) Toleration of noises. Am J Nurs 14:519–521
ISO 12913-1 (2014) Acoustics-soundscape-part 1: definition and conceptual framework. Geneva, Switzerland

ISO 18405 (2017) Underwater acoustics — Terminology. Geneva, Switzerland
ISO/TS 12913-2 (2018) Acoustics-soundscape-part 2: data collection and reporting requirements. Geneva, Switzerland
ISO/TS 12913-3 (2019) Acoustics-soundscape part 3: data analysis. Geneva. https://www.iso.org/standard/69864.html
James W (1884) What is an emotion? Mind 9:188–205
Jeon JY, Hong JY, Lavandier C, Lafon J et al (2018) A cross-national comparison in assessment of urban park soundscapes in France, Korea, and Sweden through laboratory experiments. Appl Acoust 133:107–117
Kahneman D (2000) Evaluation by moments: Past and future. In: Kahneman D, Tversky A (eds) Choices, values and frames. Cambridge University Press, New York
Kahneman D, Fredrickson BL, Schreiber CA et al (1993) When more pain is preferred to less: adding a better end. Psychol Sci 4(6):401–405
Kang J, Chourmouziadou K, Sakantamis K, Wang B, et al. (eds) (2013, March 22) COST TUD Action TD 0804. Soundscape of European cities and landscapes. Oxford
Kang J, Aletta F, Gjestland TT, Brown LA et al (2016) Ten questions on the soundscapes of the built environment. Build Environ 108:284–294
Klatte M, Hellbrück J, Seidel J, Leistner P (2010) Effects of classroom acoustics on performance and well-being in elementary school children: a field study. Environ Behav 42(5):659–692
Knudsen EI (2007) Fundamental components of attention. Annu Rev Neurosci 30:57–78
Kogan P, Turra B, Arenas JP, Hinalaf M (2017) A comprehensive methodology for the multidimensional and synchronic data collecting in soundscape. Sci Total Environ 580:1068–1077
Korte C, Grant R (1980) Traffic noise, environmental awareness and pedestrian behavior. Environ Behav 12(3):408–420
Lavia L, Dixon M, Witchel HJ, Goldsmith M (2016) Applied soundscape practices. In: Kang J, Schulte-Fortkamp B (eds) Soundscape and the built environment. CRC Press/Taylor & Francis Group
Lin LW, Weng SC, Wu HS, Tsai LJ, Lin YL, Yeh SH (2018) The effects of white noise on agitated behaviors, mental status, and activities of daily living in older adults with dementia. J Nurs Res 26(1):2–9. https://doi.org/10.1097/JNR.0000000000000211
Llorca J (2018) Virtual reality for urban sound design: a tool for architects and urban planners. In: Aceves-Fernandez MA (ed) Artificial intelligence: emerging trends and applications. IntechOpen, London, UK
Lockhead GR (2001) Context and judgment. In: Sommerfeld E, Kompass R, Lachmann T (eds) Fechner Day 2001. Pabst, Lengerich
Lockhead GR (2004) Absolute judgments are relative: a reinterpretation of some psychological scales. Rev Gen Psychol 8(4):265–272
Marks LE, Algom D (1998) Psychophysical scaling. In: Birnbaum MH (ed) Measurement, judgment and decision making. Academic Press, San Diego
Mehrabian A, Russell JA (1974) An approach to environmental psychology. MIT Press, Cambridge, Massachusetts and London
Mehta R, Zhu R, Cheema A (2012) Is noise always bad? Exploring the effects of ambient noise on creative cognition. J Consum Res 39(4):784–799
Meng Q, Zhao T, Kang J (2018) Influence of music on the behaviors of crowd in urban open public spaces. Front Psychol 9:596
Merriam-Webster (2022) Emotion. Merriam-Webster.com, Dictionary, https://www.merriam-webster.com/dictionary/emotion. Accessed 8 May 2022
Michell J (1997) Quantitative science and the definition of measurement in psychology. Br J Psychol 88:355–383
Moser G, Uzzell D (2003) Environmental psychology. In: Millon T, Lerner MJ (eds) Handbook of psychology, Personality and social psychology, vol 5. John Wiley & Sons, pp 419–445
Oldoni D, De Coensel B, Boes M, Rademaker M et al (2013) A computational model of auditory attention for use in soundscape research. J Acoust Soc Am 134:852–861

Payne SR, Bruce N (2019) Exploring the relationship between urban quiet areas and perceived restorative benefits. Int J Environ Res Public Health 16:1611

Russell JA (1980) A circumplex model of affect. J Pers Soc Psychol 39(6):1161–1178

Russell JA, Pratt G (1980) A description of the affective quality attributed to environments. J Pers Soc Psychol 38(2):311–322

Russell JA, Ward LM, Pratt G (1981) Affective quality attributed to environments: a factor analytic study. Environ Behav 13:259–288

Schafer RM (1977) The soundscape: our sonic environment and the tuning of the world. Destiny Books, USA, Rochester

Schafer RM (2012) Soundscape studies: the early days and the future. Soundscape: J Acoust Ecol 12(1):6–8

Schlittmeier SJ, Hellbrück J, Klatte M (2008) Does irrelevant music cause an irrelevant sound effect for auditory items? Eur J Cogn Psychol 20(2):252–271

Schomer P, Brown A L, de Coensel B, Genuit K, et al. (2010) On efforts to standardize a graphical description of the soundscape concept. In: Proceedings of InterNoise 2010, Lisbon, Portugal

Schulte-Fortkamp B (2002) The meaning of annoyance in relation to the quality of built environments. Noise Health 4:13–28

Schulte-Fortkamp B (2014) Improving sound quality measures through the multifaceted soundscape approach. In: Proceedings of InterNoise 2014, Melbourne, Australia

Schulte-Fortkamp B, Fiebig A (2006) Soundscape analysis in a residential area: an evaluation of noise and people's mind. Acta Acust united Ac 92(6):875–880

Schulte-Fortkamp B, Fiebig A (2016) Impact of soundscape in terms of perception. In: Kang J, Schulte-Fortkamp B (eds) Soundscape and the built environment. CRC Press/Taylor & Francis Group, Boca Raton, pp 69–88

Sellars RW (1907) The nature of experience. J Philos Psychol Sci Methods 4(1):14–18

Shield B, Dockrell J (2008) The effects of classroom and environmental noise on children's academic performance. In: Proceedings of international congress on noise as a public health problem, Foxwoods, USA

Snodgras J, Russell JA, Ward LM (1995) Planning, mood, and place-liking. In: Gärling T, Canter D (eds) Readings in environmental psychology: urban cognition. Academic Press, San Diego, pp 187–200

Snyder JS, Gregg MK, Weintraub DM, Alain C (2012) Attention, awareness, and the perception of auditory scenes. Front Psychol 3:15

Southworth MF (1967) The sonic environment of cities. MIT, thesis

Southworth MF (1969) The sonic environment of cities. Environ Behav 1(1):49–70

Steffens J, Guastavino C (2015) Trend effects in momentary and retrospective soundscape judgments. Acta Acustica united Ac 101(4):713–722. https://doi.org/10.3813/AAA.918867

Steffens J, Müller F, Schulz M, Gibson S (2019) Cognitive load influences the evaluation of complex acoustical scenarios. In: Proceedings of ICA 2019, Aachen, Germany

Stevens SS (1975) Psychophysics. Introduction to its perceptual, neural, and social prospects, Wiley, New York, USA

Sudarsono AS, Sarwono SJ, Shabrina A, Tamrin LN (2019) The soundscape dimensions of third-class hospital ward in Indonesia. In: Proceedings of ICA 2019, Aachen, Germany

Sun K, De Coensel B, Filipan K, Aletta F et al (2019) Classification of soundscapes of urban public open spaces. Landsc Urban Plan 189:139–155

Taghipour A, Pelizzari E (2019) Effects of background sounds on annoyance reaction to foreground sounds in psychoacoustic experiments in the laboratory: limits and consequences. Appl Sci 9:1872

Tarlao C, Steele D, Guastavino C (2019) Investigating factors influencing soundscape evaluations. In: Proceedings of InterNoise 2019, Madrid, Spain

Titchener EB (1922) Mach's lectures on psychophysics. Am J Psychol 33(2):213–222

To WM, Chung A, Vong I, Ip A (2018) Opportunities for soundscape appraisal in Asia. In: Proceedings of the euronoise 2018, Crete, Greece

Truax B (1984) Acoustic communication. Ablex Publishing Corporation, Norwood

van den Bosch KAM (2015) Safe and sound. Soundscape research in special needs care. Doctoral thesis, Groningen, Netherlands

van den Bosch KAM, Welch D, Andringa TC (2018) The evolution of soundscape appraisal through enactive cognition. Front Psychol 9:1129. https://doi.org/10.3389/fpsyg.2018.01129

Van Kamp I, Klaboe R, Brown AL, Lercher P (2016) Soundscapes, human restoration and quality of life. In: Kang J, Schulte-Fortkamp B (eds) Soundscape and the built environment. CRC Press/Taylor & Francis Group, Boca Raton

Västfjäll D, Slovic P, Burns WJ, Erlandsson A et al (2016) The arithmetic of emotion: integration of incidental and integral affect in judgments and decisions. Front Psychol 7:325. https://doi.org/10.3389/fpsyg.2016.00325

Welch D, Shepherd D, Dirks K, Tan MY et al (2019) Use of creative writing to develop a semantic differential tool for assessing soundscapes. Front Psychol 9:2698. https://doi.org/10.3389/fpsyg.2018.02698

Yu B, Kang J, Ma H (2016) Development of indicators for the soundscape in urban shopping streets. Acta Acustica united Ac 102(3):462–473. https://doi.org/10.3813/AAA.918965

Zadra JR, Clore GL (2011) Emotion and perception: the role of affective information. Wiley Interdiscip Rev Cogn Sci 2(6):676–685

# Chapter 3
# Soundscape: The Holistic Understanding of Acoustic Environments

**Brigitte Schulte-Fortkamp and Pamela Jordan**

**Abstract** The expanded recognition of soundscape studies has resulted in a research domain composed of various disciplinary perspectives and, at times, contradictory interpretations of what is meant by "soundscape." At its core, soundscape is a construct of human perception that factors in the entirety of an acoustic environment and the individual's responses to it. This stands in contrast to the acoustic environment, which is simply the composition of acoustic stimuli in an environment. The baseline of human perception in soundscape was recognized by the International Organization for Standardization in ISO 12913-1 (ISO 2014), which emphasizes the interrelationships between person, activity, and physical place in both space and time. The breadth of one's responses beyond auditory sensation can be much more complex than reactions to noise. The meaning an individual ascribes to sounds, individual's attitude and expectations toward the acoustic environment, socioeconomic and cultural background, and life experiences all play a role. The complexity of interrelationships between context and listener can only be understood through a multilateral, holistic approach in the field of soundscape study. This chapter traces the broad trajectory of soundscape studies to contextualize a holistic approach and concludes by highlighting various holistic research projects that sought to enhance the quality of acoustic environments and living situations.

**Keywords** Acoustic design · Community noise · ISO 12913 series · Environmental experience · Local expertise · Participatory research · Quality of life · Soundscape evaluation · Soundscape measurement

---

B. Schulte-Fortkamp (✉)
HEAD-Genuit Foundation, Herzogenrath, Germany
e-mail: bschulte_f@web.de

P. Jordan
University of Amsterdam, Amsterdam, The Netherlands
e-mail: p.jordan@uva.nl



B. Schulte-Fortkamp et al. (eds.), *Soundscapes: Humans and Their Acoustic Environment*, Springer Handbook of Auditory Research 76,
https://doi.org/10.1007/978-3-031-22779-0_3

## List of Acronyms

| | |
|---|---|
| BMVBS | Federal Ministry of Transport, Building, and Urban Affairs |
| COST | European Cooperation in Science and Technology |
| DEFRA | Department for Environment, Food and Rural Affairs |
| EEA | European Environment Agency |
| EHIA | Environmental Health Impact Assessments |
| END | Environmental Noise Directive |
| ENNAH | European Network on Noise and Health |
| ExWoSt | Experimental Housing and Urban Development |
| GSSN | Global Sustainable Soundscape Network |
| HOSSANA | Holistic and Sustainable Abatement of Noise by Optimized Combinations of Natural and Artificial Means |
| ISO | International Organization for Standardization |
| NORAH | Noise-Related Annoyance, Cognition, and Health study |
| SONORUS | The Urban Sound Planner |
| STSM | Short-Term Scientific Missions |

## 3.1 Introduction

Soundscape as a field of study has been closely intertwined with the study of community noise in daily life. For decades, assessments of noise and its effects were organized through management efforts and regulations related to community noise, including road, rail, and air traffic, industrial activity, construction and public works, and their repercussions in neighborhoods (Berglund et al. 1999). European noise policy provided harmonization of noise indicators, noise mapping, and action plans (EPC 2002). This synthesis delivered basic administrative information for noise abatement in highly noise-polluted areas and for comparisons across European countries. The Environmental Noise Directive (END; European Commission 2011) also focused on noise reduction through country-specific regulations that provided tools and essential knowledge for the design and planning of sustainable environments that would be supportive of well-being and health.

Considering the diverse action plans suggested to reduce the burden of noise, Hiramatsu (2013) asked why noise pollution is still rampant despite the development of the specialized field of noise control engineering. He also wondered if it would be more appropriate to place noise abatement within the cultural context of efficiency as a better way to convey how those involved in noise abatement understand themselves and their actions. Decades earlier, a scientific approach was proposed that incorporated altering the soundscape in collaboration with cultural contexts (Schulte-Fortkamp 1994a). The work strove to find solutions that would allow people to remain in place while reducing their daily acoustic burden. The vital element was directly involving residents and other stakeholders in the evaluation processes. As soundscape research developed, these individuals would come to be known as local experts.

Soundscape is a paradigm shift toward a multidisciplinary approach that presents a completely different understanding of the human need to solve noise problems that is related to individual perception and acceptance of an acoustic environment. Based on both human biology and social sciences, soundscape research approaches the diversity of an environment with special attention to how people experience the acoustics in a specific environment. Lifestyle and sociocultural backgrounds are additionally considered as they provide important contributions to a person's assessment of an acoustic environment.

In contrast to noise control, soundscape also shifts the understanding of any acoustic input as a "resource" by definition rather than a "waste" (Brown et al. 2011), which is explored in more detail in Sect. 3.2. When considering the built environment and modeling or analyzing dependencies, soundscape research pushes beyond sound sources to consider further contributors. Soundscape was first conceptualized by Southworth (1967) and Schafer (1977) as the (human) contextual perception of the acoustic environment. Their approach was the first time that human perception was the focus rather than the noise sources. The resulting soundscape approach revealed how people perceive sound as well as how they establish a relationship with their acoustical environment.

The ISO 12913 series subsequently defined a perceived acoustical environment as a soundscape. Erfanian et al. (2019) stated that humans and soundscapes have a dynamic bidirectional relationship. Thus, while humans and their behavior directly influence their soundscape, humans and their behavior also are influenced by their acoustic environment. Erfanian et al. (2019) emphasized that a diverse group of researchers from the areas of acoustics, environmental psychology, and auditory neuroscience have all outlined the adverse impacts of noise on well-being with regard to living standards. Therefore, to understand how the soundscape approach surpasses noise research in assessing an acceptable acoustic environment, it is necessary to trace the conceptual development and multidisciplinary origins of soundscape research. Botteldooren, De Coensel, Aletta, and Kang, Chap. 8 introduce the process of measuring the perceptions of individuals to reach a complete and holistic picture of how the community feels about the soundscape. This chapter will set the stage by introducing the origins of that holistic approach in the context of (and in contrast to) noise measurements related to community noise, which is still applied as an objective measure to protect people against noise. In contrast to noise measurements, the nature of the soundscape approach allows for a more comprehensive evaluation of community noise; however, both fields share an endpoint as well as an origin story.

## 3.2 The Concept of Soundscape

In addition to a few of his contemporaries, composer Schafer (1977) framed the acoustic world in terms of soundscape and guided the subsequent development of soundscape as a distinct field of research. His definitions and detailed conceptualizations of how people relate to the acoustic environment inspired researchers, particularly those who shared an interest in community noise, to rethink their ocular-centric understanding of place and community. Schafer pointed out that "the general acoustic environment of a society can be read as an indicator of social conditions which produce it and may tell us much about the trending and evolution of that society" (Schafer 1977, p. 7). Soundscape research has incorporated an ever-expanding range of sound-related disciplines and tools of inquiry, all with the intent to improve the acoustic environment. The eventual creation of an ISO standard for soundscape set the orientation point for the next phase of soundscape research.

### *3.2.1 Background and Development: A Holistic Approach Responding to Community Noise*

Among his many contributions to the conceptualization of soundscape research, Schafer (1977) emphasized the possibility that improving soundscapes could ultimately improve society. He pointed out three categories of research that would assist in attaining that goal: *acoustics and psychoacoustics* to reveal how sound is first interpreted by the human brain; *social studies* to indicate how sound affects and changes human behavior; and musical composition as a framework for describing the acoustic components of a soundscape and conceptualizing how to design preferred soundscapes from existing conditions (Schafer 1977). By joining science, society, and the arts in soundscape studies, Schafer introduced a new interdisciplinary field called acoustic design (Schafer 1977; Fiebig and Schulte-Fortkamp 2020). He stressed that any acoustic designer must understand the environment being studied thoroughly and must be able to draw from a multitude of professional perspectives: the researcher "must have training in acoustics, psychology, sociology, music, and a great deal more besides, as the occasion demands" (Schafer 1977, p. 206). He imagined the new discipline would incorporate "scientists, social scientists, and artists (particularly musicians), acoustic design attempts to discover principles by which the aesthetic of the acoustic environment or soundscape may be improved" (Schafer 1977, p. 271).

Along with conceptual leaps and calls for multidisciplinary collaboration, Schafer is credited with popularizing soundscape research tools that are still in use today and promoting what Hildegard Westerkamp introduced as sound-walking to explore a specific soundscape (Westerkamp 1974; Behrendt 2018). Sound-mapping had also been explored earlier by Southworth through the lens of city planning, which proved influential as well (Southworth 1969). Soundscape-mapping has since been steadily woven into soundscape research in a variety of applications. For a thorough review of its development and implementation within soundscape, see the survey of soundscape mapping by Kang et al. (2016a).

The development of outdoor public places was a shared interest of Southworth, Schafer, and others. Soundscape brought a new perspective to what had previously been the sole purview of noise management by focusing on the people burdened by increasing noise from traffic and urban activity (Schulte-Fortkamp 2001; Brooks et al. 2014). Schafer (1977) explicitly depended on the input of residents, taking into consideration their expectations and experiences as foundational research input. By doing so, he set the groundwork for the incorporation of qualitative data from "local experts" (Schulte-Fortkamp et al. 2008).

### *3.2.2 Conceptual Framework: The Role of Local Experts in Soundscape*

Local expertise in an acoustic environment signifies meaningful knowledge about an area where individuals interact over time. Their knowledge is one of the most important investigative resources available to researchers. Generally, local experts are residents or workers from the area being studied. Because they are intimately familiar with their own environment, its daily rhythms and norms, they are asked to provide their knowledge in evaluation processes such as soundwalks and various kinds of open interviews. Local expert participation sharply focuses the subsequent analysis of perceptual and acoustical data as the information provided often enhances the investigator's sensitivity to the subtle particularities of the examined area (Schulte-Fortkamp 2010). Previous experiences of those individuals are significant for comprehending their perceptions and assessments of the environment (Basturk et al. 2012) and for identifying the meaning and significance of sounds to them. Such a comprehensive understanding is essential to be able to improve the soundscape.

The acceptance of local experience over outside expertise is a more recent phenomenon, and disciplines beyond soundscape have begun incorporating similar perspectives. Fields as diverse as architectural design (Erwine 2017), engineering (Krimm et al. 2017), archaeology (Hamilakis 2013; Mills 2014; Skeates and Day 2019), automotive design (Genuit, Fiebig, and Schulte-Fortkamp, Chap. 6; Sheller 2004; Schulte-Fortkamp and Dubois 2006; Genuit and Fiebig 2014), city planning (Brooks, Chap. 4; Lippold and Lawrence 2019), and healthcare (Lercher and Dzhambov, Chap. 9) have increasingly incorporated soundscape considerations into their processes and products directly, partially in response to requirements of policy makers. Another important force driving this transition is the recent sensory turn in the humanities, which has underscored the primacy of sensory experience in defining place. For instance, phenomenology and affect theory (Merleau-Ponty 2004; Gregg and Siegworth 2010) have expanded into fields such as architectural history and theory (Norberg-Schulz 1979; Pallasmaa 2005), historical studies (Smith 2007), anthropology (Classen 1997; Howes 2008), sociology (Hollstein 2011), and psychology (Berglund and Nilsson 2006), all of which tackle the complexities of human-made spaces. Newer fields such as sound studies, media studies, and history of science have also developed robust scholarship related specifically to sound and acoustics and provide strong contextual insights (see, for instance, Howes 2018; Lingold et al. 2018; Ouzounian 2020). As a result, a professional and scholarly environment has developed in which the individual's perceptual experience is increasingly valued, and forms the core of how any particular place is defined. Local experts can be the foundational voice.

Noise consultants and researchers also have folded the contextual grounding of soundscape into their work by incorporating contributions from local experts; related projects increasingly recognize that human perception of sound is a "multi-stage process" and cannot be understood without studying the context and meaning of sound (Schulte-Fortkamp and Nitsch 1999, quoting McAdams 1993). A special issue of Acta Acustica united with Acustica (*The Journal of the European Acoustics Association*) on soundscape and *The Journal of the Acoustical Society of America* (JASA) brought together scientists and practitioners in soundscape for further collaboration to provide guidance for taking the multidisciplinary approach on which soundscape had been built (Schulte-Fortkamp and Dubois 2006; Schulte-Fortkamp and Kang 2013).

A listener's perception of the acoustic environment is equally constructed through their auditory sensation, their interpretation of these sensations, and the broader responses they have to the acoustic environment. Brooks et al. (2014) summarized the evolution of soundscape research this way: "The soundscape concept was first introduced as an approach to rethink the evaluation of 'noise' and its effects on the quality of life. Now it has evolved into something much more. Soundscape suggests exploring all the sound in an environment in its complexity, ambivalence, meaning, and context. Basically, the soundscape concept considers the conditions and purposes of its production and perception. Consequently, it is necessary to understand that the evaluation of noise/sound is a holistic approach." (Brooks et al. 2014).

### *3.2.3 Introduction of ISO 12913 Series*

By the early 2000s, there was enough interest growing and a variety of approaches to soundscape being created that an international standard for terminology and research was required. In 2014, the first part of the soundscape standard was published by the International Organization for Standardization (ISO). Common terms and concepts were defined for all practitioners: soundscape was defined as the "acoustic environment as perceived or experienced and/or understood by people, in context." Harkening back to Schafer's earlier work, an essential distinction was made between a soundscape (how a human perceives an acoustic environment) and the acoustic environment, which is defined as "sound from all sound sources as modified by the environment" (ISO 2014, p. 1).

Using this conceptual framework (see Fig. 3.1), soundscape is the "process of perceiving or experiencing and/or understanding an acoustic environment, highlighting seven general concepts and their relationships: context, sound sources,

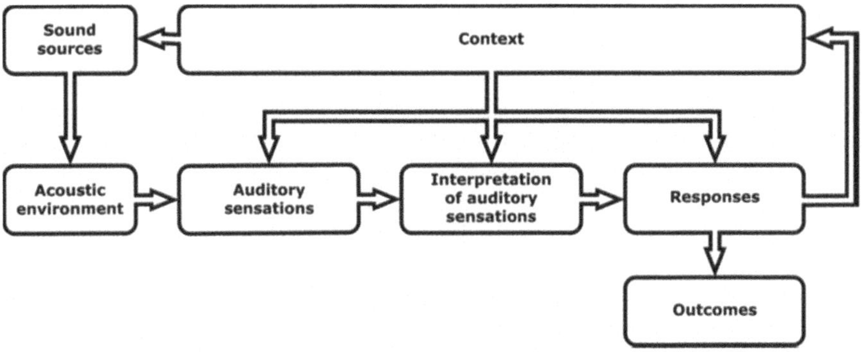


**Fig. 3.1** Diagram depicting the elements in the perceptual construct of soundscape. (From: ISO/FDIS 12931-1. Acoustics-Soundscape-Part 1: Definition and conceptual framework, April 2014 (5), p. 2)

acoustic environment, auditory sensation, interpretation of auditory sensation, responses, and outcomes" (ISO 2014, p. 1). "Context" in the standard refers to the physical place where the acoustic environment exists, but it also "includes the interrelationships between person and activity and place, in space and time and may influence soundscape through the auditory sensation, [...] the interpretation of auditory sensation, and [...] the responses to the acoustic environment" (ISO 2014, p. 1). These same distinctions will be used throughout this chapter.

As of 2022, the standard includes three approved parts and one under development:

1. ISO 12913-1 (ISO 2014) Acoustics-Soundscape-Part 1: Definition and conceptual framework. Part 1 lays out the basic definitions and the conceptual framework for soundscape studies and soundscape research.
2. ISO/ITS 12913-2 (ISO 2018) Acoustics-Soundscape Part 2: Data collection and reporting requirements. Part 2 provides many detailed descriptions of possible data collection methods and the requirements for reporting results according to the soundscape approach. Local expertise is emphasized.
3. ISO/ITS 12913-3 Part 3 (ISO 2019) offers detailed guidance on analyzing collected data related to a soundscape.
4. ISO/ITS 12913-4 Part 4 is under development at the time of writing. It will focus on how to use soundscape data and the results of specific analyses for assessing existing environmental noise situations and for determining interventions for use in urban planning and soundscape design. Primarily, this standard addresses the needs of soundscape experts (e.g., a consultant, designer, or researcher) as well as clients who are likely to commission a soundscape intervention.

Taken together, ISO 12913 Parts 1–4 present a standardized soundscape approach for conceptual, research, and design considerations alike. The ISO standards are designed to help researchers represent a specific location in its full complexity; thus, the diversity of methodologies suggested for soundscape documentation in Part 2 should come as no surprise (see ISO 2018). Universal interventions and implementation strategies for soundscape improvement simply are not possible, and such an assumption would run counter to the recognized soundscape definition in which adapting to the specific context is key. The standard makes clear that the study of soundscape first relies upon human perception and then those results should guide subsequent physical measurements.

A common approach to studying the soundscape has emerged in the application of descriptors and indicators, such as in soundscape surveys in which stakeholders are asked how closely a certain term describes the soundscape. That survey is followed by measurements recorded in the environment. Soundscape descriptors are measures of how people perceive the acoustic environment; soundscape indicators are measures used to predict the value of a soundscape descriptor. Descriptors and indicators have been applied so often that Part 2 of the ISO standard offers definitions and best practices for their utilization in soundscape research. Therein, a descriptor is defined as a "term which is used to describe the perception of any acoustic environment" (ISO 2018, 3.2). A commonly applied descriptor is annoyance. An indicator, on the other hand, is directly related to these descriptors and is "used to predict a descriptor or a part thereof" (ISO 2018, 3.4). Indicators for annoyance could include measured sound levels and sharpness, for example. The standard emphasizes that descriptors and indicators must be chosen based on the site under investigation, the stakeholders involved in the study, and the detailed context at play. Time variance should also be considered, as conditions and responses may change throughout a day or season (ISO 2018, 4.1). A survey of commonly applied descriptors and indicators applied in soundscape research by Aletta et al. (2016) pointed to the emergence of some commonly applied terms. This topic is increasingly being investigated due to the urgent need for context-sensitive operational tools, like predictive models, aimed at implementing the soundscape approach in urban planning and design (Brooks et al. 2014).

An additional requirement offered by the standard is the introduction of psychoacoustic indicators alongside acoustic ones. Studying both the sound sources and the auditory sensations they evoke is the only way that a fully representative and adequate description of the acoustic environment can be achieved (ISO 2018, 4.2). Classical acoustic indicators that are commonly found in noise management studies include continuous sound pressure levels and percentage exceedance levels. Their measurement and reporting are to be carried out according to ISO 1996-1:2016 (ISO 2016). In terms of psychoacoustic parameters, psychoacoustic loudness is emphasized as a minimum reporting requirement, but additional parameters such as tonality, sharpness, or roughness are also recommended. Some of these parameters have their own associated standards for determination and reporting, such as DIN 45692 for sharpness (DIN 2009). However, acoustic and psychoacoustic indicators describe only the sound and evoked auditory sensations. These indicators are not intended to explain the level of pleasantness or appropriateness of sound in its entirety (ISO 2018, p. 4) and should not be used as a substitute for qualitative soundscape research. Genuit, Fiebig, and Schulte-Fortkamp provide a more in-depth discussion on the application of psychoacoustic parameters in soundscape research in Genuit, Schulte-Fortkamp, and Fiebig, Chap. 6.

### *3.2.4 Comprehensive Evaluation and Measurement in Soundscape*

When conducting soundscape research, there are two types of assessments to differentiate. One is the engagement and assessment we carry out as individuals in our daily lives, for example, determining if a soundscape is too unpleasant or annoying to remain in it. The other type of assessment is a research-driven study of soundscapes and is the focus of the soundscape approach. The soundscape approach supplies a description of an acoustic environment that is gathered at the community level to achieve potential improvement of an acoustic environment, not simply to provide a description or a temporary engagement. The assessments discussed here are examples of research-driven studies.

To predict how people would perceive an acoustic environment, the underlying acoustic properties of soundscapes must be identified. Given the centrality of human perception to soundscape study, the individual acoustic environment must be evaluated, assessed, and measured in-person. Assessment through human perception is the central tenet of the soundscape approach and guides the way soundscapes should be measured. Measurements gathered from the perceptions and interpretations of individuals are best used for measuring complex soundscapes as these more holistically integrate quantities and qualities perceived by the human brain in the evaluation process of a soundscape (Berglund and Nilsson 2006).

Even though psychoacoustics offers concepts and tools that address human perceptions, tools such as psychoacoustic indicators cannot capture all relevant data to describe a soundscape. However, Genuit and Fiebig (2006) have used this logic and the importance of in-person human perceptions to position psychoacoustics as a vital component in the soundscape approach. Generally, psychoacoustics can provide a comprehensive understanding of sound quality through a detailed consideration of human signal analysis: how the human ear receives and initially processes a sound. The focus is on auditory perception in terms of hearing sensations (Fastl and Zwicker 2007). The initial reception is then interpreted by the individual through contextual influences in the environment and the personal background of the individual, which is where work with stakeholders proves essential. For instance, Genuit and Fiebig (2006) showed that it is possible to identify contributions to annoyance caused by environmental noise by using psychoacoustic parameters based on standardized procedures of measurement and analysis. Psychoacoustics offers promising tools and concepts for soundscape in its attention to predicting how people would perceive an acoustic environment, though psychoacoustic techniques and ecological acoustics need to be more tightly integrated to mediate between personal experience and group–area–society requirements and needs (see Genuit 2013; Genuit, Fiebig, and Schulte-Fortkamp, Chap. 6; ISO 2018, especially Annexes B and D).

A baseline tool of soundscape characterization through psychoacoustics is the use of binaural technology for recording the acoustic environment. Such recordings enable a listener to reexperience the acoustic environment in an aurally accurate way and mimic (though not replicate, since the listening context is different) in-situ human auditory sensations. To describe and analyze such noise measurements appropriately, psychoacoustic parameters covering several dimensions of basic auditory sensations must be applied (Genuit and Fiebig 2006).

Psychoacoustic indicators (loudness, sharpness, roughness, and fluctuation strength) are commonly used to characterize the acoustic environment (e.g., Genuit 2004). The loudness measurements conducted at Nauener Platz in Berlin clearly showed the difference in noise parameters and psychoacoustic indicators (see Sect. 3.4 and Schulte-Fortkamp et al. 2008), demonstrating how these two forms of characterization can and should be used in tandem. Masking techniques may be adopted by making use of psychoacoustic phenomena that are also explored in soundscape (Genuit and Fiebig 2006) by enhancing or introducing preferred sounds that will mask unwanted sound components or will divert the attention of the listener to other more pleasant sounds (Genuit, Fiebig, and Schulte-Fortkamp, Chap. 6). When new sounds are introduced, they should correlate with the place, human activities, and expectations so as to assure overall context coherence. Though the ISO standard offers guidance on indicator use, in practice there remains a significant gap between soundscape indicators as used by individuals (e.g., in an in situ participant survey) and instrument-based measurements by the same name (e.g., loudness). The use and understanding of named descriptors depends on the context (see the related discussion in Fiebig, Chap. 2).

Building on the possibilities of psychoacoustic analyses, Botteldooren et al. (2013) have developed a combined approach of measurement and evaluation that they term "triangulation" in which different perceptive views and measurement methods are combined. This method is promising for understanding how a soundscape works: it connects a deep understanding of the (soundscape) whole with technical measurements by using a more focused measurement approach via human participants and human-mimicking (psychoacoustic) physical measurements.

Additionally, there are aspects beyond sound variables that must be considered in any decision-making process seeking to improve the acoustic environment. For instance, visual, thermal, and general satisfaction with a place influences soundscape perception (Botteldooren et al. 2013). The professional expertise and lived experiences of people (the local experts) involved in any study also provide meaningful and essential information that relates to the place and the perception of it. Soundwalks offer an instrument for both exploring urban areas through the minds of local experts as well as incorporating the resulting specialized data into the triangulation process (see the related discussion in Brambilla and Fiebig, Chap. 7).

## 3.3 The Holistic Grounding of Soundscape

The previous discussion highlighted triangulation as a means of combining qualitative and quantitative data in soundscape research as well as multiple experiential and professional perspectives. From concept to data analyses, soundscape is inherently multidisciplinary and multifactorial with multiple perspectives, cultural factors, varying and variable research settings, and different priorities taken into consideration for research. Such a view includes the meaning of an acoustic environment, sound source characteristics, psychoacoustics, sound quality, and sound quality evaluations by individuals. Only a holistic approach has the ability to take stock of the variables and influences of so many factors, a reality that was acknowledged in the earliest days of soundscape research. Even so, soundscape has until recently remained in the orbit of noise research and mitigation efforts, which tended toward more limited methods of conceptualization and research. A review of soundscape's holistic development and the necessity for multidisciplinary research teams will be juxtaposed with noise research, and the ways these fields have approached related issues is presented in the following section.

### *3.3.1 Conceptual Background*

Soundscape is approached as a complex whole in research and analysis. Therefore, the ISO standard is focused on context and its relationship to the given circumstances, which is a holistic perspective that identifies the necessity for assessments of an acoustic environment from varying disciplinary methodologies. The ISO standard sets out a common toolbox for researchers to use in various settings but does not hardline a definition for any one soundscape approach.

Specifically, the standard does not provide a holistic definition of soundscape precisely because of the variety of possible soundscapes coupled with the variability in the perception of a soundscape among different individuals. For example, consider the many variables included in the definition provided by the adjacent field of animal bioacoustics by Erbe and Dent (2017): "Animal bioacoustics is a field of research that encompasses sound production and reception by animals, animal communication, biosonar, active and passive acoustic technologies for population monitoring, acoustic ecology, and the effects of noise on animals." With the addition of sociocultural, historical, and political factors to name only a few, a functional definition of the holistic human soundscape begins to feel ungraspable. Even without a formal definition of the holistic nature of soundscape as a research endeavor, much existing work has established the value of a holistic approach.

Since soundscape is fundamentally human perception, methodologies from different disciplines can provide new angles to understand the many facets of perception. Qualitative approaches have been drawn from the social sciences in particular, and their transformation in soundscape research creates powerful tools for a holistic perspective, such as questionnaires, interviews, recordings, and observer-based data collection. Holistic science served as an important conceptual touchstone in its emphasis on the study of complex systems, which are best understood in context and in relation to one another as much as the whole. However, holistic concepts related to perception have also been productively adopted from psychology.

Herranz-Pascual et al. (2010) linked the field back to psychology, observing that "the study of environmental perception has its roots in the most important psychological traditions, particularly in the holistic concepts that define the Gestalt. The key idea of the Gestalt, following the dynamic systems approach, is that everything is more than the sum of the parts" (Herranz-Pascual et al. 2010, p. 3). Gestalt theory makes a strong case for a soundscape as more than the sum of its sounds, and the psychological understandings of human perceptions more broadly are applicable to soundscape work in many ways.

Berglund and Nilsson (2006) defined four categories of soundscapes stemming from psychological theories: (1) perceived affective quality, (2) restorativeness, (3) overall perceived quality, and (4) appropriateness. Basturk et al. (2012) emphasized that research into soundscapes encompasses the entire context of the physical and psychosocial environment of the setting and the participants. In that context, psychology-based research tools would be necessary to evaluate such a nested condition. Basturk et al. (2012) identified listening walks as an important tool of investigation, which has its origins in Schafer's early research (see Sect. 3.2.1) but draws from the participatory research methods of social sciences as well. Vaughn and Jacquez (2020) further described participatory research as emphasizing direct engagement of local priorities and perspectives. Cargo and Mercer (2008, p. 14) made the important point that "participatory research can be defined as an umbrella term for research designs, methods, and frameworks that use systematic inquiry in direct collaboration with those affected by the issue being studied for the purpose of action or change." The foundational premise of participatory research methods is the value placed on genuine and meaningful participation—methods that offer "the ability to speak up, to participate, to experience oneself and be experienced as a person with the right to express yourself and to have the expression valued by others" (Abma et al. 2019, p. 127).

Returning to a soundscape's composition, the identification of its many components and the human responses to them poses a significant challenge that researchers have approached from different disciplinary angles. Kull (2006) sought a classification system for any acoustic environment by offering a scale of expected acoustic characteristics in various environments, ranging from urban to natural settings. Though every soundscape is unique, the entire acoustic environment is composed of natural and human-made sounds. Kull's system would capture their origin while allowing a description of how the contributions of sound sources vary as well as the role of human expectation in environmental assessment. Lercher and Schulte-Fortkamp (2003) expanded the scope of sound source considerations to include contributions from local geography, climate, wind, water, people, buildings, and animals, which requires a much broader disciplinary perspective. Soundscapes do more than simply describe the sound level or audibility of ambient and intrusive sounds from natural or human-made sources alike. The individual's response to a soundscape was found to be key in describing the components of any soundscape.

Clearly, the meaning a listener ascribes to sounds, the composition of diverse sound sources, and the listener's attitude and expectations about the acoustic environment are of primary importance in soundscape work. Moreover, the lifestyle and societal norms of an individual are also imperative contributions to one's expectations in an environment; thus, a consideration of participant sociocultural background is important to understand the assessment of any acoustic environment as well (Schulte-Fortkamp 1994b). With such strong influences from multiple sources, the measurement of perception in an acoustic environment is a heterogeneous field of research.

To address the many facets of soundscape, researchers must be versed with psychosocial and sociology-based research methods. These methods include different forms of observation, interviewing techniques with a low level of standardization (such as open-ended, unstructured interviews, partially or semistructured interviews, guided or narrative interviews), and the collection of documents or archival records (e.g., from libraries, public repositories, or private collections). Despite their differences, such qualitative approaches share common territory within the *interpretive paradigm* of social reality. This model of knowledge, originating from the social sciences, is concerned especially with the meaning that one ascribes to experience, and thus it is fundamentally applicable to studying the individual's response to soundscape. In addition, an individual's reality is dynamic and based on their subjective experience of the world around them.

Hollstein (2011) discussed different qualitative research methodologies compared with quantitative research methods that are particularly relevant for related soundscape work. She emphasizes, for instance, that social reality is always meaningful: an individual is constantly organizing his/her actions in response to this meaning. The meaning of a soundscape does not exist apart from its context or specific frame of reference for an individual (Hollstein 2011). Moreover, short-term responses (emotionally as well as behaviorally) might change the context and ultimately influence long-term actions. The ISO standard describes these as "an overall, long-term consequence facilitated or enabled by the acoustic environment" (ISO 2014, p. 3).

Data collection and analysis are embedded in the interpretive paradigm of the soundscape approach and must be calibrated appropriately in terms of research motivation and the participants chosen for study: Marsden summarized the issue well: "Several aspects need to be considered in choosing the method of data collection. First of all, it needs to be clarified what aspects of social relations will be studied and how relations and networks are to be theoretically conceptualized." (Marsden 1990). A key question in this respect is whether the research will be concerned with existing relations (e.g., network practices) or with participants' perceptions of such relations (e.g., network orientations and assessments) (Hollstein 2011, p. 410).

### *3.3.2 Pioneering Soundscape Collaborations Across Acoustic Research Fields*

Though soundscape research is inherently complex, there have been an array of projects that investigated soundscape using a multidisciplinary—not to mention multinational—perspective that are important to note. The EU European Cooperation in Science and Technology (COST) Action TD0804 on Soundscape of European Cities and Landscapes was a prominent project with the goal of "providing the underpinning science and practical guidance in soundscape" (COST 2009, p. 1). The project created a vibrant and productive international network of 52 participants from 23 COST countries and 10 participants from outside of Europe, who have continued research after the conclusion of the initial COST project. The COST Action served as the first training ground for more than 150 young researchers and practitioners through 14 Short-Term Scientific Missions (STSM). Five training schools were also associated with the Action, including three Soundscape and Psychoacoustic Training Schools in Aachen, Germany between 2010 and 2012, a 2010 Summer School on Soundscape in Ljubljana, Slovenia, and the 2013 Winter School on Soundscape in Merano, Italy. The Action has gone on to influence both national and international policy and practice through involvement with and support of international and national policy bodies such as the EU and UK Department for Environment, Food and Rural Affairs (DEFRA). The Action also stimulated a number of soundscape projects in Sweden, Germany, the United Kingdom, Spain, Belgium, and Portugal, including the Brighton White Night project (Lavia et al. 2016a).

Some of the most important impacts of the COST Action have been in promoting health and sustainability, attracting investment, conveying cultural uniqueness and diversity, and enhancing quality of life through increased awareness of the importance of soundscapes in daily life. One important topic was identifying the links between the perception of imissions and adverse health effects, which can be seen in Fig. 3.2. In addition, there have been collaborations within other networks, including the Global Sustainable Soundscape Network (GSSN) funded by the US National Science Foundation, four other COST Actions, several EU projects including Holistic and Sustainable Abatement of Noise by Optimized Combinations of Natural and Artificial Means (HOSSANA), The Urban Sound Planner (SONORUS), a number of additional EU networks such as the European Network on Noise and Health (ENNAH), and national networks such as UK Noise Future network.

Another important result from the COST Action was the identification of three types of soundscape studies. This classification clarified the relationships and potential benefits from the involvement of different stakeholders, specialists, and other individuals (Kang et al. 2013; Lercher and Schulte-Fortkamp 2013); levels can be seen in Fig. 3.3 as described by Lercher and Schulte-Fortkamp (2013).

Type I is strictly related to individual experiences and preferences; Type I studies are dominant in the field at the time of writing. Type II studies are based on collective assessments and group understanding; these are less common and mostly involve too few participants to contribute significantly to research knowledge. Type III studies relate to soundscape applications primarily for planning and political actions. While Type II studies could also serve this purpose, they have not yet been integrated into Environmental Health Impact Assessments (EHIA) or action plans triggered by the END at the time of writing. The different types of studies do not have to be pursued in isolation. Soundscape assessments could be fostered by funding Type I and II or Type III studies together as "tandem studies" to gain broader

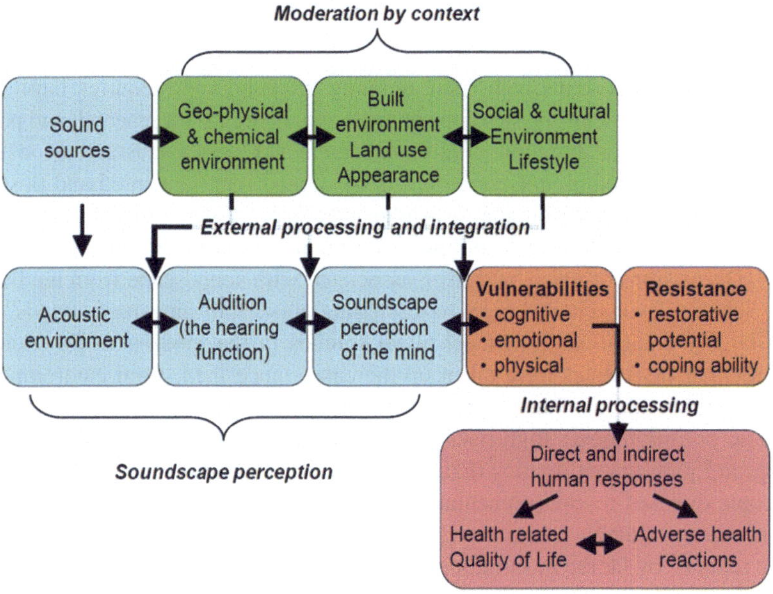


**Fig. 3.2** Integrated diagram depicting multilevel analysis of causal links between perception of imission and adverse health effects. From: Lercher and Schulte-Fortkamp (2013)

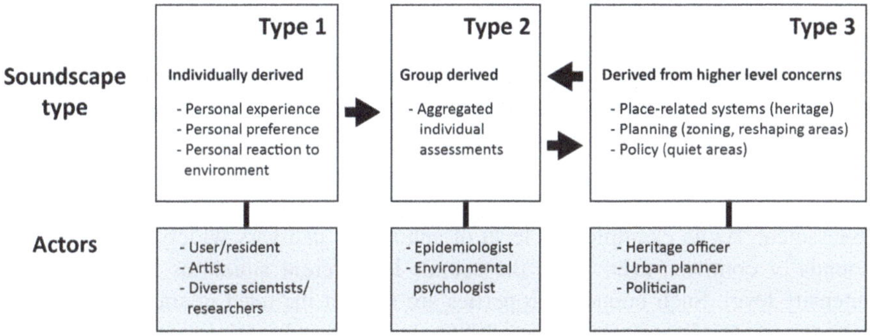


**Fig. 3.3** Diagram depicting the types of soundscape studies and their main actors. From: Lercher and Schulte-Fortkamp (2013)

insight for judgments and decisions in conservation, planning, and reshaping residential and restoration areas (Lercher and Schulte-Fortkamp 2013, p. 120).

There are many additional examples of soundscape studies applied to specific research priorities. For further projects and discussion related to soundscape and health, see Lercher and Dzhambov, Chap. 9; for those related to architecture, see Siebein and Siebein, Chap. 5; and for those related to urban planning, see Brooks, Chap. 4.

### 3.3.3 Noise Management and Soundscape

Understanding the origins of holistic thinking in soundscape requires familiarity with noise management. As one of the most dominant arenas for research and policy related to soundscape, the topic of noise management requires substantial consideration. Broadly speaking, noise research is concerned with unwanted and possibly harmful sound in an environment. Noise management "seeks to maintain low noise exposures, such that human health and well-being are protected" (Berglund et al. 1999). Discussions of noise have been associated with soundscape from the beginning. Schafer devoted many pages to discussing noise, the "unwanted sounds" that were largely emanating from mechanical sources in the modern world (Schafer 1977). Other disciplines have taken up the cause since then, from environmental management to city planning, transportation policy, and health regulations.

Noise pollution poses high risks to human health. A briefing developed by the European Environment Agency (EEA) presented updated estimates of the numbers of people exposed to environmental noise pollution in Europe. It also provided a new summary of the measures being used in Member States to manage noise. Road traffic remains by far the most important source of environmental noise: at least 100 million people in the EU are exposed to levels of traffic noise that exceed the European Union's indicator of noise annoyance (EEA 2017).

In addition, studies have shown that today most people are typically exposed to environmental noise levels between 50 and 60 dB, averaged over 24 h (or Lden; see Lercher and Dzhambov, Chap. 9 for a detailed explanation of this Day, Evening, and Night noise level measurement). Noise has only broadened as a concern since Schafer (1977) initially addressed the issue. Such noise measurements are typically carried out through quantitative measurement campaigns according to minimum code requirements; however, typically only the accepted upper health limits for exposure are considered.

Perception and context must be key considerations beyond decibel levels in noise assessment. As an example, the level of annoyance can vary widely with different sounds or contexts, even when the sounds in different situations have the same intensity level. Such nuanced properties are not yet included in standard acoustic measurements (Genuit and Fiebig 2006). In EHIA, taking into account only the acceptable upper limits for sound exposure has led to an attitude by administrations and policy makers that noise sources can be added to the maximum sound level allowed. During recent decades, this attitude has resulted in undesirable noise exposure spreading from urban centers into suburban and rural areas through development and transportation projects. In addition, the times of undesirable exposure have spread from daytime into night times. The available options are steadily diminishing for any restoration of consistent daily quiet time for rest and relaxation, and there are increasing concerns for human health and environmental quality of life as a result.

One of the evident limiting factors of simple noise mitigation efforts is the non-holistic approach, which is illustrated in recent research into the relationship between sound exposure, sleep, and health. The understanding of the impact of noise exposure on sleep stems mainly from experimental research in controlled environments. Field studies conducted with people in their normal living situations are scarce, thus ignoring important contextual information. Most field research on sleep disturbance has been conducted for aircraft noise alone (Horne et al. 1994; Passchier-Vermeer 2001). Some field studies have examined the effects of road traffic and railway noise as well (Griefahn and Spreng 2004).

A highly related study that incorporated a more holistic approach was the Noise-Related Annoyance, Cognition, and Health study (NORAH). Published in 2016, NORAH was one of the most important and well-funded studies conducted at the time, yet it was not able to confirm expectations regarding the noise burden from different sources that included aircraft, traffic, or railway noise. The most important result was simply to recognize that more investigations are needed to understand the reason for permanent disturbance through noise (UBA 2016). Ultimately, support for large, holistic noise studies has proven to be difficult to obtain. Further reflection on noise and health measurement is discussed in Lercher and Dzhambov, Chap. 9. Additional connections between psychoacoustics and noise research are discussed in Genuit, Fiebig, and Schulte-Fortkamp, Chap. 6.

#### 3.3.3.1 Noise Management

Because of the limited scope of recent noise management approaches, outputs such as strategy papers, guidelines, and directives have advocated for more perception-oriented and repeatable assessment procedures in future work. One major objective of these efforts is the protection of quiet areas (e.g., residential and hospitals areas) and at sensitive times (e.g., average sleeping hours). The soundscape approach has been introduced through hundreds of publications and an impressive number of international and European projects designed to improve the platform for future noise policy along these lines. Some key examples are reviewed here.

Noise control measures and strategies are used to reduce or eliminate unwanted sound when and where possible. Several policy mechanisms have sought to address noise and its impacts on populations, particularly in Europe. For instance, in April 2017 at the Noise Conference in Brussels, the World Health Organization (WHO) outlined the latest findings on health implications related to noise; additionally, the European Commission presented findings concerning the evaluation of the EU Directive aimed at reducing noise pollution in the Union. While addressing the implementation of the END in accordance with Article 11 of directive 2002/49/EC, the report of the Commission to the European parliament and the council pointed out that noise pollution continues to constitute a major environmental health problem in Europe (EC 2011).

According to the WHO (2018), noise pollution leads to a disease burden that is second only to air pollution among the environment-related causes in Europe (EC 2011). The EEA further stated: "What we learn from two rounds of noise mapping assessments implemented in accordance with the END is that road traffic noise, both inside and outside urban areas, is the most dominant source affecting human exposure above the action levels defined by the END. The impacts and affects resulting from this noise exposure vary depending on which levels the population is exposed to." (EEA 2014). The official discussion concerning the new format of the END left more flexibility for conducting or interpreting soundscape studies: "Exposure to excessive noise can also be addressed via urban planning policy, as suggested in the 7th Environmental Action Program. While competence in this area lies with the Member States, the Commission will stimulate and encourage activities to mitigate excessive noise in urban areas, for example by facilitating the exchange of good practices, as well as supporting research and innovation in this field." (EC 2011).

The European Commission recognized that the most dominant source of noise affecting people was road traffic noise and acknowledged the need for a deeper understanding of how people react to noise from sources beyond road traffic (EPC 2002; EC 2011; EEA 2017). While research in noise control has provided information about health issues related to noise, no guidance has been formally provided for affected people on how to deal with the noise burden more broadly. Fidell (2003, p. 5) summarized the issues as follows: "[N]o systematic explanations are available for large differences in annoyance prevalence rates in different communities with the same noise exposure [...] accurate predictions of the prevalence of annoyance in communities exposed to change in noise levels [...] remain elusive...[T]he prediction of benefits from costly measures intended to mitigate noise exposure cannot be made with confidence, and [...] regulatory policies intended to balance conflicting societal interests remain largely arbitrary and poorly supported by technical analysis."

Fidell's warning opens the discussion for further approaches regarding annoyance measurements and noise controls. Clearly, research restricted to laboratory settings will not give representational guidance to inform needed innovations in a community. A holistic approach is necessary and the interface between noise management and soundscape should be explored urgently. When it comes to noise management and improving the soundscape for local residents, one possibility would be a sensitive merging of soundscape and community noise assessment frameworks, both of which aim to capture a full understanding of a specific location.

#### 3.3.3.2 Common Interests Between Noise and Soundscape Assessment

One of the most significant distinctions to draw between noise and soundscape is how each treats the sounds studied. Noise concentrates on unwanted sound, making noise a subjective term (Schafer 1977, p. 273). In contrast, the soundscape approach treats all sound in an environment as resource rather than waste. Despite the difference, quality of life is fundamental to both fields with the goal of improving acoustic environments. EU Directive 2002/49/EC introduced a huge leap in recognizing the impacts of noise on populations as well as ways to mitigate its effects: "At EU level, Directive 2002/49/EC relating to the assessment and management of environmental noise is the key legislative instrument for protecting citizens from excessive noise pollution caused by road, rail and airport traffic, as well as by large industrial installations. Its purpose is twofold: (1) to define a common approach intended to avoid, prevent or reduce the harmful effects of environmental noise and (2) to provide a basis for developing measures to reduce noise emitted by the major sources" (EC 2011).

Following EU Directive 2002, noise maps and action plans most often shape common noise protection policies today. Unfortunately, predictions resulting from these mechanisms very often fail to define the real burden caused by noise and fail to consider the history of that problem. In fact, the insights provided by the community, the experience and knowledge that residents provide for the action plans, should be a primary assessment criterion for noise impacts. How to underpin community-based strategies successfully and respond to local needs remains unresolved. Recognition that community perception has at least equal relevance to physical measurements is only a first step, but a significant one, for researchers and policy makers when it comes to urban and community planning.

With similar motivations to this directive, the ISO 12913 series was developed in order to enable a broad international consensus on soundscape and to provide a foundation for communication across disciplines and professions with an interest in soundscape. Both initiatives seek to enhance quality of life and to bring the perception and appraisal of local individuals into the process of change. Both approaches focus on enhancing life quality as a major concern.

What falls under "life quality" can be quite broad, as Torigoe and Nagahata have demonstrated. Concerning the 2011 earthquake in Japan, they argued that communities could be better prepared for natural disasters by integrating acoustic warning systems. They also considered how humans contribute to devastation when they fail to think through the delicate balance of the acoustic environment affected by human-made constructions (Torigoe 2012; Nagahata 2012). Furthermore, regarding environmental shifts, pollution, and disasters, they discussed local and global perspectives on the value of archiving and appreciating acoustical environments that are often taken for granted. Their hope was to make soundscape an integral and consistent part of contemporary society, especially during a time when everyday auditory quality, particularly related to airport noise, has become a major issue of public debate (Torigoe 2012). Their ambition was boosted by the finding that access to high-quality acoustic environments may positively affect well-being, quality of life, and environmental health through some restorative or health and well-being promoting mechanism (Lavia et al. 2016b; van Kamp et al. 2016; WHO 2018).

Soundscape restoration has been offered as a possible approach to promote well-being and is a topic for which soundscape and noise share a great deal of common ground. Two types of restoration were discerned as part of the COST Action discussed earlier (Lercher and Schulte-Fortkamp 2013): Type 1 refers to the restorative effect of having direct access to a high-quality acoustic environment; Type 2 refers to the effect that availability (knowledge) of a high (better) quality acoustic environment has on a person who otherwise is subject to adverse effects of noise. Type 2 also includes the availability of a quiet place or access to nearby green areas. Epidemiologic evidence is limited on the intrinsic positive value of areas with high acoustic quality such as green areas, wilderness, or water; yet for restoration by way of mediation, several studies (e.g., de Kluizenaar et al. 2013) have shown that access to quiet near one's home can reduce annoyance at home and also have a beneficial effect on sleep quality and blood pressure. Temporary respite from exposure to unwanted environmental noise at home can mitigate the negative effects of home-centered noise on health and well-being. Different features of the immediate physical environment work together. For example, access to green space in the immediate vicinity of one's dwelling can contribute to a sense of quiet as much as having a quiet side of a dwelling. A need for quiet space in the neighborhood has also been found for those who live under noisy conditions (e.g., high traffic noise equivalent levels) as well as those who are noise sensitive (Booi and van den Berg 2012). There is still a need, and thus an opportunity, for soundscape research to advance our understanding of the process by which these different mechanisms operate.

Meanwhile, applications of the soundscape approach have been used for evaluations of cities and for city planning to address noise impacts on well-being. For instance, the need to improve acoustic environments in cities has led to increased interest in correcting or minimizing noise pollution in urban environments (Herranz-Pascual et al. 2010). Within noise management, the soundscape approach can help to bridge the needs of the people with noise measurement values. Effectively and sustainably reducing the number of people highly annoyed by noise is only possible with further scientific research into developing appropriate methods and assessing the effects of noise. Wherever noise maps are drawn up and action planning is proposed, the soundscape approach can offer support for reaching informed decisions that promote soundscape quality and provide long-term improvements to the quality of life. Noise maps employed within the soundscape approach can help to obtain a deeper understanding of noise reactions and to reliably identify perception-related hot spots. Genuit and Fiebig (2006) have additionally proffered psychoacoustic maps as particularly promising in areas where the noise levels are marginal below the noise level limits, offering additional help for interpretation and identification of required noise abatement measures.

### *3.3.4 Multidisciplinary Case Studies*

Fiebig and Schulte-Fortkamp, Chap. 11 lay out a series of important soundscape case studies for reference, but a few examples will be mentioned here for their incorporation of fundamentally holistic approaches that are also examples of acoustic design. A trail-blazing example of collaboration and multiparty integration is the soundscape project of "Nauener Platz" in Berlin, Germany. The project "Nauener Platz—Remodeling for Young and Old" was part of the research program "Experimental Housing and Urban Development (ExWoSt)" of the Federal Ministry of Transport, Building, and Urban Affairs (BMVBS) (Schulte-Fortkamp et al. 2008). The existing urban park on Nauener Platz was slated for a community-focused redesign with the goal of improving the soundscape of the traffic-burdened area. Community involvement was integrated at every stage of the redesign project and relied on residents as the experts in evaluation procedures. Their knowledge, based on their extensive experience coupled with their daily expectations for the Nauener Platz, was recognized as one of the most important resources available. Therefore, residents were involved in the entire redevelopment project, from conceptual brainstorming to construction feedback, through different engagement approaches. Public hearings were held about the intention to redesign the place, and workshops were organized to gain access to the different social groups for their individual expectations. Attention was paid to gender and age, and collaboration was intentionally interdisciplinary. Soundwalks enabled the attendees of several working groups to develop a certain social bond through the development of "their" new place while also providing unparalleled insights. A diverse series of design strategies were developed with community input, from zoning different activities across the site according to soundscape cues to the introduction of noise-mitigating barriers and the installation of new sound sources played in custom-designed pods. Moreover, the project served as a training ground for researchers in qualitative soundscape research in community settings. The project was awarded the 2012 European Soundscape Award by the EEA and the UK Noise Abatement Society. Over 10 years later, the park is still being used as planned.

In contrast to the local scale of Nauener Platz, a citywide approach to noise interventions using soundscape principles was laid out in Brighton and Hove in the United Kingdom (Easteal et al. 2014). There, the City Council and The Noise Abatement Society worked together on a series of demonstration projects (Lavia et al. 2016b), including West Street Story, West Street Tunnel, and Valley Gardens, to tackle the challenge of noise control during nighttime hours. The West Street Story project was the first night noise soundscape intervention pilot study: a three-dimensional, curated, ambient audio installation that was inserted into a clubbing district. The project resulted in better crowd behavior and reduced the need for a police presence in the area, which was confirmed through observational and body language analysis on-site as well as through video footage. The West Street Tunnel project was a follow-up experiment in a pedestrian subway, which had been closed due to antisocial behavior and noise. Curated sounds were added to the tunnel, and they proved helpful in minimizing public disorder and increasing feelings of safety among those passing through the tunnel. In the Valley Gardens project, different kinds of soundscape analyses were conducted in the area to incorporate considerations of the acoustic environment into broader urban regeneration efforts. The interventions from these three projects helped residents to feel safer, suffer less from noise pollution, and experience increased social cohesion through citywide collaboration (Kang et al. 2016b).

As a final example, "The Soundscape Approach for Early-Stage Urban Planning: A Case Study" demonstrated relevant and replicable methods by which planners and urban designers created the potential for good soundscapes in areas undergoing development (De Coensel et al. 2010). The focus was on stakeholder involvement at an early stage to guarantee the participation of all involved stakeholders and respected measures. This too was a commissioned project. In contrast to the others described, participants were invited rather than participation being open to the general public. The results were used to inform the conceptual design phase of the planning project, illustrating the means through which soundscapes can be enriched in practical and self-sustaining ways (Lavia et al. 2016b, p. 274).

## 3.4 Soundscape as a Paradigm Shift in Noise Control

Soundscape research represents a paradigm shift in noise evaluation through its interdisciplinary and holistic approach to acoustic environments. An important distinction between the two fields is their different focus for outcomes (Brown et al. 2016). An effective means to see this shift is through a consideration of soundscape in relation to environmental experience (Schulte-Fortkamp 2010). Environmental experience originated as a human-centered, more holistic approach in environmental studies and sociology (De Coensel et al. 2010; Herranz-Pascual et al. 2010) and has shown promise when applied to soundscapes as well.

### *3.4.1 Soundscape Applied to Evaluating Environmental Experience*

There is potential for soundscape research to contribute directly to environmental assessments; so far, however, soundscape studies primarily have been conducted at small scales (e.g., parks and gardens), and triangulation has not yet been fully exploited as a tool of investigation (Botteldooren et al. 2016). A related, larger scaled project that focused on environmental experience specifically also sought to further disseminate the soundscape concept and the soundscape approach in general practice with US national parks. In 2009, local actors and stakeholders in communities, parks, and wildernesses were advised to consider sociocultural, aesthetic, and economic effects when evaluating the impact of human-made sounds on both the natural soundscape and visitor experience (Fristrup 2009). This project clearly demonstrated how soundscape research can be employed toward environmental assessments from the start.

A critical element to such success is the establishment of robust and effective communication between stakeholders and researchers from the beginning (De Coensel et al. 2010; Schulte-Fortkamp 2019). Such a platform was critical to the success of the Nauener Platz project. The stakeholders had meetings once a month to discuss the project and their environmental experiences. A platform for communication can be a facilitated discussion, a scheduled meeting, a workshop, online meetings, or informal community events. The method of communication depends on the context of the community and what would work best for its members. In addition to the involvement of different disciplines, it is important to define areas that will build the basis for communication through further development, such as through economics, noise policy standards, combined effects, common protocols, cross-cultural studies, education about soundscape, combined measurement procedures, and perceptive parameters that include the characteristics of sounds and cross-cultural questionnaires.

The ongoing approach to the standardization of soundscapes and for psychoacoustics and noise management has provided big steps toward comprehensively evaluating environmental experience and ultimately enhancing quality of life (Botteldooren et al. 2013). Transforming public open places into cocreated spaces, for instance, requires the integration of multiple sociocultural contexts, multistakeholder perspectives and the diversity of needs, incentives for the guaranteed participation of different groups, and assorted cooperation capabilities (Mačiulienė et al. 2018). The soundscape approach has already mapped a way to undertake such work.

### 3.4.2 Environmental Experience as a Conceptual Model for Soundscape

The original conceptual model of environmental experience proposed by Herranz-Pascual et al. (2010) defined the leading factors in judging environmental experience (Fig. 3.4). Critically, the model bridged experience in both noise evaluations and soundscape approaches with great promise for soundscape applications.

According to Herranz-Pascual et al. (2010), there are multidimensional patterns that focus on activity and environmental experience via perception and valuation. These can be grouped into the following categories: community and person, person–place interaction, and place. This structure conclusively meets the central premise of the soundscape approach, which concentrates attention on limited areas where residents can offer their expertise. Related research utilizing the soundscape approach has made an important contribution in changing reactions to noise from different sources, as Lavia et al. (2016b, p. 248) found in their soundscape project: "By discussing possible options for soundscape management and design with stakeholders (such as residents, citizen groups, or transport authorities), planning technicians (architects, engineers, urban planners, consultants involved), and

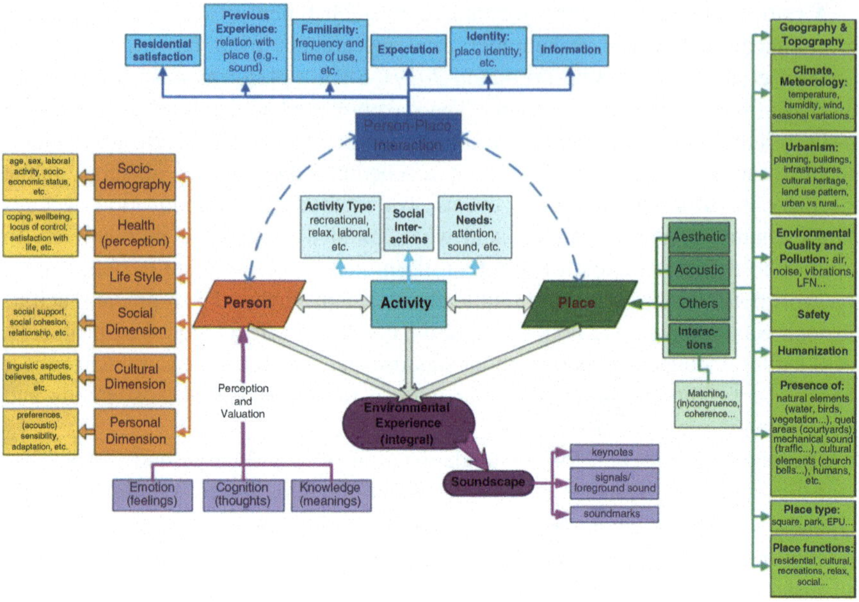


**Fig. 3.4** Proposal of a conceptual model about environmental experience. (From: Herranz-Pascual et al. (2010))

decision makers (local authorities, for example), light might be shed on the best applicable solutions and on the user's expectations."

As a natural extension from environmental experience research, careful selection of sites for soundscape surveys is essential. Locations for soundscape study generally are one of two types: preselected for research by mechanisms such as governmental taskforces or neighborhood associations with a vested interest in the findings, or freely selected by the researchers themselves for a myriad of reasons. For instance, the study at Nauener Platz was commissioned by the German BMVBS for the express purpose of improving the design solutions for the Nauener Platz through community involvement. A nearby project at the Berlin Wall Memorial, on the other hand, was an example of research questions driving the location of study: the research site was carefully selected due to its specific soundscape characteristics, their possible relationship to the Berlin Wall and its neighboring community through time, and the availability of various stakeholders to participate (Jordan 2019, 2021). Soundscape research mostly falls into the second, freely selected category.

Given the evident potential to improve quality of life for stakeholders, further effort should be made to integrate approaches from adjacent efforts into soundscape work, such as multisectorial EHIA, with attention to sustainable development, environmental zoning, citizen involvement, and the preservation of quiet areas. The totality of soundscape must be distinguished from the limited notion of a quiet zone, however (Kang et al. 2016b). Consideration of "sensitive areas" and the design of "supportive environments" requires new insights into existing annoyance data and new integrative research strategies. There is a common consensus on the necessity for additional research parameters for soundscapes beyond A-weighted sound pressure level measurements: psychoacoustic parameters can greatly contribute to effective assessments of environmental sound. An integrated evaluation procedure is needed as well, pushing soundscape beyond simple noise level reductions by accounting for the concerns and well-being of the community. Since soundscape is meant as an intervention to enhance the quality of acoustic environments and living situations, greater awareness of appropriate evaluation procedures must be fostered (Torresin et al. 2019). The development of an environmental experience model could be a valuable addition to the many possible integrated evaluation procedures.

## 3.5 Summary

The significant expansion of soundscape studies in the past decades has created a rich body of work with diverse approaches. An important differentiation among these studies is between commissioned soundscape projects versus those freely chosen by researchers. A distinction in research must also be drawn between studies that examine existing conditions and their supporting sounds as an end-goal of analysis and those studies that seek to intervene in an existing soundscape with the goal of long-term improvement. Participant engagement is also an important consideration: some projects directly involve the people concerned in the soundscape process (i.e., by sharing their local expertise, their stories, their expectations), while other projects scrutinize existing areas without directly involving locals in the same ways.

No matter the category of approach in the research, participatory research methods in soundscape work are clearly essential to achieve two goals: knowledge and real-world action realized in a democratic way. Vaughn and Jacquez (2020) have eloquently pointed out that "such engagement allows research to benefit from the collective wisdom of both researchers and communities which in turn creates more meaningful findings translated to action." Abma et al. (2019, p. 127) also confirmed this: "The foundational premise of participatory research methods is the value placed on genuine and meaningful participation—methods that offer the ability to speak up, to participate, to experience oneself and be experienced as a person with the right to express yourself and to have the expression valued by others."

With the participatory methods of soundscape supported to such an extent, and with the depth of research already conducted, it is clear that the soundscape concept and the soundscape approach are becoming firmly entrenched within policy and design applications. It is important that the respective local actors and stakeholders are properly advised on sociocultural, aesthetic, and economic effects in addition to the use of more recognized measurement strategies. Wide-ranging discussions have revealed the need for a platform for stakeholders to communicate and share common decisions as the field goes forward (Kang and Schulte-Fortkamp 2016). This is especially true today because many disciplines are already drawing from or collaborating with soundscape research. For instance, it will be necessary to connect soundscape to smart growth principles that are relevant to new urbanism and addressing the needs of citizens in modern cities. Soundscape analysis and application has proven to be a successful method of providing an improved acoustical environment for urban dwellers. The intersection and similarities between soundscape goals and urban smart-growth principles serves as an endorsement of soundscape's ability to address a significant portion of the smart growth agenda while balancing technical innovations with environmental protection. This is just one example of how holistic thinking through a soundscape approach is positioned for expanded relevance across the built environment.

At its most essential, an acoustic environment is constructed in its entirety by the people who use it and their interactions with the space. Yet the soundscape is composed of these interactions alongside how the acoustic environment is perceived—it must remain clear that soundscape involves human perception and not simply physical measurement. Thus, the strategies used to understand any particular soundscape must be adapted to these perceptual and conditional singularities. The soundscape approach is holistic in accounting for both the components within and the interrelationships between a given context and the people concerned.

The term soundscape has entered general parlance due to the expansion of related research efforts that include community planners, designers, laypersons, and even those primarily interested in management of the acoustic environment through environmental noise control. Though noise and soundscape share some points of origin, the general acknowledgment that soundscape is a construct of human perception clearly puts soundscape in a different area than noise control. In fact, soundscape research represents a paradigm shift from noise control policies toward a new, multidisciplinary approach that focuses on perception. The multidisciplinary approach involves cooperation between the humanities and social sciences to address the diversity of soundscapes across cultures with more attention paid to how people experience acoustic environments. Therefore, it is clear that modeling or analyzing soundscape dependencies with the built environment requires consideration of the sensitivities, visual aesthetics, and geography as well as social, psychological, and cultural aspects.

Soundscape standardization and the available standards with regard to community noise in psychoacoustics have provided a big step toward enhancing the quality of life for people in a variety of contexts. Although there are many opportunities for synthesizing soundscape research with other fields, much more waits to be done to improve the health of communities and the acoustic environment in developed areas.

**Compliance with Ethics Requirements** Brigitte Schulte-Fortkamp declares that she has no conflict of interest.

Pamela Jordan declares that she has no conflict of interest.

## References

Abma T, Banks S, Cook T, Dias S et al (2019) Participatory research for health and social well-being. Springer Nature Switzerland AG, Cham

Aletta F, Kang J, Axelsson O (2016) Soundscape descriptors and a conceptual framework for developing predictive soundscape models. Landsc Urban Plan 149:65–74

Basturk S, Maffei L, Masullo M (2012) Soundscape approach for a holistic urban design. In: Proceedings of the Association of European Schools of Planning (AESOP) 26th Annual Congress, pp. 1–3

Behrendt F (2018) Soundwalking. In: Bull M (ed) Routledge companion to sound studies. Routledge, New York City, pp 249–257

Berglund B, Lindvall T, Schwela DH (eds) (1999) Guidelines for community noise. World Health Organization, Geneva

Berglund B, Nilsson ME (2006) On a tool for measuring soundscape quality in urban residential areas. Acta Acust United Acust 92(6):938–944

Booi H, van den Berg F (2012) Quiet areas and the need for quietness in Amsterdam. Int J Environ Res Public Health 9(4):1030–1050

Botteldooren D, Andringa T, Aspuru I, Brown L et al (2013) Soundscape of European cities and landscape – understanding and exchanging. In: Kang J, Chourmouziadou K, Sakantamis K, Wang B, Hao Y (eds) COST action: TD-0804 – soundscape of European cities and landscapes. Soundscape-COST, Oxford, pp 36–43

Botteldooren D, Andringa T, Aspuru I, Brown AL et al (2016) From sonic environment to soundscape. In: Kang J, Schulte-Fortkamp B (eds) Soundscape and the built environment. CRC Press, Boca Raton, pp 17–41

Brooks B, Schulte-Fortkamp B, Voigt KS, Case A (2014) Exploring our sonic environment through soundscape research & theory. Acoust Today 10(1):90–99

Brown AL, Kang J, Gjestland T (2011) Towards standardization in soundscape preference assessment. Appl Acoust 72(6):387–392

Brown AL, Gjestland T, Dubois D (2016) Acoustic environments and soundscapes. In: Kang J, Schulte-Fortkamp B (eds) Soundscape and the built environment. CRC Press, Boca Raton, pp 1–16

Cargo M, Mercer SL (2008) The value and challenges of participatory research: strengthening its practice. Annu Rev Public Health 29(1):325–350

Classen C (1997) Foundations for an anthropology of the senses. Int Soc Sci 49(153):401–412

de Kluizenaar Y, Janssen SA, Vos H, Salomons EM et al (2013) Road traffic noise and annoyance: a quantification of the effect of quiet side exposure at dwellings. Int J Environ Res Public Health 10(6):2258–2270

De Coensel B, Bockstael A, Dekoninck L, Botteldooren D et al (2010) The soundscape approach for early stage urban planning: a case study. In: Proceedings of The 39th International Congress on Noise Control Engineering, Inter-Noise 2010, pp. 1–10

Deutsche Industrial Norms (DIN) (2009) 45692:2009-08 Measurement technique for the simulation of the auditory sensation of sharpness. Beuth Verlag GmbH, Berlin

Easteal M, Bannister S, Kang J, Aletta F et al (2014) Urban sound planning in Brighton and Hove. In: Proceedings of the Forum Acusticum 2014, pp. 1–6

Erbe C, Dent ML (2017) Animal bioacoustics. Acoust Today 13(2):65–67

Erfanian M, Mitchell AJ, Kang J, Aletta F (2019) The psychophysiological implications of soundscape: a systematic review of empirical literature and a research agenda. Int J Environ Res Public Health 16(19):3533

Erwine B (2017) Creating sensory spaces: the architecture of the invisible. Routledge, New York City

European Commission (EC) (2011) On the implementation of the Environmental Noise Directive in accordance with Article 11 of Directive 2002/49/EC, Document 52011DC0321 (Report from the commission to the European parliament and the council). https://eur-lex.europa.eu/legal-content/EN/TXT/?uri=celex%3A52011DC0321. Accessed 03 Apr 2022

European Environment Agency (EEA) (2014) Good practice guide on quiet areas, EEA Technical report No 4/2014. Publications Office of the European Union, Luxembourg. https://www.eea.europa.eu/publications/good-practice-guide-on-quiet-areas. Accessed 01 May 2022

European Environment Agency (EEA) (2017) Managing exposure to noise in Europe, EEA Briefing No 01/2017. Publications Office of the European Union, Luxembourg. https://www.eea.europa.eu/publications/managing-exposure-to-noise-in-europe. Accessed 01 May 2022

European Parliament and Council (EPC) (2002) Directive 2002/49/EC relating to the assessment and management of environmental noise, Final report: The Centre for Strategy & Evaluation Services LLP. Publications Office of the European Union, Brussels. https://op.europa.eu/en/publication-detail/-/publication/7febde6d-9a89-11e6-9bca-01aa75ed71a1. Accessed 01 May 2022

European Cooperation in Science and Technology (COST) (2009) COST Action TD0804 Soundscape of European Cities and Landscapes. https://www.cost.eu/actions/TD0804/. Accessed 03 Apr 2022

Fastl H, Zwicker E (2007) Psychoacoustics - facts and models, 3rd edn. Springer Verlag, Berlin

Fidell S (2003) The Schultz curve 25 years later: a research perspective. J Acoust Soc Am 114(6):3007–3009

Fiebig A, Schulte-Fortkamp B (2020) Soundscape as methodology in psychoacoustics and noise management. In: Bull M, Cobussen M (eds) The Bloomsbury handbook of sonic methodologies. Bloomsbury Academic, New York, pp 252–268

Fristrup K (2009) The acoustical status of U.S. national parks. J Acoust Soc Am 125:2716

Genuit K (2004) Sound quality in environment: psychoacoustic mapping. J Acoust Soc Am 116:2641

Genuit K (2013) The need for transdisciplinary actions - psychoacoustics, sound quality, soundscape and environmental noise. In: Proceedings of 42nd International Congress on Noise Control Engineering, Inter-Noise 2013, pp. 1–10

Genuit K, Fiebig A (2006) Psychoacoustics and its benefit for the soundscape approach. Acta Acust United Acoust 92(6):952–958

Genuit K, Fiebig A (2014) The measurement of soundscapes - is it standardizable? In: Proceedings of 43rd International Congress on Noise Control Engineering, Inter-Noise 2014, pp. 1–9

Gregg M, Siegworth GJ (2010) The affect theory reader. Duke University Press, Durham

Griefahn B, Spreng M (2004) Disturbed sleep patterns and limitation of noise. Noise Health 6(22):27–33

Hamilakis Y (2013) Archaeology and the senses. Cambridge University Press, New York

Herranz-Pascual K, Aspuru I, García I (2010) Proposed conceptual model of environmental experience as framework to study the soundscape. In: Proceedings of Inter-Noise 2010, pp. 1–9

Hiramatsu K (2013) Soundscape concept in association with noise control. In: Kang J, Chourmouziadou K, Sakantamis K, Wang B, Hao Y (eds) COST action: TD-0804 – soundscape of European cities and landscapes. Soundscape-COST, Oxford, pp 128–131

Hollstein B (2011) Qualitative approaches to social reality: the search for meaning. In: Scott J, Carrington PJ (eds) The sage handbook of social network analysis. Sage Publications, London, pp 404–416

Horne JA, Pankhurst FL, Reyner LA, Hume K, Diamond ID (1994) A field study of sleep disturbance: effects of aircraft noise and other factors on 5,742 nights of actimetrically monitored sleep in a large subject sample. Sleep 17(2):146–159

Howes D (2008) Can these dry bones live? An anthropological approach to the history of the senses. J Am Hist 95(2):442–451

Howes D (2018) The Routledge companion to sound studies. Routledge, London

ISO (2014) International Organization for Standardization. ISO 12913-1:2014: Acoustics — Soundscape Part 1: Definition and conceptual framework. ISO, Geneva. https://www.iso.org/standard/52161.html. Accessed 01 May 2022

ISO (2016) International Organization for Standardization. ISO 1996-1:2016: Acoustics — Description, measurement and assessment of environmental noise — Part 1: Basic quantities and assessment procedures. ISO, Geneva. https://www.iso.org/standard/59765.html. Accessed 01 May 2022

ISO (2018) International Organization for Standardization. ISO 12913-2:2018: Acoustics — Soundscape Part 2: Data collection and reporting requirements. ISO, Geneva. https://www.iso.org/standard/75267.html. Accessed 01 May 2022

ISO (2019) International Organization for Standardization. ISO 12913-3:2019: Acoustics — Soundscape Part 3: Data analysis. ISO, Geneva. https://www.iso.org/standard/69864.html. Accessed 01 May 2022

Jordan P (2019) Historic approaches to sonic encounter at the Berlin Wall Memorial. Acoustics 1(3):517–537

Jordan P (2021) The border of sound and silence - sonic preservation at the Berlin Wall. In: Sioli A, Kiourtsoglou E (eds) The sound of architecture: acoustic atmospheres in place. Leuven University Press, Leuven, pp 215–228

Kang J, Schulte-Fortkamp B (eds) (2016) Soundscape and the built environment. CRC Press, Boca Raton

Kang J, Chourmouziadou K, Sakantamis K, Wang B, Hao Y (eds) (2013) COST action: TD-0804 – soundscape of European cities and landscapes. Soundscape-COST, Oxford

Kang J, Schulte-Fortkamp B, Fiebig A, Botteldooren D (2016a) Mapping of soundscape. In: Kang J, Schulte-Fortkamp B (eds) Soundscape and the built environment. CRC Press, Boca Raton, pp 161–196

Kang J, Aletta F, Gjestland TT, Brown LA et al (2016b) Ten questions on the soundscapes of the built environment. Build Environ 108(1):284–294

Krimm J, Techen H, Knaack U (2017) Updated urban facade design for quieter outdoor spaces. J Facade Des Eng 5(1):63–75

Kull RC (2006) Natural and urban soundscapes: the need for a multi-disciplinary approach. Acta Acust United Acust 92(6):898–902

Lavia L, Witchel HJ, Kang J, Aletta F (2016a) A preliminary soundscape management model for added sound in public spaces to discourage anti-social and support pro-social effects on public behaviour. In: Proceedings of the DAGA 2016 Conference, pp. 1–4

Lavia L, Dixon M, Witchel HJ, Goldsmith M (2016b) Applied soundscape practices. In: Kang J, Schulte-Fortkamp B (eds) Soundscape and the built environment. CRC Press, Boca Raton, pp 243–301

Lercher P, Schulte-Fortkamp B (2003) The relevance of soundscape research to the assessment of noise annoyance at the community level. In: Proceedings of the 8th International Congress on Noise as a Public Health Problem, p. 16

Lercher P, Schulte-Fortkamp B (2013) Soundscape of European cities and landscape: Harmonising. In: Kang J, Chourmouziadou K, Sakantamis K, Wang B, Hao Y (eds) COST action: TD-0804 – soundscape of European cities and landscapes. Soundscape-COST, Oxford, pp 120–127

Lingold MC, Mueller D, Trettien W (eds) (2018) Digital sound studies. Duke University Press, Durham

Lippold M, Lawrence B (2019) Soundscape as a tool for urban planning. In: Proceedings of the 23rd International Congress on Acoustics, pp. 7035–7042

Mačiulienė M, Skaržauskienė A, Botteldooren D (2018) Developing a digital co-creation assessment methodology. Contemp Econ 12(4 special issue):399–408

Marsden PV (1990) Network data and measurement. Annu Rev Sociol 16:435–463

McAdams S (1993) Recognition of sound sources and events. In: McAdams S, Bigand E (eds) Thinking in sound: the cognitive psychology of human audition. Clarendon Press, Oxford, pp 146–198
Merleau-Ponty M (2004) The world of perception. Routledge, London
Mills S (2014) An auditory archaeology. In: Mills S (ed) Auditory archaeology. Left Coast Press, Walnut Creek, pp 75–101
Nagahata K (2012) What should the soundscape community do when listening to the soundscapes of Fukushima in soundscape. J Acoust Ecol 12(1):15–19
Norberg-Schulz C (1979) Genius loci: towards a phenomenology of architecture. Rizzoli, New York City
Ouzounian G (2020) Stereophonica: sound and space in science, technology, and the arts. MIT Press, Cambridge, MA
Pallasmaa J (2005) The eyes of the skin: architecture and the senses, 2nd edn. John Wiley & Sons, Ltd, Chichester
Passchier-Vermeer W (2001) Effects of noise. In: Bistrup ML (ed) Health effects of noise on children and perception of the risk of noise. National Institute of Public Health, Denmark, pp 47–68
Schafer RM (1977) The tuning of the world. A.A. Knopf, New York
Schulte-Fortkamp B (1994a) Geräusche beurteilen im Labor. VDI-Verlag, Düsseldorf
Schulte-Fortkamp B (1994b) Loudness judgements on the background of subjective experience with environmental sounds. In: Proceedings of Inter-Noise 1994, pp. 791–796
Schulte-Fortkamp B (2001) Soundscapes in the sense of reaction to sound and vibration. J Acoust Soc Am 109(5):2345
Schulte-Fortkamp B (2010) The tuning of noise pollution with respect to the expertise of people's mind (keynote lecture). In: Proceedings of Inter-Noise 2010
Schulte-Fortkamp B (2019) The soundscape standard – its development and challenges. J Acoust Soc Am 145(3):1752
Schulte-Fortkamp B, Dubois D (2006) Preface to the special issue: recent advances in soundscape research. Acta Acust United Acust 92:I–VIII
Schulte-Fortkamp B, Kang J (2013) Introduction to the special issue on soundscapes. J Acoust Soc Am 134(1):765–766
Schulte-Fortkamp B, Nitsch W (1999) On soundscapes and their meaning regarding noise annoyance measurement. In: The Proceedings of INTER-NOISE 99/ The 1999 International Congress on Noise Control Engineering, pp. 1387–1392
Schulte-Fortkamp B, Volz R, Jakob A (2008) Using the soundscape approach to develop a public space in Berlin perception and evaluation. J Acoust Soc Am 123:3808
Sheller M (2004) Automotive emotions. Theory Cult Soc 21(4–5):221–242
Skeates R, Day J (eds) (2019) The Routledge handbook of sensory archaeology. Routledge, New York City
Smith MM (2007) Sensory history. Berg, Oxford
Southworth MF (1967) The sonic environment of cities (Thesis). Massachusetts Institute of Technology, Dept. of City and Regional Planning, Boston
Southworth MF (1969) The sonic environment of cities. Environ Behav 1:49–70
Torigoe K (2012) Insights into the global composition taken at three stricken places in Japan. J Acoust Ecol 12(1):9–14
Torresin S, Albatici R, Aletta F, Babich F et al (2019) Assessment methods and factors determining positive indoor soundscapes in residential buildings: a systematic review. Sustainability 11(19):5290
Umweltbundesamt (UBA) (2016) Umweltbewusstsein in Deutschland 2016. Resource document. Bundesministerium für Umwelt, Naturschutz, Bau und Reaktorsicherheit (BMUB). https://www.umweltbundesamt.de/sites/default/files/medien/376/publikationen/umweltbewusstsein_deutschland_2016_bf.pdf. Accessed 03 Apr 2022

Van Kamp I, Klaboe R, Brown L, Lercher P (2016) Soundscapes, human restoration and quality of life. In: Kang J, Schulte-Fortkamp B (eds) Soundscape and the built environment. CRC Press, Boca Raton, pp 43–68

Vaughn LM, Jacquez F (2020) Participatory research methods – choice points in the research process. J Particip Res Methods 1(1):1–14

Westerkamp H (1974) Soundwalking. Sound Heritage III(4)

World Health Organization (WHO) (2018) Environmental noise guidelines for the European region. WHO Regional Office for Europe, Copenhagen. https://www.euro.who.int/en/publications/abstracts/environmental-noise-guidelines-for-the-european-region-2018. Accessed 01 May 2022

# Chapter 4
# Soundscape and Urban Planning

**Bennett M. Brooks**

**Abstract** The soundscape technique can provide an improved quality of life in our communities. Urban planning is a key vector in the process of actualizing the soundscape theory and implementing holistic improvements in the acoustical environment on a larger scale. This chapter presents the basic concepts and principles of urban planning and urban design, which is the bridge between urban planning and architecture. The basic concepts of community noise control and a variety of options are discussed, which include the need to combine perceptual and objective criteria. Also presented are the principles of smart growth, a recent movement of urban planning. Smart growth is a method of integrating planning with other disciplines to achieve a better quality of life by creating neighborhoods of viable residential and commercial mixed use that promote walkable, livable, and enjoyable residential areas. Smart growth design is the result of attention to the needs of occupants and users, or the "local experts" of the community. With their common goals and methods, smart growth integrates soundscape principles with the other design disciplines to achieve a harmonized acoustical environment. The chapter closes with a "toolbox" of techniques from which the practitioner may draw to apply to a specific real-world urban planning or design situation and to develop sustainable solutions.

**Keywords** City planning · Noise ordinances · Quality of life · Smart growth · Urban design

## 4.1 Introduction

The soundscape technique is a powerful tool for understanding the acoustical environment. Urban planning is the method by which our communities shape themselves and prepare for the future. How then are these two, seemingly very different,

B. M. Brooks (✉)
Brooks Acoustics Corporation, Pompano Beach, FL, USA
e-mail: bbrooks@brooksacoustics.com



B. Schulte-Fortkamp et al. (eds.), *Soundscapes: Humans and Their Acoustic Environment*, Springer Handbook of Auditory Research 76,
https://doi.org/10.1007/978-3-031-22779-0_4

disciplines to interact? How can the soundscape tool be employed to improve our communities and the quality of life for those who work, live, and play there? This chapter answers these questions, describes methods for the practitioner, and provides guidance for interested parties to formulate and then to achieve their goals. Following a review of relevant literature, it can be used as a working handbook for the application of the soundscape technique to the real issues facing urban planners and others who manage and mold our cities.

### 4.1.1 Research Versus Application in Soundscape and City Planning

At this time, the integration of the soundscape technique with urban planning principles and practices is in its very early stages. As city planning and the way in which residents live, work, and play evolves to account for the desires and needs of the populace, there will be an increasing need for the benefits of soundscape work. Current research within the soundscape discipline focuses mainly on academic studies of individual spaces or methods of data collection and analyses, and there are few examples of applications on a large urban or city-wide scale. These applications would come under the Type III designation studies relate to soundscape applications primarily for planning and political actions (see Schulte-Fortkamp and Jordan, Chap. 3).

Improving the soundscape benefits individuals and communities in terms of health and well-being (Schulte-Fortkamp 2017). Additionally, many current objectives, such as livable streetscapes with pedestrian-friendly districts and neighborhoods that mix commercial, civic, cultural, educational, and recreational activities, align directly with the goals of properly implemented soundscape design projects. There are clear opportunities to incorporate the soundscape approach directly within city planning efforts at various scales, from neighborhoods to districts and city masterplans.

It is imperative that soundscape researchers and university-based professionals closely engage with their faculty counterparts in the urban and regional planning department of their university, to develop approaches which integrate soundscape into the standard planning curriculum and practicums. The purpose is to foster awareness of, and action in, collaborative soundscape projects among current planning professionals through continuous learning and outreach programs, and among students, the future professionals, through their class and research work. In addition, soundscape researchers and practitioners can engage government-based planning professionals in workshops, congresses, and symposia to promote the development of soundscape practice (Steele et al. 2012, 2020).

One purpose of this chapter is to encourage soundscape researchers to study the interaction between the acoustical environment and urban planning and to develop the research methods that can be used to collaborate more fully with the urban planning process. Another purpose is to provide soundscape practitioners with the tools by which they can apply soundscape principles to planning and development projects of varied sizes and scopes.

### 4.1.2 Objective Design Standards Based on Perceptions

Urban planning, urban design, and architecture are design disciplines that utilize form and function, as shaped by creativity, to serve a practical purpose. Designs must meet accepted standards. Subjective design standards are not effective and objective standards have to incorporate the needs of the community. This approach highlights the importance of understanding the perceptions and expectations of the stakeholders in context and recognizing the community identity when developing project and community standards.

Broad stakeholder participation is important across disciplines, including in soundscape and planning projects. There is evidence that using soundscape principles and practices is an excellent way to gain the acceptance of disparate groups of stakeholders (see Siebein and Siebein, Chap. 5). The *diplomacy of soundscape* may be the most important facet of the process. Understanding the acoustical environment and bringing a community together in a shared goal can be a rewarding achievement. The basic goal is to *always engage more users/stakeholders* of the community whenever possible. This provides a more varied, deep, diverse, and expansive pool of perceptions from which broader themes and trends may be discovered. In this endeavor we seek to find consensus with multiple interests and develop bonds between community members for a common purpose. The integration of stakeholder opinions leads to the desired result of what may be called the diplomacy of soundscape (ISO 12913-1 2014; ISO 12913-2 2018).

The task is to filter a specific design concept through the medium of stakeholder perceptions. To do this, many questions must be considered and answered. How are user sensibilities identified and integrated to be implemented into a project? How is the presence of sound part of the project vision? What is the effect of sound on the prestige, image, cost, and constructability of a project within a neighborhood or district? Will the people like it? How do we maneuver through all the political and social constructs to achieve a building or a city? Soundscape methods fill a gap in the process to evaluate the acceptability and potential beauty of sound in the space and environment.

### 4.1.3 Soundscape and City Planning: Actualizing Positive Outcomes

This chapter expands the current state of soundscape research with the goal of moving toward more universal applications, including the setting of standards, norms, and policies that enhance the livability of an entire area of interest. The traditional means for enhancing livability are based on noise control engineering principles and practices that treat sound source emissions (Brooks et al. 1995; Lyon and Brooks 2003). Most of the referenced work to date that connects the soundscape technique to city planning and design focuses on specific development projects or smaller urban areas, such as individual streets, parks, and university campuses. While the results of these studies are vitally important, the limited extent cannot be applied to larger scale areas such as neighborhoods, districts, or the city as a whole (Coelho 2016; Steele 2018).

The receiver aspects of the source–path–receiver system are known as "immissions" and are treated in a limited way in most noise control and city planning approaches. The soundscape technique greatly expands the understanding of the receiver component of the system, which can achieve better, more balanced results. Second, the whole idea of sound as a positive resource rather than simply a negative aspect of an environment supports this approach (Schulte-Fortkamp and Fiebig 2016).

The soundscape approach can provide the means to actualize positive outcomes in environmental acoustic analyses. The following sections are organized to provide a detailed look at the soundscape process from the urban planning point of view. "The ultimate goal here is for the soundscape tool to be recognized as so powerful, so effective, and so influential that private developers, architects and urban planners will understand that they must use it, or risk the failure of their project" (Brooks et al. 2014, p. 39). Practitioners have known how to change the acoustical experience for many years. Yet, the same planning and design errors keep reoccurring. Sound is not typically treated in a holistic way. This is because sound is something that we cannot see, we must experience it, and for planners and designers that is not always common or easy.

Knowledge of how the sonic environment is perceived, offered by application of the soundscape method, opens a window of opportunity for improved planning and design. The soundscape technique now gives planners and designers the tools to achieve their objective of an improved quality of life for the city's inhabitants.

## 4.2 Urban Planning Principles

In large part, no city plan or land-use development project, large or small, will happen without the participation of an urban planner. Depending on the scale of the project and the jurisdiction, the planner may be involved directly, peripherally, or even tacitly. It is incumbent on soundscape researchers and practitioners who seek to achieve positive acoustical outcomes through urban planning and design to understand and adopt the concepts and language of the urban planner.

### 4.2.1 Planning Overview

Urban planning is both a technical and political process that is truly interdisciplinary as it seeks to develop a physical area. This process is best described by John Levy (2017), a widely accepted authority, who states that planning is a highly political activity, and inseparable from the law. The ultimate arbiter of many a planning dispute is the court. And for every case that comes to court, many planning decisions are conditioned by what the participants in the process think would be the decision if the matter did come to court. This is particularly important as often large sums of money, both public and private, are involved and planning decisions imply economic and financial issues delivering large benefits to some and large losses to others. Thus, discussions are frequently held about the role of government and authorities and how to draw the line between public needs and private rights, between political decisions and market rules. Urban planning is not simply a matter of architectural design or meeting requirements and regulations.

Usually, unanimous agreement is not found on precisely what constitutes the public interest, and compromises need to be sought. For example, to shut down a facility to enhance what one person sees as environmental protection might mean unemployment for the next person. "Planning, like politics, is in large measure the art of compromise" (Levy 2017, p. 99).

All land-use decisions affect the community in many ways, such as road traffic, air and water quality, the need for utilities, public services and finances, and opportunities for social interactions. The practitioner who seeks to build a city based on soundscape principles must be able to navigate the complexities of the political and financial environments. The second challenge for the soundscape practitioner is to acknowledge and address the needs of the community in a fully holistic manner. To do this, the soundscape practitioner must embrace the existing urban planning process. A key urban planning tool for understanding these complexities and community interconnectedness is the *Comprehensive Plan*.

### 4.2.2 Comprehensive Plan

A *Comprehensive Plan*, sometimes called a *Master Plan*, is the basic guide for the development of a community. Such a plan is typically commissioned and proposed by a governing body of the area, such as a local government, neighborhood association, or university. A key feature is that the Comprehensive Plan includes the entire community. Often the Comprehensive Plan specifies controls on land use, such as whether the land is designated for residential, commercial, or industrial use. Various legal and administrative means are used by the governing jurisdiction to control how privately owned land is used, such as zoning regulations and conditional use permits (Levy 2017).

The goals of a Comprehensive Plan usually include three primary public concerns: health, safety, and public welfare. Today, the development of a Comprehensive Plan is typically a participatory process that includes public officials, municipal staff, and citizens of the particular community under study. The process can vary but will often include elements that start with a research phase and clarification of community goals and objectives. Then the plan is formulated, implemented, and finally reviewed, assessed, and possibly revised based on the lessons learned, or on the changing needs of the community. In short, the Comprehensive Plan should recognize and address the wide variety of infrastructure and social needs of the community and should be guided by soundscape principles to be a useful roadmap to a successful urban environment.

### *4.2.3 Urban Design Process*

Urban design bridges the disciplines of planning and architecture. Examples of modern planned cities that incorporated elements of urban design include Seoul (South Korea), Singapore (Republic of Singapore), Zurich (Switzerland, master plan by architect Le Corbusier), Brasilia (Brazil), Washington D.C. (USA, by architect Pierre Charles L'Enfant), Chandigarh (India), Shanghai (China), and Medellin (Colombia). In many of these cities, the traditional infrastructure design was enhanced or modified to facilitate transport, include green surroundings, and improve residential livability (planningtank.com/blog/top-planned-cities-in-the-world).

Notable examples of urban designs through history include the grand boulevards, uniform building facades, and public spaces in Paris by Baron Haussmann and in Central Park in New York City by Frederick Law Olmstead and Calvin Vaux. Both projects were conducted in the mid-nineteenth century. Haussmann's design concept (see Fig. 4.1) was to create grand tree-lined boulevards to control traffic flow and create a sense of rhythm and order. He shaped the skyline and created vistas focused on public buildings and gardens over a 17-year period. Following Kang (2018), New York's Central Park was designed to be a welcoming and democratic public space to bring together the diverse people of the rapidly growing population of the time.

Urban design differs from architectural design in scale, time frame, and complexity. Whereas architecture usually focuses on an individual building, urban design works over a larger area. A single building may take several years to construct, while an urban design implementation may span decades. Urban design addresses many complex and interconnected variables, including transportation, utility services, pedestrian orientation, and neighborhood identity.

**Fig. 4.1** Haussmann's Paris design. A view down the Seine from Notre Dame cathedral (Levy 2017)

Urban design also differs from urban planning. An urban design project site may be designated in a city and can be a subset of a larger planning process. Importantly, urban designers are not as likely to be involved in the political process as are the planners. The fundamental unit of urban design is the neighborhood. The neighborhood concept was developed in the 1920s to include residences, retail shopping, schools, and small parks. Although neighborhoods as groups of residents clearly existed prior to that time, formal planning of these living areas with their integrated services had not been done (Levy 2017).

Levy (2017) proposes a common procedure for urban design related to four phases: analysis, synthesis, evaluation, and implementation. Urban design studies begin with an *analysis* of the gathered information on land use, population, transportation, utilities, topography, the character of the site, the structure of the neighborhoods, and the economics of the business areas. In the *synthesis* phase, solution concepts are proposed that address the stated problem. Given the often-conflicting demands and necessary trade-offs, specific schematic designs are considered, and then preliminary plans are developed. An *evaluation* of these plans is then done considering the original problem and design goals, including whether the solutions are adequate and can be readily implemented. Finally, financing and construction strategies are devised for *implementation*.

A good urban design varies depending on the context, much as a soundscape is best when it meets the needs of its specific stakeholders. The general intent of urban design is to improve the quality of daily life through better designs that are coordinated with local needs, uses, and capabilities. This may be accomplished through the elimination of barriers and the creation of opportunities for people to move around the city in a free, safe, and pleasant way.

Moreover, Levy (2017) suggests that a successful urban design could be judged on the following criteria as unity and coherence, minimum conflict between pedestrians and vehicles, protection from the elements, ease of orientation, compatible land uses, places to rest, observe and meet, creation of a sense of security and pleasantness. In this context, Levy remarked that urban design is not an exact science since there is always the element of personal taste. One person's peace and tranquility will be another person's boredom and sterility (Levy 2017). This statement is a prescient description of how the perception of the sound environment should be included in an urban design.

### 4.2.4 Soundscape Connection to Urban Planning and Design

Although the traditional view of the urban planning and design processes would suggest the perceived soundscape should be treated as an important factor, in general practice very little is done to manage or to improve the acoustical conditions. Most of the attention is spent on transit, efficiency of mobility, universal access, beautification, controlled expansion/growth, and economic viability among other concerns. Only recently has attention begun to widen into the sensory aspects of city experience (Woolworth 2013).

Until the advent of the soundscape technique, urban planners and designers did not have a formal system to evaluate human perceptions of an acoustical environment. There is currently a palpable lack of awareness and no grounded methodology in the urban planning discipline concerning the sense of hearing and sound. This is true both in the academic realm and within urban planning practice. Noise issues are generally addressed using pre-soundscape processes, in which sound is considered only as something to be mitigated and controlled, and not as a resource (see Schulte-Fortkamp and Jordan, Chap. 3). The word "sound" is often not used at all by planners: "noise" is used instead to signify what one hears.

An example from urban planning is traffic noise. A typical complaint received by urban planners is the noise due to higher speed local road traffic. Traffic calming, such as narrower road passages, roundabouts, and speed bumps, is the typical planning response to this prominent source of urban sound. Despite the lack of awareness of soundscape approaches in planning practice, when introduced to it planners are enthusiastic about using the soundscape method to evaluate the acoustical environment. Greater exposure to the soundscape discipline among planners is the key.

It is important that the architectural design and urban planning communities work more closely together to create great cities. The collaboration between these disciplines must be more than a zoning and permitting checklist. The development of a city mixed-use district is a good example. The question is, how can the interests of the commercial users be met and balanced with the needs of the residents such that they coexist in harmony? The soundscape technique can be a powerful tool to bring the planning and design disciplines closer together to build livable urban environments.

### 4.2.5 The Path Forward

It is vitally important to introduce the concept of soundscape into the urban master plan process.

We must integrate soundscape thinking into any comprehensive plan as a public facing, visionary document. The key is to inspire proactive planning versus reactive measures regarding sound resources, their benefits, and impacts. Urban sound can be a valuable commodity and not just an unwanted by-product of urban life.

Further, a soundscape analysis must be included as an important element of future urban design studies. Instead of crude noise speculations being applied to an urban design, a comprehensive soundscape study can provide the precise data that the designers and decision makers need to develop viable solutions. How does a great city get built? How does sound fit into the total vision? This is the direction in which we are heading with soundscape studies and their growing role in urban planning.

## 4.3 Traditional Noise Control Practices and Urban Planning

The practice of noise control in our cities has been a concern since ancient times, beginning with royal edicts and continuing into the era of government regulations. The results of noise control efforts in our urban centers have been modest at best. The "traditional approach" of the regulation of sound sources treats sound as unwanted noise that must be eliminated or removed. Unfortunately, this approach has resulted in minimal improvements in the quality of life for urban residents, workers, and visitors. The new approach using the soundscape method recognizes urban sound as a resource (see Schulte-Fortkamp and Jordan, Chap. 3) that can enhance the lives of those in urban environments.

### 4.3.1 Noise Control from the Nineteenth Century Onward

Informal reports (apocryphal) state that U.S. President Thomas Jefferson (1801–1809) had the berms around the White House residence built to reduce the noise from horses and carriages on nearby streets. Early urban designers recognized that broad boulevards and open spaces created less noise than narrow city streets.

Early scientific efforts to describe noise compared to sound were begun in the nineteenth century. In New York City (NYC) at the turn of the twentieth century, a wealthy socialite, Mrs. Rice began a crusade against what she believed to be excessive noise in the city. She formed the Committee Against Noise in 1908, focusing on the incessant blasting of horns by riverboats (Keizer 2012; Prochnik 2011).

In the 1920s and early 1930s, the New York City Noise Commission studied the sounds of the city. Their findings were reported by Slocum (1931). The Commission developed a list for the classification of city noise:

- Automotive traffic
- Rail transportation
- Water transportation
- Building (construction) operations
- Street noises (vendors, taxi stands)
- Collections and deliveries
- Homes (radios, dogs, parties)
- Miscellaneous (fire dept, aircraft, restaurants, factories, sirens, amusement halls)

The study of urban sound was revisited in New York City in the 1960s, by the Mayor's Task Force on Noise, culminating in a report "Toward a Quieter City" in 1970. Again, construction and transportation noise were deemed to have the greatest impacts (NYMTFN 1970).

## *4.3.2 Current Noise Policy and Soundscape*

The implementation of noise control regulations, codes, and ordinances of a given jurisdiction is driven by the policy that is determined by a governing authority. However, soundscape methods are not yet incorporated in these policies to maximum benefit. There is a need to fully incorporate the soundscape technique for better compatibility between environmental noise control approaches and improved quality of life.

### 4.3.2.1 Noise Policy Status in Europe

Following the EU Directive of 2002 (END 2002), dB(A) noise maps and action plans have shaped common noise protection policy; however, predictions based on these maps very often fail to define the real perceptual burden borne by the populace. There remain questions as to how individual noise complaints get handled locally in the context of the continent-wide directives.

Noise maps do not accurately nor fully describe the positive and negative aspects of sound in communities. The experience and knowledge that residents bring to the "action plans" should become the main assessment criteria for noise and its impact. The question that soundscape practitioners should pose to the regulators and politicians is how to underpin the strategy of the community and respond to needs arising out of the given conditions and established structures (Coelho 2007).

#### 4.3.2.2 Noise Policy in the United States

There is no unified noise policy in the United States. There are 50 states with 50 different kinds of noise criteria. In addition, noise policy is a mixture of federal, state, and local laws, ordinances, and regulations. These regulations are almost all based on source noise emission, not on receiver immission (Brooks et al. 1995; Lyon and Brooks 2003).

Federal regulations primarily govern interstate transportation issues such as for aircraft and highways. State and local regulations primarily govern local stationary sources, which fall into the categories of industrial, commercial, and residential sources. State and local noise ordinances generally follow a policy of *source noise emission control* because the common law system provides means to control individual or corporate behavior (emissions), and the purported receiver retains individual property rights and the right to the quiet enjoyment of the home. The exception to this is the definition of the *plainly audible* nuisance, described in more detail in Sects. 4.3.4.4 and 4.3.4.5.

### *4.3.3 Traditional Methods for Characterizing Soundscapes*

Traditional characterizations of the acoustical environment rely primarily on measurable physical sound parameters, such as A-weighted level in decibels, dB(A), and time- and frequency-weighted variations. Characteristic sound parameters can include the spectral content, sound level averages and statistics, and sound level time history (Brooks 2006; Beranek 1992).

The spectral content of sound could be broadband or tonal and could be dominated by high or low frequencies. Environmental sound levels may be described using the A-weighted scale. The A-weighting filter mimics human hearing sensitivity and is used for assessing the impacts of sounds on people. Sound level measurements which apply A-weighting are designated by the symbol "dBA" or "dB(A)." The C-weighting scale may be used to emphasize more bass (low frequency) intensive sounds; alternatively, the unweighted frequency response (Z-weighting) has zero weighting with no frequency biases.

Typical outdoor A-weighted sound levels are:

- 25 dB(A) pasture, no wind
- 35 dB(A) whisper
- 40 dB(A) small town residence
- 50 dB(A) wind in trees
- 55 dB(A) light traffic @ 100 ft (30 m)
- 60–70 dB(A) conversation
- 75 dB(A) busy city street
- 85 dB(A) heavy truck @ 50 ft (15 m)

Sound level statistics are useful to describe sound in the environment. For a particular test period, sound levels may fluctuate due to the variation of sound source signals received at that location. Sound level fluctuations may occur due to the transient presence of natural sources, such as birds, insects, and rustling leaves, and also due to the presence of anthropogenic sound sources, such as machinery, road traffic, and aircraft operating in the area. Often, A-weighted levels or sound spectra may be used to compare statistical levels. Typical parameters used to describe outdoor environments include: the sound level that is exceeded during 90% of the sample time period, $L_{90}$; the sound level exceeded 10% of the time, $L_{10}$; and the equivalent energy average during the sample period, $L_{eq}$.

Statistical characterizations can be applied as follows:

- $L_{90}$ background ambient sound
- $L_{10}$ typical road traffic sound
- $L_{10}$–$L_{90}$ effect of intrusive noise
- $L_{eq}$–$L_{50}$ smoothness or jaggedness of sound variations

A way of using sound averages to characterize an outdoor environment is the Day–Night level, notated as $L_{dn}$, or DNL. For this characterization, A-weighting of the frequency spectrum is assumed. This level is defined using the measured daytime average sound level, $L_{eq}$ (Day), over the 15 daytime hours of 07:00 to 22:00, and the measured nighttime average sound level, $L_{eq}$ (Night), over the 9 nighttime hours of 22:00 to 07:00, per the following equation:

$$L_{dn} = 10\log\left[\left(15/24 \times P_{day}\right) + \left(9/24 \times P_{night}\right)\right]$$

where:

$P_{day} = 10^{(Leq\ day/10)}$
$P_{night} = 10^{[(Leq\ night\ +\ 10)/10]}$

Note that the actual measured average nighttime sound level has a penalty of 10 dB imposed, as the time period of 22:00 to 07:00 is during normal sleeping hours. Typical day–night sound levels are characterized by the U.S. Environmental Protection Agency as:

- DNL 50 rural/small town
- DNL 55 suburban
- DNL 60 urban
- DNL 65 noisy urban
- DNL 70 very noisy urban
- DNL 75–80 downtown metropolis

Detailed temporal (time history) sound test results are given in the form of a Time History Chart, which shows the change in sound level over time for each test record. Time history analysis of sound data can be very helpful for understanding the character of the tested acoustical environment. Simply stated, the sound level time history indicates the sound level that is measured at any given moment of time during the test period.

The sound time history is represented by a chart showing how the measured sound levels vary with time. A steady sound (e.g., a constant fan) will appear to be more of a flat line on the chart; variable sounds of shorter duration (e.g., passing vehicles) will appear as a series of peaks and valleys on the chart. This is illustrated for the sound levels near a hospital emergency generator measured in a residential area, using a 1-s sample period, as shown in Fig. 4.2 (top). This time history shows a sonic environment with a variety of sound sources. Another example of a sound time history taken over a period of several days, using 1-h sample period averages and statistical parameters, is shown for the front of a residence in a city entertainment district, as shown in Fig. 4.2 (bottom). Some features to note in this time history include the high maximum levels during a 1-h period due to the passing of loud vehicles, the drop-off in sound levels during the very early morning hours each day, and the steady entertainment sound levels during the prime entertainment hours on Friday and Saturday evenings.

It is important to recognize that these traditional methods for characterizing the sound in outdoor environments provide the objective, physical acoustical basis for analyzing that environment using the soundscape methods. Correlations between the physical acoustical measurements and the perceptual soundscape measurements can beneficially inform the urban planning and design processes.

### *4.3.4 Noise Control Ordinances*

The traditional means that many jurisdictions use to control the community acoustical environment focus on noise ordinances. Urban planners and other officials are usually familiar with these legal tools. Therefore, it is important for soundscape researchers and practitioners to be conversant with the potential range and breadth of noise ordinance provisions.

A noise ordinance is a law, written and enforced by a state or a local jurisdiction, that usually pertains to the amount (volume or level) of noise, the duration of noise, and sources of noise that affect the inhabitants of a community. Basically, a noise ordinance defines which sounds may or may not be acceptable at any given time so that residents can live comfortably within a community. This effect can be called the acoustical quality of life. A city or county noise ordinance is usually intended to have provisions that are in effect at certain times of the day. For example, certain provisions may only apply during daytime hours when most people are awake and other provisions may apply during nighttime hours when most people sleep. Perceived violations of a noise ordinance may be reported to police or local officials.

In legal terms, noise may be defined as unwanted sound. Of course, whether a sound is considered wanted or unwanted is a subjective judgment by an individual. Sounds that may be addressed by a noise ordinance not only include those commonly produced by residents but also include industrial and commercial facilities.

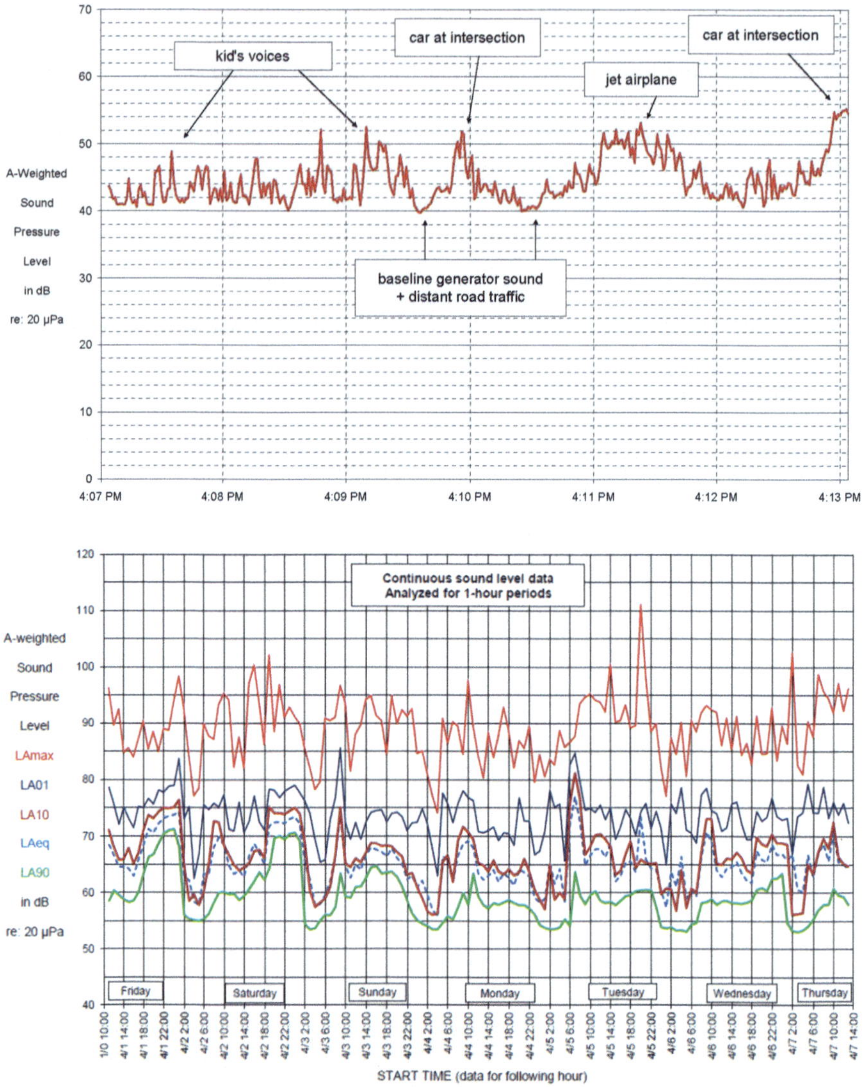


**Fig. 4.2** Sound time history showing the relationship between noise from a hospital emergency generator and ambient background sounds in a residential area (top). Sound time history near a residence in a city entertainment district (bottom)

Examples of potential sound sources that a noise ordinance may address are person(s) shouting, barking dogs, loud music, power tools, cars or motorcycles with excessively loud engines, fireworks or explosives, mechanical equipment in commercial buildings, manufacturing plants, and power-generation facilities. An effective noise ordinance should be designed to keep a community's residents comfortable in their own homes and prevent businesses and industries from interfering with residents and each other.

#### 4.3.4.1 Acoustical and Non-acoustical Measurement-Based Noise Ordinances

Noise ordinances can be based on objective, measured data. These data can be measured acoustical parameters, as described in Sect. 4.3.3, or based on other physical measurements, such as distance or time of day. The preferred approach to developing noise ordinances or regulations for stationary commercial and industrial potential noise sources involves the use of measurable and verifiable sound level limits and assessment procedures. Developers of businesses and industries must have a target for which they can plan and be able to design their operations to be economically viable. The *acoustical measurement* is the first feature of a successful noise ordinance, the second feature is *effective enforcement*, and the third feature is *community consensus.*

One of the most effective means to protect the population from adverse noise exposure from commercial and industrial sources is through the application of *objective* (physical) acoustical measurements for comparison with accepted, community-based criteria. These objective methods involve the measurement of environmental sound pressure levels at a specific location and offer the following features:

- Objective testing provides the community with a transparent and fair representation and assessment of the existing and desired environmental sound climates.
- Objective testing allows the community to determine through collective due process the environmental sound climate that it deems to be reasonable.
- Objective testing provides the community with an enforceable standard for environmental noise or unwanted sound.
- Objective testing removes the limited subjective feature present in some existing ordinances that base a noise violation upon a single individual's opinion of "nuisance" or "annoyance." The opportunity afforded by the soundscape technique elevates the subjective responses to a much wider group, which may be analyzed in an ordered and scientific manner.
- Objective testing affords equal protection under the law to all citizens.
- Objective testing affords protection of a citizen's legal property rights, such as the "quiet enjoyment of one's home."
- Objectivity is required to pass the strictest judicial review.

For the reasons listed, objective acoustical tests are strongly recommended as the primary methods by which communities can control unwanted sound, or noise, from commercial and industrial sources. Objective testing methods are most likely to provide successful remedies for the majority of harmful or offending environmental noise exposures. In addition, defined noise limits provide planning agencies and business developers with predetermined criteria that can be used to judge the acceptability of a proposed plan of development. This places great importance on the selection of criteria that will be acceptable to the community.

To guarantee effective enforcement, the legislative and executive branches of government should have a clear idea of the skill level, training, and anticipated workload of the enforcement personnel. Several types of enforcement authorities can be identified. These authorities could include the local police, state police, noise control officers, health departments, planning and zoning boards, and citizen-initiated civil court cases.

A consensus process is strongly recommended for developing a noise ordinance or regulation, which includes listening to and accounting for the interests of all the stakeholders in the community. These stakeholders may include the resident citizens, business owners and employees, community development organizations, city administrative and legislative officials, enforcement officers, and legal counsel. Soundscape techniques are perhaps the most efficient way to gather information about community perceptions and to apply those findings to build a consensus. Forums that include soundwalks, workshops, public hearings, and in-depth interviews of the "local experts" may be the most effective means for determining the interests of the community as a whole and how to address community concerns. Once the soundscape-driven consensus process has been completed, legislation to control noise pollution, based on measurements by an enforcing authority, may be drafted. It is important from a legal framework that the legislation be categorized by the nature of the sound source or sources.

Under certain circumstances, "plainly audible" *non-acoustical measurement standards* may be appropriate as the basis for a noise ordinance. Although the term "plainly audible" is used without the application of any quantitative acoustical measurements, the use of the plainly audible test can be cost-effective when compared to the cost of performing more sophisticated acoustical measurements by an enforcement authority such as the local police. Experience has shown that such a law can be enforced in a fair and equitable manner using that test.

Additional appropriate standards may include other non-acoustical standards, such as time-of-day restrictions, zoning and area-based restrictions, minimum-distance restrictions, and specific bans or prohibitions. In addition, the methods and guidance provided can include administrative noise laws, such as the enforcement of a state law prohibiting a motorcycle owner from modifying the original muffler. This type of law can be enforced by having parking meter authorities examine the mufflers of cycles parked in public spaces and issuing repair notices when appropriate. Another administrative law example would be the definition of permissible time periods for garbage collection or routine lawn maintenance. Such administrative laws can be effective since they do not require noise measurements or evaluation of perception of the noise source while it is in operation.

#### 4.3.4.2 Non-measurement-Based Noise Ordinances

Noise laws may be categorized as either quantitative (measurement-based) or qualitative (non-measurement-based). Qualitative, non-measurement-based laws contain provisions that rely on a determination that noise is "excessive" or a "nuisance" or "disturbing." They generally have no specific measurable aspect, except possibly for time-of-day limitations. One advantage of qualitative laws is that they can be applied to many special, limited-duration, or difficult-to-quantify situations such as noisy parties, barking dogs, or generally loud behaviors in public. For enforcement, they also require no special equipment or training. The primary disadvantage of these laws is that the determination of a violation usually depends upon judgment of the enforcing official, which may or may not be successfully challenged.

The legal record on so-called "nuisance ordinances" based on subjective opinion is mixed when they lack appropriate reference to acoustic properties. Some nuisance ordinances have been rejected by the state courts and then upheld at the appellate level, while others have been found to be unconstitutional in federal court.

In some cases, communities have been subjected to expensive financial penalties and protracted legal proceedings for attempting to enforce subjective nuisance ordinances. Thus, nuisance ordinances are not recommended but instead ordinances that rely on objective testing and decibel-based acoustics metrics are preferred. However, in some jurisdictions, local noise regulations have successfully included both objective and subjective provisions, and there are circumstances where just a subjective approach is useful. In some cases, subjective provisions can complement and supplement objective, measurement-based ordinances.

#### 4.3.4.3 Advantages and Disadvantages of Measurement-Based Ordinances

Quantitative acoustical measurement-based laws have specific numerical provisions that generally are in terms of sound level limits in decibels. These laws usually fall into one of two categories: source-specific limits or property-line limits.

Source-specific limits are often used to control the noise produced by sources that can be measured either in normal operation or by using specific standardized measurement procedures (e.g., automobiles, motorcycles, trucks, snowmobiles, motorboats). Source levels of many vehicles, such as cars, trucks, trains, and aircraft, are regulated by the federal government, and the state or local governments may not be able to further regulate those sources.

Property-line limits are based on the land uses or zoning of the "sound emitting" property and of the "sound receiving" property. As the name implies, these laws are often enforced through measurements made at the property line of the receiving property. However, it is not necessary for a noise emitter and a noise receiver to be adjacent to each other (i.e., to share the same boundary line). They may be separated by some distance. In any case, the offending sound must be clearly attributable to the alleged offender and must also constitute a dominant feature of the sound environment. These provisions are not always clearly specified in noise ordinances but should be so stated as to leave no confusion.

Quantitative, acoustical measurement-based laws provide certain benefits from having the statutory limits specified in terms of decibels. The existing source levels can then be measured with the proper instruments and methods and then compared with the sound level limits mandated by law. Trained personnel can objectively measure whether a specific sound source produces levels that exceed the permitted limits.

Those who are subject to limitations on their noise may determine for themselves whether they comply by conducting straightforward testing. Specific noise level limits provide the violating emitter with the opportunity to mitigate the problem by applying standard noise control engineering methods. They may modify their operations such that their noise emissions are below the noise limit target at a property line or other relevant receptor site. Alternatively, the emitter may be required to alter the hours of operation during which excessive noise is produced, depending on the language in the ordinance.

Specific noise level limit criteria provide developers that are creating new businesses and industries with the opportunity to build a facility that they know will meet with community approval. The planning and zoning authorities in the locality or state can make compliance with objective noise standards a condition for approval of the proposed plan of development. The authorities may demand that the developer demonstrate, using engineering methods, how their plans will comply with the appropriate noise standards. The authorities may also mandate penalties if the developer fails to comply. Such penalties could include the rescinding of a certificate of occupancy or the issuance of a shutdown notice. Therefore, objective noise standards provide a community with a great deal of control over the future sound environment, which can be applied in a fair, equitable, and predictable manner.

Objective noise measurement ordinances and regulations also benefit from the well-established standard test methodologies and procedures that have been developed for this purpose. Such methods are described in ANSI S12.9 Part 3 (2013) and ASTM E1503 (2014). There are several disadvantages to enforcing the established objective standards. For enforcement of existing noise violations, instruments must be purchased and maintained. Enforcement staff must be trained. Development, implementation, and at least initial prosecution of violations will likely require expert assistance. In addition, during development of a project, selecting the appropriate sound level metric(s) and determining the values to use for the permitted maximums are likely to require consultations between acoustical experts and those who write the laws. Clearly, considerable expense can be incurred when enforcing compliance with regulations. Enforcement personnel would also have to understand the instruments used and the potential sources of methodological error that might invalidate the measurement.

#### 4.3.4.4 Determining Audibility

The audibility, sometimes termed detectability, of a target sound depends on the frequency content (pitch or character) and the sound levels (loudness or amplitude) of two sound categories: the target sound one is attempting to identify as audible and all other background or masking sounds. If the sound levels of all frequencies of the target are well below the levels of all frequencies of the masking sound, then the target sound will not be audible. However, the human auditory system is well adapted to picking out a single source of sound in the presence of considerable masking sound, particularly targeted sounds that are tonal or periodic in nature. The target sound must have comparable or greater sound levels than the masking sound at only a few frequencies to become audible. For example, we have no trouble hearing the melody played by a piccolo within a very loud marching band; the piccolo's high frequencies easily stand out audibly from the lower frequency sounds produced by the rest of the band instruments.

When the target sound is variable and close to the level of the masking sounds, the target will not necessarily be audible. In these types of complex situations, however, one person may hear the target and another person, listening at a different time, may miss the target. In addition, they may have different preconceived notions of how the target should sound. However, the term "plainly audible" implies that these complex situations are not under consideration. By definition, anyone listening for the target should be able to hear it if it is plainly audible.

The most relevant aspect of audibility is that the performance of people listening for one sound in the presence of other sounds is predictable. Well-established methods have been developed to predict how audible a target sound of known character will be in the presence of a masking sound of known character (Rossing 2015).

#### 4.3.4.5 Advantages and Disadvantages of Non-measurement-Based Ordinances

The requirement that a suspect sound source be "plainly audible" at a particular location to be in violation places this type of law between the subjective nuisance type and the objective quantitative type of noise law. On the one hand, though the "plainly audible" provision has a defined location, it does not include a measurable sound level. However, unlike for the qualitative, nuisance-type laws, the decision that a sound is audible is not as subjective as is deciding whether a sound is disturbing.

It may be determined through due process if the observing person, such as a police officer, is fair and objective in reporting hearing perceptions. In addition, the hearing acuity of that person may be objectively determined by a hearing test. Therefore, the ability of a person with normal hearing to hear a specific sound in the presence of other background sounds is more the result of human physiology than of subjective human judgment or opinion.

There are several advantages to the "plainly audible" test for noise ordinances:

- No special instrument or knowledge of acoustics required
- Unlike the quantitative noise laws, enforcement officials, those subject to the requirements of the ordinance, the public, juries, and adjudicating officials need not acquire an understanding of acoustics or of the instruments used for enforcement.
- Expectation of consistent judgments from person to person
- Extensive laboratory experiments as well as limited fieldwork support the conclusion that determination of a source's audibility should not vary from person to person.
- Easy to understand
- After a brief description, most people should be able to understand what it would mean to experience a source of sound that is plainly audible. One of the most consistent problems encountered when working on noise issues with the general public is the difficulty people have in developing an understanding of what a sound level, expressed in decibels, means in terms of personal experience.

There are also a number of serious disadvantages to the "plainly audible" test for noise ordinances. These disadvantages may preclude the use of this test for an effective ordinance:

- "Plainly audible" not a valid design target
- There is a public need to establish design targets, guidelines, and/or procedures for commercial and industrial developments to ensure creation of a facility that will be compatible with their neighbors. Reasonable targets, established through a soundscape method-based understanding of the area, are crucial for the design and building of the business.
- "Plainly audible" may not mean disturbing
- Whether or not a sound is disturbing when it is plainly audible depends upon what criteria are used to determine disturbance. For example, it is likely that for some individuals, simply hearing a particular source of sound may be disturbing to them.
- However, the development of most quantitative laws and regulations that are known at this time use several underlying assumptions concerning the definition of disturbance. One assumption is based on the particular source itself. For example, a legally operating truck during daytime hours may be plainly audible, but since the operator has an expectation to use the public roads, and the public expects to see and hear them, such an audible sound would not be considered disturbing. However, a late-night neighborhood house party that is audible in an otherwise quiet area could be considered disturbing.
- Another assumption in the attempt to quantify the definition is that a disturbance may be considered based on the generalized or averaged reactions across the community, such as the percent of people who are likely to complain, or the percent of people who are likely to be highly annoyed. In legal terms, the test is whether or not the sound would disturb a "reasonable" person. In any case,

because of the variability in individual reactions and, hence, in the overall range of the community reactions, more factors than audibility (a sound's level) will determine whether or not a sound is considered to be disturbing.

- Time variable sound levels
- In most cases, both the target sound (e.g., a nightclub) and the masking or background sounds will vary with time. Hence, whether or not the target is plainly audible may depend on at what time and for how long one listens. For example, does the provision mean that the target sound is plainly audible at any time between 11:00 p.m. and 7:00 a.m. or must be plainly audible for some critical percentage of the night or most of the time under consideration?
- Location (spatially) variable sound levels

- Both the target and masking sounds are likely to vary from location to location. The target's audibility should be determined at the noise-sensitive receptor site. For example, consider several locations located 100 feet from the target source. One location 100 feet away could be adjacent to a busy street with high traffic-produced sound levels, while another location could be set back from the street in a much quieter residential location. It should be noted that these last two disadvantages (time and spatial variability of source sound) also need resolution for quantitative (measurement-based) property-line type noise laws.

### 4.3.5 The Need for Perception-Based Understanding of the Acoustical Environment

Traditional urban noise control methods create the need for alternate approaches that account for community stakeholder perceptions, such as the soundscape technique, rather than top-down regulatory practices. The soundscape-based approach is vitally important as it is much more likely to garner community support and acceptance of solutions than dictatorial top-down controls. An objective standard (decibel-based) noise ordinance should match the local community's expectations and manner of living, working, and playing, and should also bring comfort and wellness to the occupants. These expectations may be reflected in the results of a soundscape study of the area.

Before the emergence of the soundscape technique, there were early pioneers in community noise research and applications. Kryter (1970) and Beranek (1992) contributed to the idea that perception was an important factor in the assessment of noise from aircraft. Sutherland (1968) and Miller (2014) conducted similar studies. Schultz (1982) and Schomer (1983) developed the use of sound metrics and formulas, which would correlate to the predicted reactions of a community to various levels of noise.

Building upon this early work, the importance of perception in noise control becomes clear. Soundscape methods provide the framework for applying perception measurements to the task of controlling noise in a community.

#### 4.3.5.1 Soundscape Principles

The principles of the soundscape technique are described in great detail elsewhere in this volume. A brief review of these principles will be provided here. The soundscape technique combines the physical and perceptual evaluation of conditions at a location. The physical conditions alone are called the *shallow soundscape* (Hiramatsu et al. 2009). These conditions may be characterized by the familiar acoustical metrics of community sound studies. Some of these acoustical metrics are described in Sect. 4.3.3.

The listener's perception of the soundscape is through the lens of their "Acoustic Biography" of past experience. The acoustic biography of the listener is formed by their aesthetics, semantic values and identity, and symbolic view of their surroundings, which are based on their individual and societal psychologies. These factors all influence the interpretation of one's environment. With such complexity, sound sources can take on complex meanings. This can be termed the "Acoustic coloration" of the larger environment.

Due to these many influences, a soundscape evaluation must include a combination of acoustical factors and other sensory, aesthetic, geographic, social, psychological, and cultural modalities across space, time, and society. This *deep soundscape* approach can be a challenge to study comprehensively. Although procedures have been standardized for soundscape evaluation, the process is not "one size fits all" when applied (Hiramatsu et al. 2009).

The soundscape research methodologies used to evaluate the acoustic environment call for interdisciplinarity (Davies et al. 2013; Schulte-Fortkamp 2014). It is important to capture the complexity of the environment beyond the A-weighted sound level. Features that describe important aspects of the environment may include psychoacoustic parameters, such as loudness, roughness, sharpness, and tonality. Ideally, these parameters are measured with binaural devices using standard metrics, and they can be used to explain annoyance from, or acceptance of, environmental noise in greater detail (see Genuit, Schulte-Fortkamp, and Fiebig, Chap. 6).

Subjective conditions are measured through individuals in the affected population who are designated as *local experts*. Several techniques are available, including soundwalks, narrative interviews, and group workshops. Interaction with local experts sharply focuses the subsequent analyses of acoustical and perceptual data on their specific concerns. Once the primary data are collected, they are analyzed to match the physical sound criteria to perceptual descriptors. This will generate comparative data between individuals and among the members of a group. Realistically, the multidimensionality of human perception *cannot* be simplified to singular numbers. The attitudes, expectations, and experiences of the local experts must be considered, as the knowledge that individuals have about the area in which they live may be the most significant factor in data collection. These diverse research strategies rely on interdisciplinarity and on consideration of the needs of the local population, including the more noise-sensitive and vulnerable groups, any cultural aspects that may be associated with the area, and the possible relevance of natural soundscapes or existing quiet areas.

An example of a successful application of the soundscape technique is the Nauener Platz (Berlin, Germany) renovation project (Schulte-Fortkamp et al. 2008). For this project, a run-down, inhospitable, and little-used public park space was transformed into a safe, welcoming, and delightful community asset. In addition to the local experts familiar with that space, the stakeholders and contributors to the renovation included architects, acoustical engineers, environmental health specialists, psychologists, social scientists, and urban planners and developers (CSI 2022).

#### 4.3.5.2 Soundscape Studies

Three types of soundscape studies have been defined based on the scope of the action for the study. Type I studies focus on the individual, Type II studies focus on a group of individuals, and Type III studies focus on higher level concerns on a larger scale. These concerns could include issues of planning and zoning, heritage sites, and existing policy. Clearly, the urban planning process is best served by a Type III study. However, the methods for reaching conclusions on urban issues involve working with groups of individuals (Lercher and Schulte-Fortkamp 2013).

Triangulation is a technique used in soundscape studies for the validation of data by comparing and cross-verifying three components of data collection: people, context, and the acoustical environment (see Botteldooren, De Coensel, Aletta, and Kang, Chap. 8). One measurement of the people component of the study is the soundscape questionnaire or survey, such as might be taken on a soundwalk that is led by the investigator. The context component may be expressed by narrative interviews or through a community workshop. Although these procedures may be organized by the investigator, the leading voices will be the local experts who are the users of the space. The final component, sound analysis, is collected by the acoustical instruments that measure related aspects of the physical environment (Schulte-Fortkamp and Fiebig 2016). With this wide variety of collected information, the investigator can integrate contextual and subjective variables and fully account for people's expertise.

## 4.4 Smart Growth Principles

*Smart Growth* is a recent movement in the urban planning process (EPA 2006). It is a comprehensive approach to land use development that incorporates specific design principles. Smart growth encourages development in existing urbanized areas and creates more compact development in non-urbanized areas to limit suburban sprawl. It promotes the proximity of jobs, shopping, and services to residential areas and provides more transportation options while ensuring access to natural areas (e.g., see Smart Growth Partnership 2022).

Smart growth seeks to enhance the quality of life by attracting a diverse population of residents and businesses to an area and then keeping them there. The quality of life is also enhanced by creating pedestrian-oriented walkable neighborhoods, diverse housing, and an array of locally owned businesses and services. Having resources nearby minimizes car trips and maximizes walking, bike-riding, and person-to-person interactions. The human scale features of smart growth, although perhaps broader, align with those promoted by the "new urbanism" approach to urban design. This is compact land-use planning at its best and includes residential, commercial, office, public, and recreational uses.

One objective of smart growth is to promote economic development. This may be done by attracting businesses and jobs to locations near housing, infrastructure, and transportation options. It seeks to promote economic development in industry clusters, expand access to education, training, and entrepreneurial opportunities, and to support the growth of local businesses, including sustainable natural resource-based businesses, such as agriculture, forestry, clean energy technology, and fisheries.

Another objective of smart growth is to create livable communities. This approach seeks to support the revitalization of city and town centers and surrounding neighborhoods by promoting development that is compact, conserves land, protects historic resources, and integrates uses (mixed-use). Smart growth encourages the remediation and reuse of existing sites, structures, and infrastructure rather than new construction in undeveloped areas. Importantly, smart growth creates pedestrian-friendly districts and neighborhoods that mix commercial, civic, cultural, educational, and recreational activities with open spaces and homes. Examples of plans for livable communities include those for the Margate City Center in Florida and the Pompano Beach Fishing Village in Florida (Figs. 4.3 and 4.4).

Smart growth principles promote alternative transportation. The objectives are to maintain and expand transportation options that maximize mobility, reduce congestion, conserve fuel, and improve air quality. Means toward these objectives include prioritizing rail, bus, boat, rapid and surface transit, shared-vehicle and shared-ride services, bicycling, and walking, and investing strategically in existing and new passenger and freight transportation infrastructure that supports sound economic development consistent with smart growth objectives. Examples of alternative transportation modes are shown in Fig. 4.5.

A significant objective of smart growth in urban planning is to design *good streets*. Really good streets are shaped (by buildings), comfortable (shaded), connected (to destinations), safe (low speed traffic with parked cars separating sidewalks from traffic lanes), and memorable (interesting visually) (Dover and Massengale 2014).

Another smart growth objective is to create a range of housing opportunities. Actions include to support the construction and rehabilitation of homes to meet the needs of people of all abilities, income levels, and household types. Other actions are to build homes near jobs, transit, and services, to foster the development of housing, particularly multifamily and smaller single-family homes compatible with

**Fig. 4.3** Plan for Margate, Florida City Center (Timothy L. Hernandez, AICP, New Urban Communities)

a community's character and vision (context), and to provide new housing choices for people of all means.

Smart growth also seeks to preserve open space, natural resources, and the environment by protecting or restoring environmentally sensitive lands, natural resources, agricultural lands, critical habitats, wetlands and water resources, and cultural and historic landscapes. A smart growth approach will increase the quantity, quality, and accessibility of open spaces and recreational opportunities.

### 4.4.1 Soundscape Techniques as an Integral Part of Smart Growth

Soundscape methods are a logical and appropriate tool to accomplish smart growth goals. It is a perfect fit to use soundscape studies to develop smart growth urban plans as these disciplines share objectives and priorities (Brooks and Schulte-Fortkamp 2019). The methods can include *soundwalks* as part of the plan evaluation. Interviews, workshops, and questionnaires can be part of the data collection and analysis procedures while applying ISO 12913-2 (2018) methods to determine user/resident preferences and priorities.

**Fig. 4.4** Site of the Pompano Beach planned Fishing Village development (top). Plan for Pompano Beach Fishing Village (bottom). (Timothy L. Hernandez, AICP, New Urban Communities)

**Fig. 4.5** Outline of alternate transportation modes (top), electric powered scooters for short distance transport (middle), (fortlauderdale.gov), free small bus intra-district transport (bottom) (ridefreebee.com)

There are several key planning and project development sound issues that may be addressed by specific perception and psychoacoustics data (see Genuit, Schulte-Fortkamp, and Fiebig, Chap. 6). These typically include:

- Construction sound—temporary or ongoing
- Transportation sound—correlated to vehicle type and speed
- Entertainment sound—locations relative to housing and hours of operation
- Other relevant sources

The way to achieve success with soundscape research and smart growth is to work with the key players, in addition to other stakeholders, to incorporate these principles. This should start at the vision phase of the project or city-wide development and continue through implementation of the plan.

The greatest success will be achieved if there is a basis of common knowledge and fluency among the players about soundscape and smart growth. The key players include:

- City Planners
- City Commission/Planning Board
- City Attorneys and Enforcement Officials
- Politicians
- Developers/Land Use Attorneys
- Architects

An accepted urban planning/development technique is the "scoring" of a planned development by a team of third-party reviewers. In the scoring process the reviewers will rate the project on a scale of 0 to 100 on how well they achieve smart growth objectives. Reviewers who are familiar with smart growth principles can introduce soundscape measurements and analysis methods as part of the evaluation process. Recommendations can be made to include ISO 12913-2 (2018) techniques into the project, which could range from project charettes to city code reviews to Master Plan revisions. Once soundscape methods are introduced and accepted as part of the development scoring process, it can be expected that soundscape considerations will be included in future development plans.

As the integration of soundscaping into the urban planning process is in 2022 in its early stages, there is a need to promote and recognize sound as an important resource and contributor to the quality of life. This integration won't happen by itself, as soundscape is not yet a part of the standard urban planning vocabulary. How do we make soundscape essential to different industries? How do we fit soundscape into the "big picture" of city/urban development? Acoustical designers can take the lead to promote this powerful and effective technique.

## 4.5 Quality of Life: The Importance of Local Experts

### 4.5.1 *Scale of Soundscape Studies*

The tools traditionally used in dB(A) mapping or other systems of measurement are not necessarily applicable for city-scale soundscape studies. This is an especially important consideration for acquiring information and using it to inform city planning ordinances or measures originating from local governments. The input and expertise of the local inhabitants of the neighborhood, district, and city are required to develop and to gain the acceptance of new or revised government mandates.

How to proceed? Unfortunately, there are not many examples of fully fledged soundscape studies with design interventions. There is no doubt that incorporating soundscape methodology into the urban planning or urban design process can demonstrate the range of possibilities and the various advantages of this approach. Indeed, when quality of life is the question, the soundscape method is indispensable! However, some limitations could arise. Different ways to conduct soundscape studies exist that vary in their scope and relevance, and some soundscape procedures may provide greater assistance to the urban policy decision-making process than others. Some soundscape procedures are easier to conduct and to be understood in the context of a given urban planning/design context. For example, which soundscape data-gathering process is more expedient and relevant for a given urban design project: soundwalks, workshops, or detailed personal interviews? The current approach is to treat each urban planning and urban design effort individually and to determine on a case-by-case basis which soundscape procedure will be the most effective in collecting appropriate data, communicating results, and informing the decision process.

### 4.5.2 *Application to Quality-of-Life Issues*

The historical development of the connection between soundscape methods and urban planning can inform the selection of the methods to be applied to a particular urban planning or urban design project, and the importance of the stakeholders (De Coensel et al. 2010; Schulte-Fortkamp and Brooks 2018). The users of the physical area encompassing the soundscape study must be considered along with their expectations. For example, are they living on a farm in an urban center? The initial quality-of-life assessment may need to be reassessed as stakeholder needs and expectations become clearer using questionnaires or group discussions.

There is a need to have an established concept for "quality of life." The issues and targets must be identified in the early stages of any design program. It is important to recognize that the motivation for the project, that is, what and who are the drivers of the program, will influence public perception. Does this effort entail the creation or revision of an existing land-use Master Plan? Is this an exercise in developing a new ordinance? Is it an individual building project? For each situation there is a need to define goals and targets. To establish quality-of-life goals, the stakeholder group must be balanced: people from all sides and viewpoints of the issue must be included. As always, there is the need to balance residential and commercial interests, as well as other potential factors specific to certain locations, such as tourism and historical relevance. Once goals are established there must be a clear path to enhancing the quality of life through sound in urban centers, suburbs, and exurbs. The following section describes a framework for defining the path toward the successful implementation of an urban planning or design project.

## 4.6 How to Implement the Toolbox

The aim of this section is to provide a guide, or "toolbox," that can be used by the practitioner to integrate the soundscape technique into the urban planning process. When using the term *urban planning*, it is including the more limited scope of *urban design*.

The holistic approach of the soundscape technique is well-suited to the urban planning and design processes, which are intended to take the concerns of a wide range of stakeholders into account. Our urban environments encompass many factors that have both benefits and impacts on the populations within them. The variety of sound sources and the perceptions of those sounds by groups of individuals in context are but one perceptual component, as important as it may be for the quality of life. Efforts to increase the positive aspects of sound perception into the design and implementation of our urban environments should be the objective of planners, officials, and other stakeholders. This toolbox can help those efforts.

The planning process has the following basic framework:

1. Identify the physical area that encompasses a project, neighborhood, district, or jurisdiction in the plan under consideration, which we will call the *soundscape space*.
2. Identify the means of developing and implementing a comprehensive plan: design best practices, project development permitting, district rules (zoning) definitions, city ordinances.
3. Identify the stakeholder populations.
4. Establish the existing acoustical/perceptual conditions in that urban space.
5. Determine what the ideal or improved conditions could be for that space, balance stakeholder interests, and create targets to move closer to those improved conditions.
6. Create, modify, or revise existing plans/rules to enable the ideal/improved environments.
7. Implement the proposed plan.
8. Conduct tests to determine the acoustical/perceptual outcomes.

In integrating the soundscape technique into the urban planning process, it is important to recognize the need to triangulate the various components of that process, including experiences, expectations, ordinances, and economic and political factors. For more information about triangulation techniques, see Botteldooren, De Coensel, Aletta, and Kang, Chap. 8. The following sections will look at each of the steps in this process and the tools that can be used to address specific issues.

### 4.6.1 Identify the Soundscape Space

The soundscape space is usually predetermined by forces that are external to the soundscape practitioner. Determine the scale of the project from neighborhood to district to city-wide. The purpose or goal for this area may be called the soundscape "program." There may be a development project that is deemed to require a beautiful or at least pleasing sound environment. There may be the need to balance competing sound generation and receiving interests in a particular city district. There may be a desire to improve the soundscape environment city-wide. Once the physical boundaries of the soundscape space are defined, the "program" can be defined, and the soundscape process can begin. Discussions with soundscape program drivers can help to define the soundscape space.

### 4.6.2 Identify the Methods for Implementation

Determine the desired means to improved quality-of-life outcomes such as a code, ordinance, standard, master plan, best practice, or a combination of means. If possible, it is important to incorporate soundscape studies into the Comprehensive Plan for the city, county, or jurisdictional entity involved. This gives the acoustical environment a place at the planning table with other important quality-of-life considerations and may be the best way to ensure that the sound in the community is treated seriously at the outset of any urban design or development project.

The implementation means on an individual project level may be driven by best practices of architecture, engineering, and urban planning or by restrictive means such as a municipality-issued conditional use permit (CUP), special permits, or similar. For a city district, such as a downtown entertainment district or zoning overlay, the implementation means could be practical rules for siting and operating residences and businesses. For city-wide soundscape spaces the implementation means could be zoning rules and enforceable municipal ordinances. Discussions at the outset with the program drivers (officials, planners, and developers) will help to define and determine the needs of the specific soundscape space in question, as an individual project, district, or city-wide endeavor. The outcomes of those discussions will lead to development of appropriate means for implementation.

### 4.6.3 Identify the Stakeholder Populations

The stakeholders will be those who work, live and/or play in the soundscape space and are affected by its environment, and those that have the responsibility to design, monitor, and/or control the outcomes in that space. These could include residents, business owners and operators, visitors and guests, and municipal officials such as planners, inspectors, enforcers, administrators, and regulatory bodies. There should be a wide net to seek out stakeholders. This can be done by announcements of workshops or other forms of public hearings, and through other community networks to draw participants to the soundscape program. A public workshop can be a useful tool to identify and attract local experts and other stakeholders to the process at the outset of the program, and to explain the relevance of soundscape data in designing effective solutions. This provides the opportunity to bring together the public and the design and planning disciplines for meaningful discussions and informed decision-making.

### 4.6.4 Establish the Existing Conditions: Context

The existing conditions may be established through physical acoustical measurements, psychoacoustic measurements, and perceptual measurements. These measurements should be done in recognition of the contextual relationships of the locations and stakeholders, in terms of both the soundscape and design sense. Just as we calibrate the physical sound-measuring equipment, sound level meters, and analyzers, so we must calibrate the soundwalk process. This would include determining the time and place of the soundwalk as well as recognition of any implicit sample bias in the participants.

### 4.6.5 Determine the Improved Conditions

Once the existing conditions are established, work with stakeholders to develop a sense of what would constitute ideal or at least improved conditions that balance the expectations of the various stakeholder sectors, using workshops and potentially virtual reality presentations. This process should recognize the potential need for conflict resolution between stakeholder sectors, which is focused through a best practice (don't design to the minimum) and "standard of care" lens.

Based on the existing condition data, discussions with program drivers should be followed by a public workshop to develop the vision for the ideal/improved conditions. This discussion should be guided by the established expectations of stakeholders. Discussions should be tempered by recognizing that stakeholders may not be aware of the possibilities for an improved acoustical and perceptual experience:

it may be necessary to manage stakeholder expectations to achieve a positive result. Discussions of relative loudness (e.g., explaining that a 10 dB reduction will be perceived, on average, as half as loud) can introduce the topic. Virtual reality presentations can be very useful for this approach. Soundwalks in already improved or target locations can also be part of the toolbox to help make the process as tangible as possible.

Determine how the quality of acoustical life may be treated at different residential economic levels, from subsidized housing to market rate to luxury housing. Determine if the outcomes are driven by private development, public regulation, or a partnership, and if the outcomes will be resource driven. Determine the aesthetic considerations for outcomes, such as the desire to make beautiful sounds within the city, structure, and people context. Determine what could constitute "stakeholder happiness." Determine the amount of flexibility which could occur during shifts in stakeholder perceptions and attitudes (Bieletto-Bueno 2017; Djimantoro et al. 2020).

### *4.6.6 Create and Enable the Means Toward Improved Environments*

This is the opportunity for the design aspect of the program to be brought forward. For an individual development project, the creation of a path to positive soundscape outcomes will encompass actual architectural, engineering, and urban planning designs, and/or project-specific conditional uses or special permits. For districts it may be the crafting of siting rules or district ordinances. For city-wide soundscape space, it could be crafting new ordinance language or zoning regulations.

Discussions may be held with program drivers to create a preliminary design or to draft legal language and then with a small subset of stakeholders (executive committee) to formulate details of the implementation plan. When the plan is fully formulated, a public workshop may be used to present and refine the plan.

Design principles and practices for the project should be applied as indicated. If a noise control ordinance or zoning regulation is required, use the guidance of Sect. 4.3 to develop the appropriate ordinance type, develop the legal outline, and craft the detailed language.

### *4.6.7 Implementing the Plan and Testing*

To implement the plan for the soundscape space the governing body for the project (owner/developer/design team), district, or city must adapt and approve the preferred means to drive the improvements into its set of designs, rules, and laws. Present the final design or ordinance, which has been developed from the measured data and tailored to the specific soundscape space to the governing bodies for acceptance and approval, in continuous consultation with program drivers and stakeholders.

To test the results, repeat the measurements that were conducted in step 4 of Sect. 4.6 (Sect. 4.6.4) for existing conditions to determine the amount of improvement post-implementation. Conduct physical sound test surveys and soundscape perception surveys in the appropriate post-build locations and conditions, to demonstrate and document any changes that occurred.

## 4.7 Soundscape Case Studies

Soundscape case studies can illuminate some of the issues to be addressed and the methods that can be used to achieve solutions. Most of the studies conducted have been focused on spaces that are limited in scale. Few studies have been conducted on a city-wide scale.

Soundscape projects of interest include those outlined in the Catalog of Soundscape Interventions (CSI 2022). The curators of the catalog state that the aim is to contribute to the conservation and design of positive and meaningful acoustic environments. They believe that novel methods and interventions can create better soundscape experiences.

Other soundscape case studies that focus on specific areas include those for the downtown beer garden neighborhood in Jamestown, Rhode Island and for a senior living development in Simsbury, Connecticut (Brooks and Paoletti 2014). The Jamestown study used a variety of methods, including physical sound surveys and stakeholder workshops. The unexpected result in Jamestown was that the residents opposed limitations on the beer garden sound emissions as this area defined their sense of place and community identity. The Simsbury study included physical sound surveys and detailed narrative interviews with residents of the subject area. The attitudes toward specific sound sources, including a fire station siren, a police firearms training range, and hunting packs of coyotes, depended on the residents' perceptions of the positive or negative impact of these sources on other facets of their lives, in other words, the context. The fire station, despite being the highest sound level source by far, was considered by all to be a beneficial life safety feature of the community and so was universally accepted. Other sources were accepted or not accepted depending on the residents' attitudes toward them, regardless of their sound level.

One study that treated a city-wide soundscape was that for New Orleans, Louisiana (Woolworth 2013). This study took a holistic approach to the issues of urban sound. It identified the main issues to be addressed, including low frequency sound, simplification of the methods for determining sound violations, the lack of focused enforcement resources, but also to consider importance of cultural events. The New Orleans study highlighted the relevance of recognizing and balancing stakeholders' needs and determining that the path forward would include a comprehensive social, cultural, and political process on a large scale.

## 4.8 Summary

This chapter described the current state of research and practice in the study of soundscape as applied to urban planning. Suggestions are made for a path forward to make soundscape considerations an integral and indispensable part of the urban planning and urban design processes. This will be necessary to achieve improvements in the quality of life experienced by all the people within our urban environments.

Physical acoustical measurements are very helpful, but now we must include the perceptual measurements embodied in the soundscape method to make further progress. The record of traditional, simple noise source emission control alone has been, at best, one of limited success. A more effective approach is to apply the current standardization of soundscape measurements and analysis, in concert with the available standards in psychoacoustics and sound management, to support our efforts to improve the quality of life.

A vital part of these efforts is to connect soundscape to smart growth principles, congruent with new urbanism, to address the needs, acceptance, and future of modern cities. Smart growth serves the economy, community, and the environment to enhance the livability in modern cities, and an essential feature of livability is the acoustical experience. Soundscape analysis and application is a proven method of providing an improved acoustical environment for urban dwellers, thus addressing a significant portion of the smart growth agenda. The intersection and similarities between soundscape goals and urban smart growth principles are the enhancement of the quality of life by balancing technical innovations and environmental protection. Moreover, integrating soundscape and smart growth advances the collaboration and co-creation between all stakeholders in the community. The application of soundscape tools to the urban planning process is the most promising approach available to create acoustical designs for better environments. This is our best opportunity to make a real difference.

**Acknowledgments** The author would like to thank those whose inspiration and assistance made this chapter possible. Foremost, I gratefully acknowledge the wonderful ideas, contributions, and encouragement that were given by my colleague and fellow practitioner in the built-environment, Pamela Jordan. Her cogent insights, pointed questions and considered solutions, expressed during the many hours of thoughtful and lively discussion about this topic were invaluable in developing the text and are greatly appreciated. I would like to thank my Smart Growth Partnership colleagues, Brandon Johnson, Bridget Huston, and Tim Hernandez, for their urban planning knowledge and the guidance that they graciously shared. I wish to express my sincere thanks and appreciation to Peggy Walton who edited my chapter in an exceptional way and encouraged me to organize my ideas most effectively. Finally, I am very thankful to Brigitte Schulte-Fortkamp and Andre Fiebig for the opportunity to contribute to this volume, for their expertise, and for their patience and kindness shown to me.

**Compliance with Ethics Requirements** Bennett M. Brooks declares that he has no conflict of interest.

## References

ANSI S12.9 Part 3 (2013) Quantities and procedures for description and measurement of environmental sound – part 3: short-term measurements with an observer present. Acoustical Society of America, American National Standard

ASTM E1503 (2014) Standard test method for conducting outdoor sound measurements using a digital statistical sound analysis system. ASTM standards

Beranek LL (1992) Noise and vibration control engineering: principles and applications. In: Beranek LL, Ver IL (eds) Criteria for noise and vibration in communities, buildings and vehicles. John Wiley & Sons, Inc, Hoboken

Bieletto-Bueno N (2017) Noise, soundscape and heritage: sound cartographies and urban segregation in twenty-first-century Mexico City. J Urban Cult Stud 4(1):107–126. https://doi.org/10.1386/jucs.4.1-2.107_1

Brooks BM (2006) Traditional measurement methods for characterizing soundscapes. J Acoust Soc Am 119:3260. https://doi.org/10.1121/1.4786099

Brooks BM, Paoletti D (2014) Soundscape and architecture—what is your vision? J Acoust Soc Am 135:2147. https://doi.org/10.1121/1.4876007

Brooks BM, Schulte-Fortkamp B (2019) Soundscape and smart growth. J Acoust Soc Am 146:2872. https://doi.org/10.1121/1.5136969

Brooks BM, DuBois TJ, Hoover R, Maling G et al (1995) A global vision for the noise control marketplace. Noise Control Eng J 44(3):153. https://doi.org/10.3397/1.2828398

Brooks BM, Schulte-Fortkamp B, Voigt KS, Case AU (2014) Exploring our sonic environment through soundscape research and theory. Acoust Today 10(1):30–40

Coelho JLB (2007) Community noise ordinances. In: Crocker MJ (ed) Handbook of noise and vibration control. John Wiley & Sons, Inc, Hoboken. https://doi.org/10.1002/9780470209707.ch130

Coelho JLB (2016) Approaches to urban soundscape management, planning and design. In: Kang J, Schulte-Fortkamp B (eds) Soundscape and the built environment. CRC Press, Boca Raton, pp 197–214

CSI (2022) Catalogue of Soundscape Intervention – Nauener Platz, https://soundscape--intervention.org/ex-nauener-platz. Last accessed Sept 2022

Davies WJ, Adams MD, Bruce NS, Cain R et al (2013) Perception of soundscapes: an interdisciplinary approach. Appl Acoust 74(2):224–231. https://doi.org/10.1016/j.apacoust.2012.05.010

De Coensel B, Bockstael A, Dekoninck L, Botteldooren D, et al. (2010) The soundscape approach for early stage urban planning: a case study. In Internoise 2010, Proc., pp. 2333–2342, Lisbon

Djimantoro MI, Martokusumo W, Poerbo HW, Sarwono RJ (2020) The historical soundscape analysis of Fatahillah Square, Jakarta. Acoustics 2(4):847–867. https://doi.org/10.3390/acoustics2040048

Dover V, Massengale J (2014) Street design: the secret to great cities and towns. Wiley, Hoboken

END (2002) European Directive 2002/49/EC 2002 of the European Parliament and of the Council of 25 June 2002 (2002) relating to the assessment and management of environmental noise. Official Journal of the European Communities L 189/12, Brussels

EPA (2006) This is smart growth, U.S. Environmental Protection Agency, https://www.epa.gov/sites/default/files/2014-04/documents/this-is-smart-growth.pdf. Last accessed 4 Sept 2022

Hiramatsu K, Torigoe K Dubois D, Schulte-Fortkamp B (2009) The concepts of soundscape. Are there shallow soundscapes and deep soundscapes? Internoise 2009, Proc., Ottawa, pp. 4084–4090

ISO 12913-1:2014 Acoustics-Soundscape-Part 1: Definition and conceptual framework. Geneva. https://www.iso.org/standard/52161.html

ISO/TS 12913-2:2018 Acoustics-Soundscape-Part 2: Data collection and reporting requirements. Geneva. https://www.iso.org/standard/75267.html

Kang T (2018) How Central Park was designed as a park for the people. https://www.centralparknyc.org/articles/peoples-park-design. Last accessed 4 Sept 2022

Keizer G (2012) The unwanted sound of everything we want. A book about noise. Public Affairs, New York

Kryter KD (1970) The effects of noise on man. Academic Press, New York and London. https://doi.org/10.1016/C2013-0-11015-X. last accessed 7 Sept 2022

Lercher P, Schulte-Fortkamp B (2013) Harmonising. In Kang J et al (eds) Soundscape of European cities and landscapes, pp. 118–127. Available online: http://soundscape-cost.org/documents/COST_TD0804_E-book_2013.pdf. Last accessed 4 Sept 2022

Levy JM (2017) Contemporary urban planning, 11th edn. Routledge, New York

Lyon RH, Brooks B (2003) The economic impact on America's products without a new national noise policy. Noise Control Eng J 51(3):166–171

Miller LN (2014) NCAC anthology in noise and vibration. EBM print on demand, distributed by the National Council of Acoustical Consultants. ISBN13: 1110000149369. https://ncac.com/an-ncac-anthology-in-noise-and-vibration-by-laymon-miller/. Last accessed 5 Sept 2022

NYMTFN (1970) New York Mayor's Task Force on Noise, Anderson, NH, Chair, Toward a Quieter City. Library of Congress TD893.3 N7 N48, Open Library ID: OL18172337M, distributed through the NY Board of Trade

planningtank.com/blog/top-planned-cities-in-the-world. Last accessed 7 Sept 2022

Prochnik G (2011) In pursuit of silence. Listening for meaning in a world of noise. Anchor books, New York

Rossing TD (ed) (2015) Springer handbook of acoustics. Springer, Science+Business Media, LLC, New York

Schomer PD (1983) A survey of community attitudes towards noise near a general aviation airport. J Acoust Soc Am 74:1773. https://doi.org/10.1121/1.390262

Schulte-Fortkamp B (2014) The challenge of interdisciplinarity in soundscape. J Acoust Soc Am 135:2148. https://doi.org/10.1121/1.4876009

Schulte-Fortkamp B (2017) Soundscape and noise management – a matter of life quality. Keynote, Internoise 2017, Proc., Hong Kong

Schulte-Fortkamp B, Brooks BM (2018) Urban planning integrating the soundscape approach. Internoise 2018, Proc., pp. 3551–3556, Chicago

Schulte-Fortkamp B, Fiebig A (2016) Impacts of soundscape terms of perception. In: Kang J, Schulte-Fortkamp B (eds) Soundscape and the built environment. CRC Press Taylor & Francis Group, Boca Raton, pp 69–88

Schulte-Fortkamp B, Genuit K, Fiebig A (2008) Developing a public space in Berlin based on the knowledge of new experts, Internoise 2008, Proc., Shanghai

Schultz TJ (1982) Community noise rating. Applied Science; Elsevier, 239–245, ISBN: 978-0853341376

Slocum SE (1931) Noise and vibration engineering, D. Van Nostrand, New York

Smart Growth Partnership (2022). http://www.smartgrowthpartnership.org/. Last accessed 4 Sept 2022

Steele D (2018) Bridging the gap between soundscape research to urban planning and design practice: How do professionals conceptualize, work with, and seek information about sound? PhD Thesis, McGill Univ, Montreal

Steele D, Luka N, Guastavino C (2012) Constructing ideal soundscapes: a practical study on closing the gaps between soundscape studies and urban design. Acoustics 2012, Proc., pp. 2163–2168, Nantes

Steele D, Kerrigan C, Guastavino C (2020) Sounds in the city: bridging the gaps from research to practice through soundscape workshops. J Urban Des 25(5):646–664. https://doi.org/10.1080/13574809.2019.1699399

Sutherland LC (1968) Sonic and vibration environments for ground facilities – a design manual. NASA-CR-61636, Doc ID 19760066879

Woolworth DS (2013) New Orleans sound ordinance and soundscape evaluation and recommendations. New Orleans, Louisiana for City Council of New Orleans, By Oxford Acoustics, Inc., 2013, Project No. 11-01883-01

# Chapter 5
# Architectural Soundscapes: Theories, Methods, and Practice

**Gary W. Siebein and Keely M. Siebein**

**Abstract** This chapter describes 5 levels and 19 elements of soundscape theory that can become integral elements in the design of architectural spaces. The chapter is focused on the transformative steps that can be taken to translate soundscape data and analysis into the physical form of a building where sound is conceived as a generator of form and is not necessarily a result of form or a series of elements added to the form. The links between architectural theories and soundscape theories are used to illustrate the basis of the levels and elements of the architectural soundscape design theory. Case studies of recent research, practice, and architectural studio classes are presented to illustrate how sound can become a part of the conceptual structure of buildings in addition to one of the functions to be accommodated in buildings. Soundscape principles are presented as tangible architectural interventions in buildings and are explored as one way to translate soundscape theory into a contributor to the initial generation of architectural space.

**Keywords** Architectural acoustics · Acoustical design · Hospital acoustics · Architecture

## 5.1 Introduction

In *Consilience: The Unity of Knowledge* (1998), Edward Wilson discusses the work of artist Piet Mondrian and the union of art and science. He states that at first Mondrian sought to imitate nature as exemplified by the painting *Wood with Beech Trees* (https://tinyurl.com/2j2798hx, 1899). The artist began with objects that look familiar: a series of trees forming a wooded area. The second phase is when art attempts to humanize nature, to make it geometrical by comparing and emphasizing form and structure as exemplified by Mondrian's paintings *Study for Blue Apple*

G. W. Siebein (✉) · K. M. Siebein
Siebein Associates, Inc., Gainesville, FL, USA
e-mail: gsiebein@siebeinacoustic.com; ksiebein@siebeinacoustic.com



B. Schulte-Fortkamp et al. (eds.), *Soundscapes: Humans and Their Acoustic Environment*, Springer Handbook of Auditory Research 76,
https://doi.org/10.1007/978-3-031-22779-0_5

*Tree* (https://tinyurl.com/2apwsuz2, 1908) and *Woods Near Oele* (https://tinyurl.com/35ufewkj, 1908). In the study *Tree 2* (https://tinyurl.com/4dpxumav, 1912), Mondrian was intensifying nature by finding elemental patterns and projecting feelings onto nature. He was drawing from the power of nature by creating analogies and calling into play all the sensory qualities of the composition—the shade, dampness, smell, sheltering of the tree—as in *The Flowering Apple Tree* (https://tinyurl.com/mr355mxa, 1912).

Wilson specifically mentions the painting *Farm Near Duivendrecht* (https://tinyurl.com/36d22e9h, 1916), in which the relative reality of the view of the farmhouse and its reflection in the water are contrasted with the abstracted pattern of the tree branches, which are not realistic portrayals at all but are abstracted renditions of the trees. This sequence of studies over a 21-year period is increasingly abstracting the whole. The tree is gradually becoming a metaphor for a tree—where the reality of the tree is gradually transformed into a symbol or interpretation of itself. The form is abstracted in increasing degrees.

By 1913 with the *Oval Composition* (trees) (https://tinyurl.com/4cjsjvyz), a new reality is revealed—a creative expression—where the "objective" information is transformed into a more conceptual structure so that the metaphor can move fluidly from one context to another, still providing the conceptual structure of the original tree, but allowing it to reveal a previously hidden conceptual structure, which gives it a new expression with the *Oval Composition No. 11* (https://tinyurl.com/4xuk8s9z, 1913–1914). Wilson (1998) concludes that this form of abstract interpretation is the link between science and art. He cites Rothstein who in *Emblems of the Mind: The Inner Life of Music and Mathematics* states:

> We begin with objects that look similar. We compare, find patterns, analogies with what we already know. We distance ourselves and create abstractions, laws, systems using transformations, mapping, and metaphors. This is how mathematics grows increasingly abstract and powerful. He continues that this is also how music obtains much of its power, with grand structures growing out of details. We pursue knowledge that is universal in its perspective but its powers are grounded in the particular. We use principles that are shared but reveal details that are distinct. (Wilson 1998, p. 239)

Architectural soundscapes often use details that are shared and universals that are distinct. In other words, one can share acoustical data, measurements, questionnaire responses, and other data and practical information. Measurements are conducted in accordance with standards, but the transformations, interpretations, metaphors, and expressions that convert the raw data into an architectural soundscape will be distinct in a similar way that the work of art is distinct from the original object.

By analogy, this projects the essence of an architectural soundscape. For example, the same acoustical consultant may work on the design of 11 different performing arts facilities with different architects, programs, and clients as shown in Fig. 5.1. All the rooms may have a similar reverberation time, but they are very different rooms architecturally and acoustically.

Siebein proposed five levels (Siebein 2013a) and seven elements (Siebein 2012; Siebein et al. 2006) of soundscape theory as can be applied to the design of natural,

**Fig. 5.1** Photographs of 11 different performing arts spaces with acoustical design by the authors. Each space has similar reverberation times and different architectural expressions. (Photographs by the authors)

urban, and architectural soundscapes beginning in the early 2000s. The five levels include inspiration, planning, conceptual structure, tectonics, and details (Siebein 2013a). These are discussed in greater detail in Sect. 5.2.1.

The seven elements are described by Siebein et al. (2010b). Elements one through five include identifying the acoustical communities involved in the soundscape, developing taxonomies of the sounds, mapping the acoustic itineraries, identifying acoustical rooms, and identifying the acoustical calendar in the soundscape. Each element has quantitative, objective, measurement aspects to them; however, these elements can also have interpretive, expressive, and design-oriented components to them as well. As a result, the sensitive soundscape designer will transform the analytic information in the process to create integrated architectural sonic interventions. They may want to immerse themselves in the physical/sonic/textural context of the project so that the expression can evolve from the essence of the context. This is perhaps where the point of departure from analytical work occurs, and the work becomes the expression of an individual designer for a specific project on a unique site for a client.

### *5.1.1 Defining an Architectural Soundscape*

An architectural soundscape is an interesting concept that allows one to link the way that sounds combine with forms that enclose the sounds (e.g., people, terrain, equipment, wildlife) that make sounds and the people and wildlife that perceive the sounds. This can be compared to a landscape, for which the etymology was shown by Stilgoe (2018) to be derived from words that indicate the area where the water and land come together, which is shaped by the waves, water, tide, currents, and the shifting sands of the littoral zone along the coast. The idea of a soundscape in constant flux, changing moment by moment, as the activities within the soundscape and those that surround the soundscape mix, is an interesting idea that can open doors to ways that architectural soundscapes can be conceived.

It is worthwhile to ask a few questions of an ontological nature about what the essence of an architectural soundscape may be that distinguishes it from other soundscapes and forms of inquiry. Kuhn (1996) argues that advancements in science are often not a result of development by accumulation, whereby one set of ideas builds on those that existed prior to its identification. It consists of a series of revolutions that required that the scientific imagination be transformed to accept what were previously thought to be incompatible ideas. This often occurs through crises identified through unexplained anomalies in several situations. In many cases cited by Kuhn, there were also concomitant changes in basic beliefs or values that accompanied the new base of scientific investigation. Design and the essence of architectural soundscapes offer possibilities of this type that can help advance thinking and design in the art and science of architecture and acoustics.

Southworth (1967, p. 2) considered the soundscape as the quality and types of sounds and their arrangements in space and time in relation to visible form, activity, and physical settings. Sabine (1993) defined architectural acoustics as the science of sound as it pertains to buildings. McCleary (1988) distinguished between science, which is concerned with the nature of nature, the humanities, which are concerned with the nature of culture, and building, which is concerned with the mutual productions of people and nature. Moreover, architecture is different from a building.

Architecture embodies the conscious design, intellectual constructs, materiality, shape, form, and experience of buildings (Rasmussen 1959; Merleau-Ponty 1964; Raskin 1966; Rapoport 1969; Fitch 1975; Heidegger 1977; Banham 1984; Bachelard 1994; Frampton 1996).

Architectural theorists have often defined architecture in terms of basic elements. Semper (1989) describes four elements that were abstracted from his analysis of a primordial dwelling: the earthwork, the hearth, the framework or roof, and a lightweight, enclosing membrane. Frampton (1996) simplified Semper's elements into two basic elements: light and heavy. The lightweight tectonics of the building frame that often consist of linear elements assembled to "encompass a spatial matrix" are derived from Semper's ideas about the framework and lightweight enclosing membrane of the building. Tectonics refer to the framework that holds the building skin and the conceptual framework that holds the concept together. The heavy, stereotomic "mass and volume often formed through the repetitious piling up or cutting away of heavyweight elements" (Frampton 1996, p. 5) of the earthwork or hearth to form the base or foundation. Semper and Frampton describe an intrinsic relationship between materiality, form, and architectural space.

Unwin (2000) carries this discussion further stating that one can have an awareness of the power of walls, as instruments of the mind, to overlay a framework of a perhaps magical order on the world. Unwin is referring not only to the physical properties of the wall as a barrier to weather, its materiality, its role in holding up the building, and other practical items, but also to its symbolic or poetic meaning in creating this order.

Gartner (1990 in Frampton 1996) concludes that "the body" and the senses (e.g., hearing) are often reduced to an aggregate of needs and constraints. The human body and senses can be accommodated by methods of design grounded in behavioral and ergonomic analyses. Gartner advocates for allowing the body and its experience to participate in the constitution and realization of architectural meaning, which is a reference to the *phenomenological* school of thought. This school postulates that meaning in design can be derived from the immersion of the body through its senses, its physical and psychophysical experiences, and its conception of space. An architectural soundscape can be considered as one of the elements that could be used to shape architectural space and fill it with meaning in addition to solving the functional needs defined by building science.

Sekler (1965) states that the tectonic has an expressivity arising from the static resistance of a constructional form in such a way that the expression cannot be accounted for in terms of structure and construction alone. Similar observations could also be made about the potential for sonic flows to become expressive in ways that cannot be accounted for only in terms of physics or psychophysics. This infers that one can make a space by giving shape to sounds while reciprocally making sounds that are shaping space. The architectural soundscape and its associated tectonics can have elements that have inseparable sonic and architectural meanings and expressions. These elements reflect the duality of making and shaping space, sound, and expression.

Southworth (1967) studied the interactions between sound, visible activity, and spatial form in a sequence or itinerary in central Boston. He found that dominant visual-auditory settings (acoustic rooms) have visible forms and activities that are supported by informative and unique sounds, both visually and sonically. He concluded that the sonic environment is an important area for new design work because it can increase a person's delight and acceptance of visible form. He pinpointed large open spaces, signs, and other communication media, a sequence, network or itinerary, and small responsive spaces also called sonic niches (Pijanowski et al. 2011) or acoustic rooms (Siebein et al. 2010b), as having the greatest possibilities for designing interventions to enhance the "delight" of people living in and experiencing the soundscape. Siebein (2013a) extended this thinking to include the exploration of the perceptual form of the soundscape. This allows one to understand the types and qualities of sounds, their spatial distributions, the extent to which they identify settings, changes in the soundscape over time, and how the soundscape shapes the space that is created in an architectural design.

Rasmussen (1959, p. 236) states that hearing architecture is possible and that if one is open, (architecture) will "open up and reveal its true essence." For example, one can conceive of a church as an instrument in which the text of a Latin prayer or one of the psalms could be chanted in a slow and solemn rhythm, carefully adjusted to the time of reverberation. The text becomes a song that lives in the church in a soul-stirring manner that turns the building into a musical experience (ibid., pp. 228–229).

Schafer (1977) discusses the concept of a soundscape as a composition or orchestration of "anyone and anything that sounds." Raskin (1966) states the purpose of composition is to lead people through a sequence of planned experiences (itineraries) that reveal the structural, ordinal, and relational qualities that make a building an organic entity, distinct from all others. When this particular meaning is conveyed to people it is called architecture. Architectural character is an honest, direct expression of discerned meaning, which is derived through a creative process and evokes an awareness of purpose.

Kuhn (1996) describes a scientific theory as an instrument for discovery. The architect tries to establish "contact with the sensitivities of his observer using as his instruments the elements of structure" (Raskin 1966, p. 53). Rhythm is one of the conditioned abilities of the human mind that translates a visually perceived pattern into a pattern that is felt as if it were being heard (Raskin 1966, p. 60). The notion of the "feel" of a space or building is at least partly related to its acoustical character (Groak 1992). Groak characterizes the language of structure as if buildings can speak. He discusses the idea that the underlying structure of architecture is often imbued with symbolism and meanings that are distinct from the functional attributes of the building and its program. He further states that acoustical design is concerned with controlling the relationships between reservoirs and flows of sound energy in buildings. The reservoirs and flows are affected by the form, volume, and boundaries of spaces, and their physical disposition in the building.

### 5.1.2 Sonic Flows and Acoustical Rooms

McLuhan, cited in Schafer (1985), described how sonic flows, similar to air or water flows, can create a space. The concept of space formed by sound is described by the way that speech gives a structure or an "invisible architecture" to the boundless, directionless space of the Arctic tundra perceived by Eskimos. In other words, when people gather and speak, they form an acoustical space within a larger environment that is shaped by their sounds and their ability to listen. The sound spreads from the person speaking and decreases in level as it propagates through space. It is reflected back to the listeners by surfaces such as walls, floors, and ceilings in buildings, which increases its level. It is obstructed by walls and is reduced in level behind a wall. The sound forms a space for hearing as it moves away from the source.

The shape and location of the space can change as someone starts speaking and stops speaking, raises the level of their voice, or moves through the environment. The sounds can be heard less clearly if there are other sounds present in the environment that mask or cover up the original signal. Cities, towns, landscapes, homes, places of work, civic buildings, and natural areas are comprised of this invisible acoustic architecture that creates acoustical rooms within larger spaces. These acoustic rooms, which are formed by sounds waiting to be heard, are the basis of the soundscape. Soundscape design is the conscious design of the acoustical properties of urban spaces, buildings, and natural areas so that high-quality communications of various types can occur among the inhabitants as they gather and form acoustical communities.

## 5.2 Design Theory for Architectural Soundscapes

### 5.2.1 Levels of Architectural Soundscape Design

Siebein (2010, 2013a) identified five levels of architectural soundscapes. The five levels are inspiration, planning, conceptual structure, tectonics, and detail (Table 5.1). Architects often use metaphors to begin the design of a building. To develop the *inspiration* for a project before any design begins, architects may contemplate in detail the sounds and textures one might experience as s/he approaches the project along an itinerary that moves through the context, into the building, and through the various spaces of the building. They may focus on special moments or places within the project that have designed sensory qualities including soundscapes. The inspiration and underlying philosophy guide the design and help to set the framework within which design *planning* occurs. In addition, buildings are often planned so that special features of the site, context, and program are exploited in the scheme.

Within soundscape design, the *conceptual structure* is the underlying set of principles and ultimately geometries that form the basis for the intellectual and formal

**Table 5.1** Summary of the five levels of architectural soundscapes

| | |
|---|---|
| 1. Inspiration | The *inspiration* for a project occurs before any design actually begins. The inspiration and underlying philosophy guide the design and help set the framework within which design occurs |
| 2. Planning | *Planning* is the larger scale design ideas that organize the experience of the building along an itinerary in space and time |
| 3. Conceptual structure | A *conceptual structure* is the underlying set of principles and ultimately geometries that form the basis for the intellectual and formal aspects of the project. A conceptual structure is often derived from transformational mapping studies of a site and localized contextual influences such as sounds, weather, climate, social forces, circulation systems, traditions, historical influences, spatial systems, and so on |
| 4. Tectonics | *Tectonics* are the elements that form the architectural system that the soundscape occurs within. The tectonic elements are arranged in a conceptual structure derived from the local ecology or interrelationships between elements. It is a pattern or system that can be mapped in its literal, physical, or metaphysical dimensions. The tectonics are those elements that give a unique identity and form to a place |
| 5. Details | *Details* are the local connections among the tectonic elements that support and express the inspiration and the conceptual structure of the project. The details are also the elements that often provide weather protection, connections between structural elements and enclosural systems, and elements that allow for sonic and other environmental flows in indoor environments |

Adopted from Siebein and Siebein (2017)

aspects of the project. A conceptual structure is often derived from transformational mapping studies of a site and from localized contextual influences that could include sounds, social forces traditions, historical influences, spatial systems, circulation systems, and geographical considerations such as weather and climate. This conceptual structure is ultimately the creator of space, the shaper or giver of sound, and the "coloration" it receives from the environment whether it is indoors or outdoors.

*Tectonics* are the elements that form the architectural system within which the soundscape occurs. These elements can be the framework that holds the walls and roof up, such as the four corner posts and roof frame described in Semper's primitive hut. The tectonic elements are arranged in a conceptual structure that is derived from the local ecology or from the interrelationships between elements in a pattern or system that can be mapped in their literal, physical, or metaphysical dimensions. The tectonics are those elements that give a unique identity and form to a place. The notion of place as defined by Norberg-Schulz (1980) is physical but metaphysical as well in the way it is understood by people.

*Details* are the local connections among the tectonic elements that support and express the inspiration and the conceptual structure of the project. The details are also the elements that often provide weather protection, connections among structural elements and enclosural systems, and elements that allow for environmental flows to occur in indoor and outdoor environments.

The levels of an architectural soundscape represent the places in the design process where soundscape theory can be effectively implemented to become an integrated part of the creation of architectural space. The levels also provide a working model for architects, soundscape designers, acoustical consultants, and others involved with the design of architectural spaces (e.g., interior designers, sculptors, sound artists, ecologists, landscape architects, and urban planners) to create and design soundscapes as part of an integrated, participatory design process for architectural spaces and other interventions.

In practice, acoustical design often occurs *after* the inspiration, planning, and conceptual structure of a building are already determined. The acoustical elements of the project must then be addressed as tectonic or detail items that are inserted into a conceptual "box" determined by other factors or the acoustical elements are built as an internal or external layer in a construction assembly. For example, one may add an internal layer in an assembly to reduce sounds that will propagate from one space to another, or an external layer may be added that is shaped with a material selected to absorb, reflect, or diffuse sounds inside a room to correct the deficiencies in a design. These parts of the process are philosophical and abstract in nature and one may seek the poetic expression of the essence of the project (Heidegger 1977; McCleary 1983).

### 5.2.2 Elements of Architectural Soundscape Design

The elements of a method to study, design, interpret, and/or creatively transform the attributes of an architectural soundscape are described in this section. The seven elements of the theory originally proposed by Siebein et al. (2010b) are presented in Table 5.2. These seven elements were expanded to 19 elements as more projects were brought to fruition (Siebein 2010, 2013a). Kuhn (1996) stated that a theory is an instrument for discovery, an idea to be tested, the ground on which a study can begin. The original seven elements of the theory were intended only as a framework to expand, to grow, or to evolve with use. They were reflective exercises structured to guide the exploration of integrated architectural and acoustical designs of the built environment. A theory is meant to be questioned, investigated, broken apart, modified, and ultimately transformed as a part of the process. As more and complex projects were undertaken, additional experiments were conducted that resulted in the addition of more steps in the process and more detail to the investigations.

The elements of the proposed method to study, design, interpret, and creatively transform the attributes of architectural soundscapes are described here, summarized in Table 5.3, and shown graphically in Figs. 5.2 and 5.3.

1. Identify and involve the acoustical communities in the soundscape process. This is often accomplished by discussions with community groups involved in a particular project or proposed land use action. For example, when a large institution such as a hospital or governmental agency develops a plan for the expansion of their facilities, the institution and the architects, planners, and attorneys acting on their behalf constitute one group. The governmental

**Table 5.2** Summary of the original seven elements of soundscape theory

| | |
|---|---|
| 1. Identify the acoustical communities | A group of people linked by the importance of the communication they exchange |
| 2. Identify the ecological connections | Webs of ecological relationships that structurally connect the members of the community such as the need to hear each other and to be able to discern subtle meanings and cues from the sounds of others are identified |
| 3. Map the acoustic itineraries | Map a path where the members of each community move, create, and listen to sounds during the course of their activities; the communication that these sounds represent is one of the connective elements of a vital functioning soundscape |
| 4. Identify the acoustic rooms | Specific locations with localized sonic events that are uniquely colored or otherwise affected by the surroundings<br>In its simplest form the acoustic room is literally a room with four walls, a floor, and a ceiling whose boundaries form the physical and sonic limits of the exchanges within it. In large indoor spaces the boundaries of the acoustic room are defined more by the three-dimensional acoustical horizon of sound reflections that travel from one person to other people that vary across the room |
| 5. Identify the acoustic calendars | The sound cycles and activity cycles for each of the participants that usually result in some variations of sounds and activities over a diurnal, monthly, yearly, or other cycle |
| 6. Map sonic flows | Flow includes sounds that occur within, between, and among acoustic rooms in the soundscape; the concept that soundscapes have somewhat permeable edges offers interesting design possibilities |
| Transform analytical information into aesthetic, architectural, urban, or natural design interventions with aural components | |
| 7. Design sonic niches or sonic interventions to fit within ecological niches in the soundscape | Sounds are not made by all members of the acoustical community simultaneously, nor are all members making the same sounds. People find, arrange, and use niches in time, pitch, loudness, location, and rhythm that their sounds can inhabit to allow the sounds to be heard alone or in combination with the sounds of others |

agencies (e.g., planning commissions, city or county commissions) and others that must approve the plans constitute other groups. Neighbors who may live or work nearby, and may or may not support the project, represent other groups. The users of the facility may represent yet another group. Open neighborhood meetings, formal commission meetings, and focus group discussions are among the ways that the various groups can be identified and their contributions to the process heard. Actively seeking and encouraging the participation of all stakeholders is an essential part of the process.

2. Study the structural/ecological relationships among the acoustical communities in the soundscape, the individual sources of sound in each community, and the specific acoustic events attributed to each sound source. The first step in this process is a series of one or more initial soundwalks to determine where each acoustical community resides, what functions each performs in the larger community, and the sounds associated with each. ISO 12913 Part 2 (2018) describes soundwalks, provides instructions on how to perform a soundwalk, and pro-

**Table 5.3** 19 Elements of architectural soundscapes

| | |
|---|---|
| 1. Acoustical communities | A group of people linked by the importance of the communication they exchange. Open informal meetings, formal meetings, and focus group discussions are among the ways that the various groups can be identified and their contributions to the process heard |
| 2. Structural/ ecological relationships | Identify the individual sources of sound in each community and the specific acoustic events attributed to each sound source |
| 3. Systems ecology diagram | Study the interactions among the participants in the soundscape related to sonic exchanges/production using a systems ecology diagram to identify hidden links among community members |
| 4. Taxonomy of sound sources | An organized list or taxonomy of individual sound sources and specific acoustic events for each of the acoustical communities in the soundscape |
| 5. Document, record, and measure | It is important to understand the level, pitch, and other acoustical attributes of each of the participants in the soundscape |
| 6. Itinerary | Document the paths of the participants in the soundscape as they move between and among various acoustic rooms to give a place and time to each of the specific acoustic events |
| 7. Acoustic rooms | Locate the acoustic rooms in the soundscape and identify the salient characteristics of each |
| 8. Sonic mapping | Map the itineraries and the acoustic rooms based on their physical location as well as on the sequence of experiences of the participants |
| 9. Acoustic calendar | Identify the acoustical calendars for each of the participants in the soundscape in time and location |
| 10. Model the existing soundscape | Model the existing soundscapes and use the measurement data to calibrate the model with the existing conditions |
| 11. Verify | Verify the model calibration with qualitative and quantitative data analysis |
| 12. Transform | The qualitative and quantitative data are often transformed in various ways by skilled designers into mappings that allow aesthetic interpretations to be made. This allows the sonic forces to become generational, formal, conceptual, structural, and/or theoretical elements in the proposed soundscapes |
| 13. Sonic flows | Investigate the flows and reservoirs of new and existing sounds and how they can form acoustical spaces |
| 14. Model new design interventions | Develop models of proposed situations that will alter the existing environment using calibrated aural simulations or auralizations of the proposed situations for evaluations by stakeholders |
| 15. Evaluate | Conduct qualitative and/or quantitative evaluations of the proposed soundscapes by the stakeholders |
| 16. Iterative process of design | Continue the development and refinement of the design proposals through an iterative design, evaluate, research, revise process |
| 17. Construction documents and construction | Work during the development of construction documents and construction with continued studies and evaluations to maintain the essential soundscape elements in the completed project |
| 18. Evaluate | Evaluate the completed project after construction using sound walks through the project, controlled listening/looking tests or other methods |
| 19. Lessons learned | Formulate "lessons learned" and topics for further research and development |

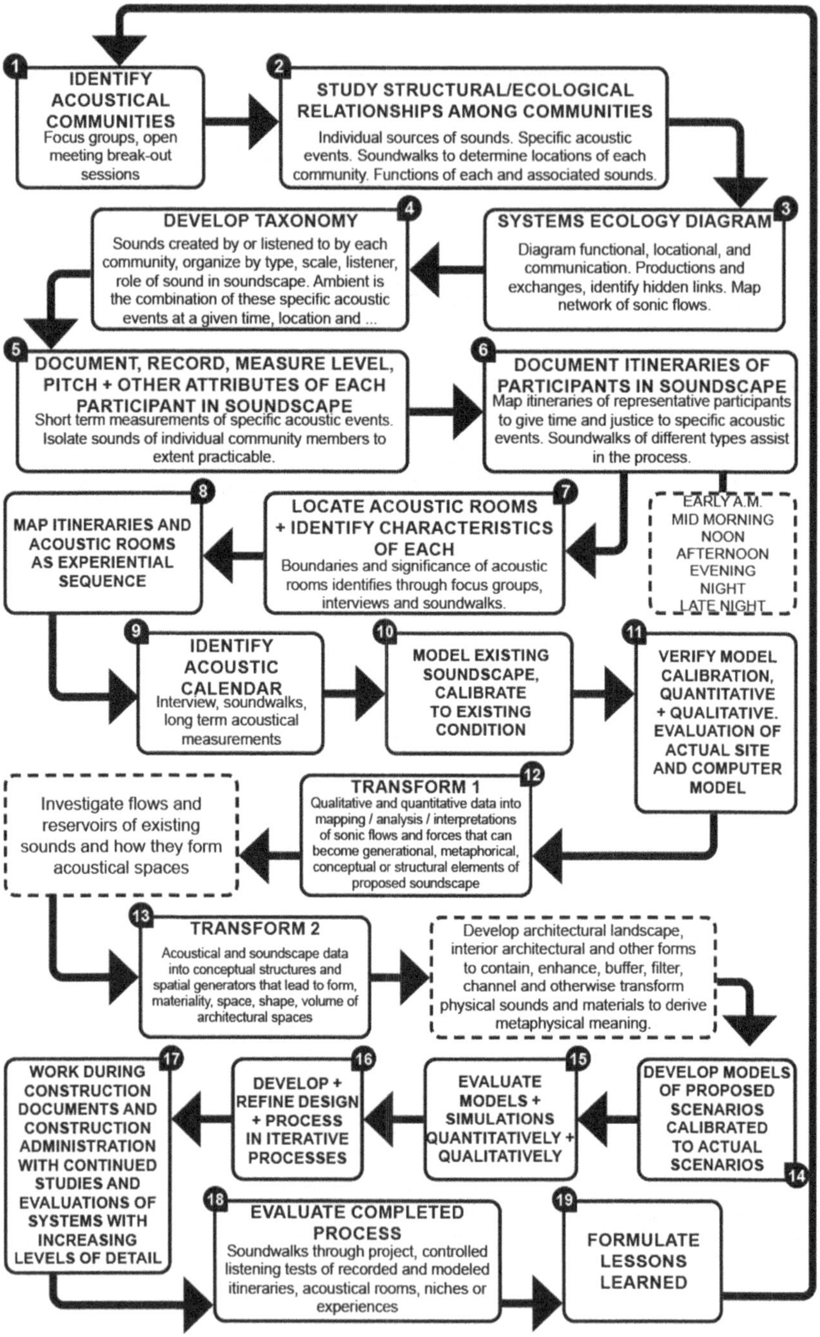


**Fig. 5.2** The 19 elements of the architectural soundscape process denoted as a single linear system

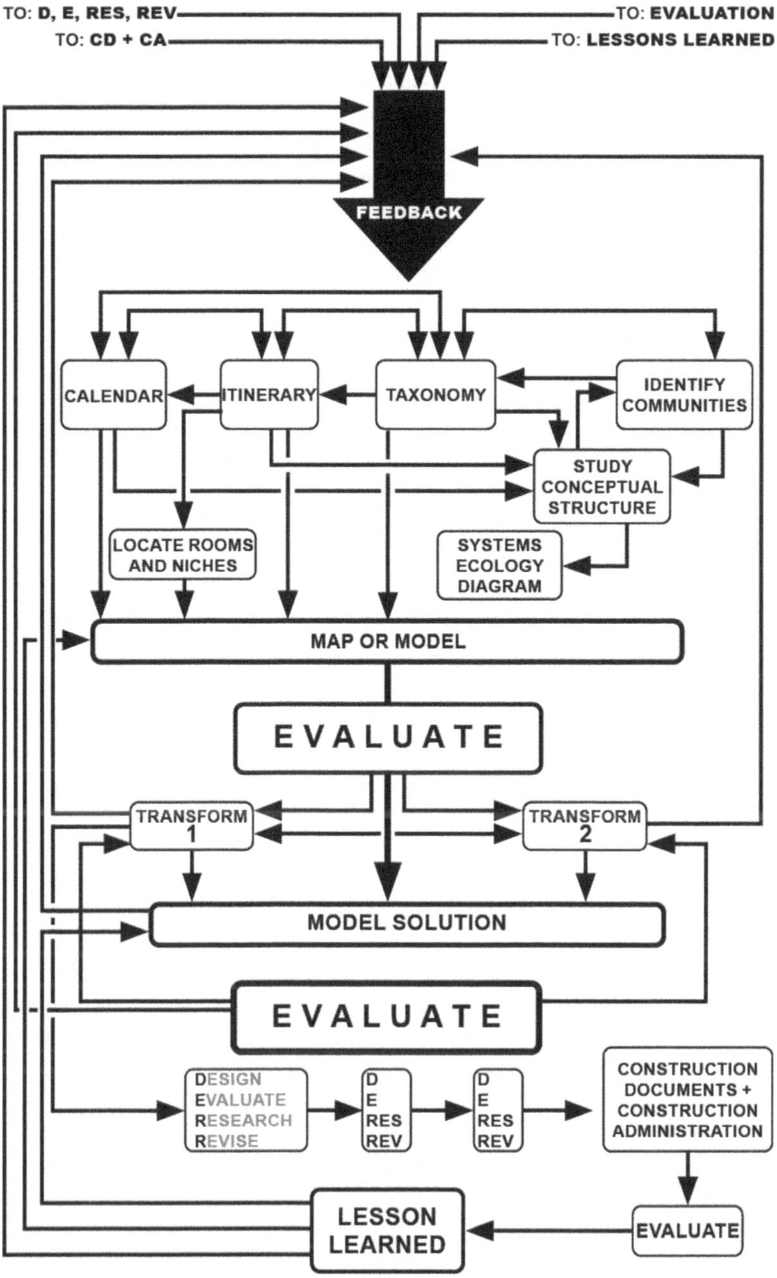


**Fig. 5.3** The feedback loops and connections between and among the 19 elements of the architectural soundscape process illustrate the complex, interactive web of information flows that dynamically connect the 19 elements over the duration of a project. Ideally, the knowledge from the interpretive mapping, modeling, and transformational elements are feeding back into both the prior data gathering elements and the later phases of design, construction, and use. Similarly, the initial and later more scientifically oriented elements feedback to illuminate the interpretive, artistic, and creative elements of the process

vides a list of qualitative questions to participants and directives for quantification. Siebein (2013b) described six types of soundwalks used for specific purposes that help to build a foundation for the project that is rooted in the context and values of the communities.

3. Construct a systems ecology diagram of the interactions among the participants in the soundscape. This is done using the methods described in Siebein et al. (2010a) and Odum (2007). Systems ecology diagramming methods can help identify hidden links among community members that perhaps they were not aware of initially. This strategy can uncover bases for consensus. The diagram can also map material, economic, energetic, sonic, and other flows among community members.
4. Develop a taxonomy of sound sources and specific acoustic events attributed to each sound source for each of the acoustical communities in the soundscape. A taxonomy is an organized list of all the members and identifies each of the sounds created by or listened to by the members as part of their daily routines. The taxonomy is organized by types of sounds, the source(s) of sound, or the listeners, and the role of each in the soundscape. Each sound is treated as a specific acoustic event that can be identified as to its source, role in the ecology of the community, level, pitch, times of occurrence, and location. This is a substantial departure from contemporary noise control or community noise regulation that considers ambient sound as a uniform, continuous sound that can be regulated by its level. The taxonomy is also used in ISO 12913 Part 2 as a means of reporting and classifying the sound sources without a value judgement. Individual designers often transform the taxonomy in a variety of ways to begin the process of design.
5. Document, record, and measure the level, pitch, and other acoustical attributes of each of the participants in the soundscape. This is usually done by short-term acoustical measurements that can, to the extent practical, isolate the sounds made by individual community members. Acoustical measurements of the overall A-weighted, C-weighted, and octave or one-third octave band flat-weighted measurements are taken as 1 second or shorter continuous equivalent sound levels. Additionally, ".wav" files are recorded at high resolution for each sound, when possible, to be used in aural simulations of the existing condition or the proposed conditions. Binaural or ambisonic recordings can yield directional and spatial information in later analysis. This measurement and recording protocol include details of sound propagation in complex environments and provides the data needed to simulate and evaluate the meanings associated with each of the sounds. Simultaneous photographic or video recording of each specific acoustic event is helpful for later analysis.
6. Document the itineraries of the participants in the soundscape as they move between and among various acoustic rooms. Mapping the itineraries of representative participants in the soundscape gives a place and a time to each of the specific acoustic events. Soundwalks of several types are conducted to assist in the documentation of the soundscape (Siebein 2013b). One type of soundwalk is for initial observation and orientation to the situation. This may be done by

the soundscape consultant alone or with one or more of the stakeholder groups. A second type of soundwalk may be conducted with one or more of the stakeholder groups present to identify the concerns and evaluations of each group as they move through their daily routines. Each group may be given questionnaires to fill out at selected locations as they experience specific acoustic events, or they may be interviewed at selected locations along the walk to elicit comments and concerns about specific situations. A third type of soundwalk, the undercover soundwalk, may be conducted by the soundscape consultant with measuring and/or recording instrumentation that is concealed so that active participants in the soundscape are not aware that documentation is occurring. This may be important when participants have control over sound sources, such as amplified entertainment equipment, because they may reduce the sound level if they know they are being measured, thereby changing the normal soundscape. A fourth type of sound walk may be conducted to collect detailed acoustical measurements or recordings of the itinerary or the acoustic rooms along the itinerary for later analysis or simulation.

7. Locate the acoustic rooms in the soundscape and identify the salient characteristics of each, including the way that sound is "colored" by its interaction with the boundaries or air space of the room. The coloration of sounds given by the rooms can be measured using impulse response techniques. The boundaries and significance of the rooms can be identified through interviews or focus group discussions with stakeholder groups or through critical listening by the soundscape consultant. In some cases, the acoustic rooms are the same as the architectural rooms. In other cases, the acoustical rooms may be smaller than the architectural rooms. Sometimes the acoustic rooms may occur between two or more architectural rooms where sounds can flow freely between the rooms.
8. Map the itineraries and the acoustic rooms based on their physical location as well as the sequence of events that soundscape participants experience. Schafer (1977) called the mapping of the soundscape as *sonography*. He noted that it is very difficult to map the psychological interpretation of sounds and their meanings, which is the ultimate goal of soundscape design and analysis. Notational systems to identify location, participant, meaning, sound level, pitch, direction, and other attributes are necessary so that the pertinent soundscape elements can be mapped, modeled, simulated, and ultimately designed into new and emerging environments.
9. Identify the acoustical calendars (sound cycles, activity cycles) for each of the participants in the soundscape and map the calendars in time and location. This is done through long-term acoustical monitoring of average sound levels in the soundscape at representative locations within each acoustic zone, along each itinerary, or in each acoustic room. Daily, weekly, monthly, or yearly patterns can be determined depending upon the needs of each project.
10. Model the existing soundscapes and use the data recorded in the previous measurement and mapping exercises to calibrate the model with the existing conditions. The models are constructed in acoustical software programs. Acoustic sources are located and assigned sound power or pressure levels, directionality,

and other parameters. Surfaces are assigned acoustic coefficients and listeners are located.

11. Verify the model calibration with qualitative and quantitative data analysis. The qualitative analysis consists of people (subjects or focus groups) taking a sound walk and critically evaluating situations they experience. They listen to recordings or other controlled aural and/or visual experiences and provide qualitative evaluations or reactions through narrative interviews, questionnaires, or other means. The quantitative analysis consists of physical, acoustical, or psycho-acoustical parameters measured through various means from the model data. Several alternative design proposals can be evaluated to determine preferences among stakeholders for variables in each scheme if desired.
12. Transform the qualitative and quantitative data into mappings for aesthetic interpretations, which allow the sonic forces to become generational, formal, conceptual, structural, and/or theoretical elements in the proposed soundscapes. This step is at the heart of the design process as data do not directly result in physical form or material. Architects, landscape architects, sound designers, exhibit designers, and others involved in the aesthetic design process transform quantitative and qualitative data into elements of a design composition, a process that is often highly individualized to the designer.
13. Investigate the flows and reservoirs of new and existing sounds and how they can individually and collectively form acoustical spaces. Contain, enhance, channel, and otherwise transform the sonic flows, reservoirs, and material elements to derive a metaphysical meaning, as described by Frampton (1996), and the poesis, or the activity in which a person brings something into being that did not exist before, described by Heidegger (1977) and McCleary (1983). The process of transforming data into forms that convey meaning as elements in a design is the point of departure from traditional acoustical consulting practice, which often addresses technical issues after the architectural and acoustical space have been defined and the associated meanings have been determined by others.
14. Develop models of proposed situations that will alter the existing environment; then develop calibrated aural simulations or auralizations of the proposed situations for evaluations by stakeholders.
15. Conduct qualitative and/or quantitative evaluations of the proposed soundscapes by the stakeholder groups.
16. Develop and refine the design proposals through an iterative process: evaluate, revise, research with the continued involvement of the design team, community members, focus groups, and other participants in the process.
17. Continue studies and evaluations during the development of construction documents and construction to maintain the essential soundscape elements in the completed project.
18. Evaluate the completed project after construction using sound walks through the project, controlled listening/looking tests of recorded or modeled itineraries, acoustical rooms, or experiences using the qualitative and quantitative methods described previously.

19. Formulate "lessons learned," topics for further research, and development postscript to enhance the design process for the next project. A summary of the process is shown is Fig. 5.2. The feedback loops between the 19 elements are highlighted in Fig. 5.3.

## 5.3 Developing Academic Case Studies

A central question that relates soundscape theory to architectural design, planning, and construction is how does one design/compose the soundscape when the building, urban environment, or natural setting does not exist at the time of the study? The aesthetic transformation of qualitative and quantitative data, along with information, and concepts from multiple influences are brought together in a composed/designed holistic soundscape that expresses the poesis of the designer/composer/artist's understandings of the project and its influences. The composer builds on the analysis, but it is not the analysis that determines the result.

Soundscape design studios were offered at the University of Florida beginning in the early 2000s that incorporated and tested design ideas through applied design work on complex building programs with acoustical issues (e.g., large performing arts centers in urban environments). Exercises explored the idea of using soundscape design as part of the conceptual structure of architecture through the transformation of "objective" acoustical and soundscape data (Siebein et al. 2019). The exercises started with the basics of a taxonomy, a calendar, and mapping of sounds on an existing site with the ultimate goal of understanding how to document a soundscape that does not yet exist. Program, site, soundscape, and mapping were transformed into three dimensions so that the space of sound became the shape of the space. A second transformation was used to organize concept plans for a school of music building on the campus of a major university. A third series of transformations were used to arrive at the actual plans for the project where the conceptual structure reads through the organization, and the tectonics and details of acoustics begin to emerge. Section sketches were developed with acoustical tectonics in mind while revealing the essence of the conceptual structure. The data gathering, analysis, and transformation process are shown in diagrammatic form in Fig. 5.4.

In another exercise, historic soundscapes were mapped to explore soundscape mapping methods. Mapping a soundscape that cannot be visited, such as an historic concert hall in a major city, led to mapping of the actual site in increasingly abstracted formats. The result was an abstracted map of the soundscape of the site that was ultimately transformed into the underlying conceptual structure for the building. Notes for the transformation analysis led to the process of abstraction that led to the concept for the building.

The class sometimes began with measurement, mapping, and modeling of sounds at an actual site. The students immersed themselves in the soundscape of the site and in the soundscape of the building program. They analyzed, drew, designed, and experienced the site. Their mappings became progressively more abstracted as the

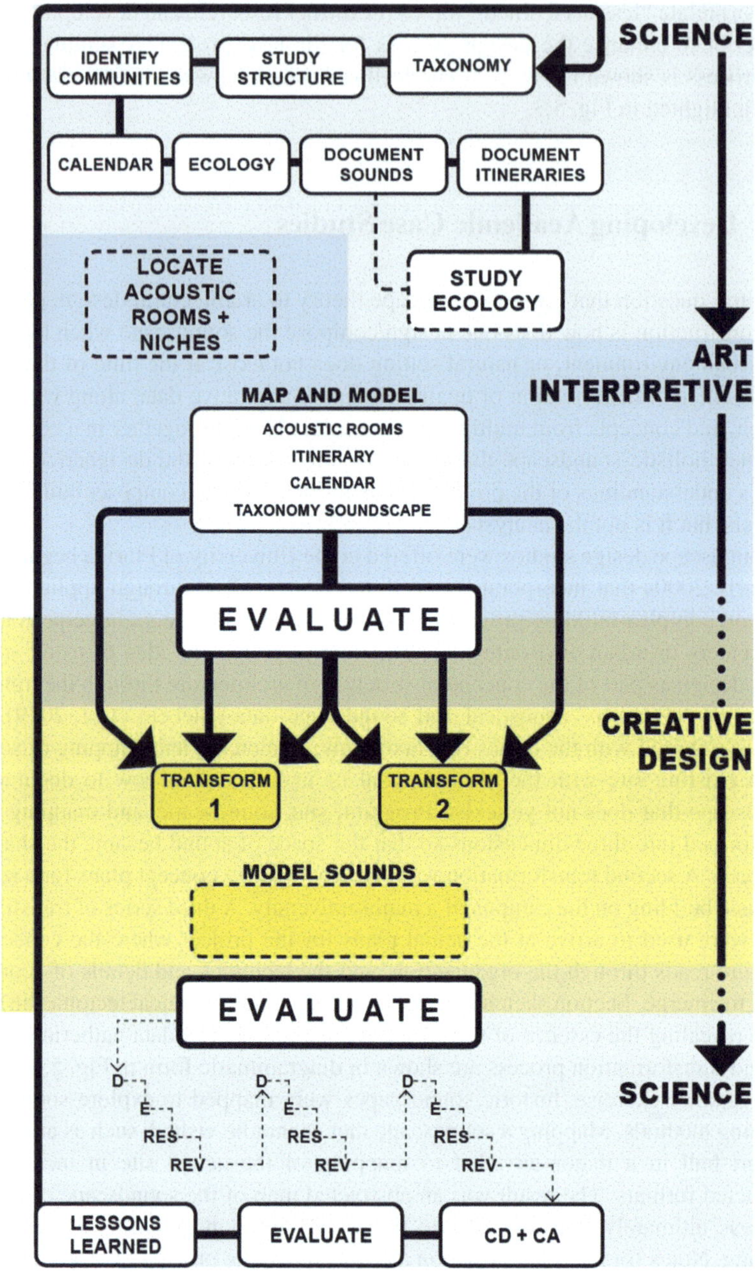


**Fig. 5.4** The 19 elements of the architectural soundscape process are shown in four general groups. The initial group of activities is scientifically based and is focused on gathering and analyzing data of different types. The second and third groups of activities are more interpretive and potentially artistically oriented. The data are evaluated and transformed into one of the five levels of soundscape design described previously. This is the domain of the soundscape, architectural, or

data became idea (or the metaphor) that was modified through the process of design. It revealed something about the nature of building and the particular program for the specific site. The connections between the actual site and sound were conceptualized as intellectual threads linking the abstracted site and sound. The mapping moved into three dimensions, sometimes using a computer and sometimes constructing a physical diagram. Taking the transformation steps from literal sound map to abstracted maps to concept model to the model of the building and site allowed the students to explore the transformation process.

The students in the soundscape design class searched, re-searched, and transformed a knowledge of acoustics into architectural ideas from the initial site map to the abstracted version. This was transformed into the layered architectural diagram where the structure of the diagram was developed from the structure of the soundscape and the forms that arose from the diagrams. The relationship of the diagram to the completed plan and three-dimensional model summarized the aesthetic, design-based process.

Schütz (2019) developed interesting soundscape design interventions and installations to create virtual extensions of space in multiple projects. This has included playing sampled and reassembled recordings of sounds made in public spaces that were designed to adjust the cognitive perception and provide perceptual masking of traffic noise by moving it to the background. She has also injected human chanting to enhance sounds (semantic transformation) and to change the acoustic perception in a gallery in New York using the concept of *spectralization*.

The contribution of architectural soundscapes to this totality can be one of functions by improving or providing suitable acoustics for a building, or by developing specific knowledge of a constituent part of the whole of architecture. However, the real potential lies in an ability to serve as a guiding light in terms of process, emerging knowledge, and aesthetic potential that could allow the architects of the future to design the sonic qualities of environments as an inherent part of the conceptual structure of the buildings and as a fundamental part of the experience of the people who work or dwell within those buildings.

This approach forms a new agenda for architectural soundscape design because it adds philosophic, theoretic, synthetic, and aesthetic dimensions to the field that demand increasing knowledge and participation in the design process. The case studies of student projects over a 20-year period illustrate the potential for soundscape information to become the conceptual structure for emerging architectural ideas in addition to informing the tectonics and details of the buildings (summarized from Siebein et al. 2019).

**Fig. 5.4** (continued) other designer. The architectural design studio classes explored the gathering and analysis of soundscape data and the transformation process as a generator of architectural form. The fourth group of activities is also somewhat scientifically based where the design intent is transformed into a building or other physical artifact through an iterative process of Design (D), Evaluation (E), Research (Res), and Revision (Rev) during the construction documents, construction, and post-construction phases of a project. The fourth step is usually only included in actual design

## 5.4 Soundscape Principles in Architectural Acoustics Research

### *5.4.1 Performance Halls*

Architectural soundscape principles have also influenced research in generating space and its underlying structure. Chiang (1994) found that responses to questionnaires about sound quality in performance halls varied among rooms, within rooms with very small distances between respondents, with the specific piece of music, and even with the specific portion of a piece of music that was evaluated. This work was expanded to include the multisensory contexts for performance spaces (Chiang 2019). Cervone (1990) also found that responses of listeners recorded on questionnaires varied not only between different halls but also among different seat location and with the music played in the same halls he studied.

The acoustical properties of performance rooms varied with the location of the listeners and with the paths by which the direct, reflected, and reverberant sound reached them in the room (Chiang 1994). Chiang developed a method to characterize the architectural properties of performance halls to represent the acoustically significant architectural features of the rooms (e.g., room length, width, balcony configurations, shell configurations, canopy configurations, and materials). The results of that study and the questionnaire studies by Cervone were used by Mahalingam (1995) to develop a computer program that designed a performance hall based on optimizing the qualitative response of listeners given the architectural shape, dimensions, and materiality of the room. That program has evolved into an expert system for architectural design of performance spaces based on acoustical performance (Mahalingam 2019).

Kwon (2006) developed an orchestral impulse response technique using an electronic sound source that simulated each of the primary instrumental groups in an entire orchestra. Multiple loudspeakers were placed on the stage of a 1800-seat performance hall to simulate the locations of the instruments. Acoustical measurements taken with the orchestral impulse response were compared to measurements made with an omnidirectional dodecahedral loudspeaker typically used for conducting acoustical tests in these rooms. The impulse responses measured at specific seats in the hall were very different for the two sound sources. The direct sounds in the orchestral impulse responses lasted for over 20 ms at many locations in the room due to the spatial separation of the loudspeakers. Reverberation times, early decay times, and early-to-late temporal energy ratios were relatively similar between the two sets of measurements. There were significant differences in center time and some early-to-late temporal energy ratios taken at increased dividing times between the early and late portions of the integrated impulse responses. Evaluations by listeners of the simulated sounds during playback in the actual performing arts hall and in listening tests conducted over headphones in the laboratory also showed significant differences for recordings made with the two sound sources.

Tsaih and Shin (2011) performed similar studies in a large auditorium constructed in a renovated church and in a band rehearsal room. This work was extended by Lokki and Patynen (2016) to create an electronic model of an orchestra with loudspeakers configured to faithfully duplicate the frequency response, loudness, and directional characteristics of each individual instrument in an orchestra.

### 5.4.2 Classrooms

Siebein (2013a) and Shin (2012, 2019) developed case studies for soundscape evaluation of open plan, collaborative learning environments. Their studies showed that when the actual communication paths used in the school were evaluated much higher Speech Transmission Index (STI) values were achieved in teaching groups compared to generic classrooms in which a teacher stands at the front of the room and speaks to children who are seated in rows of seats that are spaced evenly from the front of the room to the back. This was especially evident given the relatively high ambient sound levels in the open, active learning environments. The team spent over 1 week observing and mapping the communication paths and teaching methods used in the classrooms before any acoustical documentation was undertaken. Sound levels from students actively engaged in conversations with each other and with teachers were relatively high at many times in the building, especially when observed from outside the localized communication and learning groups. However, when sounds were measured at the locations of specific listeners, much lower levels of ambient noise and much higher levels of the teacher's voice were found. This was due to the creation of *acoustical rooms* within the larger space. Groups of students and the teacher sitting in small clusters with no more than 2–3 m from each other formed an acoustical enclosure with their heads and bodies that separated them from the other sounds and activities in the room. Engel et al. (2019) used a questionnaire and acoustical measurements to assess comfort and perception of acoustics in a classroom at multiple locations with real and fake absorbent materials in the room, and average sound pressure levels taken for the duration of the entire class and at the moment the survey was answered were compared.

### 5.4.3 Courtrooms

Von Crawford (2012) found similar results in his study of 12 court rooms. Specific communication paths were identified from observations of hearings in the rooms. Communication paths among the judge, defendant, defense attorney, prosecuting attorney, witness, the public, and the court reporter were identified. These communication paths created a matrix of 36 source–path–receiver combinations. When acoustical evaluations were made for the overall room using either the room average or one set of measurements, many of the acoustical criteria for the court rooms were met; however, when the evaluations were based on specific source–path–receiver communication channels, over 33% of the individual communication paths did not meet the criteria. Tsaih and Shin (2011) found similar results in her investigation of acoustical qualities of band rehearsal rooms as did Park (2012) in his studies of worship spaces and natural areas.

If one can discern the conceptual structure of the soundscape, one can design interventions to enhance desired elements and reduce, buffer, or mitigate those elements that are not desired (Siebein, Kwon and Smitthakorn, 2006). Furthermore, this conceptual structure can be recognized and then transformed into aesthetic dimensions. This allows sound to become an inherent part of the creation of architectural space and allows for soundscape interventions within larger environments. Computer modeling methods can be used during design to intelligently assess options for improvements in constructed spaces and for measurement and evaluation protocols employed after construction.

### 5.4.4 Music Rehearsal

Sound quality in music rehearsal spaces has been examined in a variety of studies. Pirn (1973) demonstrated the need for room volume and ceiling height in music rehearsal rooms to help dissipate excessive loudness. Egan (1988), Harris (1998), Beranek and Vér (2006), and many others have recommended reverberation times for music rehearsal rooms. Wenger (2008) published derivatives of the precedent work as recommendations for ceiling height, room volume, and floor area per musician and included suggestions for having some sound absorbing materials and some sound diffusing materials in music rehearsal rooms.

Gade (2011) interviewed orchestral musicians about sound qualities that were important to them on stage. Hearing each other was identified as the primary acoustical attribute of stage enclosures that was important to the musicians interviewed. He also took acoustical measurements on stage and identified the measurements that were related to the ability of musicians on stage to hear each other. The quantification is called "Support" and consists of two measures: ST1 and ST2 are 10 times the log of the reflected sound energy that arrives from the stage enclosure from other musicians compared to the direct sound.

Tsaih and Shin (2011) studied the soundscape of music rehearsal. Specific guidance about the type, locations, and amounts of acoustical materials to achieve the sound qualities musicians think are important during rehearsals were not generally available at the time of Tsaih's study. Tsaih and Shin (2011) used qualitative discussions and narrative documentation of discussions with music directors to identify the qualities that music directors of university performance groups think are important to hear during rehearsals. The initial set of questions to discuss with the music instructors and students was determined by focus group discussions with small groups of instructors and students and by observing and recording multiple rehearsal sessions in different rooms with very different architectural and acoustical attributes.

The responses from music directors were statistically similar to the judgments of student performers about what they listen for during rehearsals and were also correlated with the physical design of rehearsal rooms. The study identified important architectural features and the related acoustical metrics that were correlated with the ability to hear each other, play in time, and monitor intonation, dynamics, and articulation.

The acoustical qualities associated with hearing each other and other more subtle musical attributes varied not only among rooms but also within rooms at different locations. Therefore, communication paths between individual performers were studied as well as paths between each student and the instructor. Different qualities could be identified with a high degree of certainty between those seated near the source, where the direct sound and diffracted paths are important in the communication path, and those seated farther away from the source, where diffuse reflected sounds from the room are important constituents of the communication path. This research demonstrated the need for a system of diffuse reflectors to provide sound reflections across the room to allow musicians to hear each other and to allow the instructor to hear the musicians.

### 5.4.5 Worship Spaces

Park (2012) found similar results in a soundscape study of worship spaces. He gave questionnaires to five groups of participants in the worship services: the celebrant, the music director, the choir, the congregation, and the sound system operator. These groups had different roles in the service and were in different locations in the room, and there were distinct differences in the questionnaire results from each group within each space. The differences in the questionnaire data were related to differences in the composition of the sound field that each group heard. Interestingly, Park (2012) conducted both the acoustical measurements and the questionnaire study for two modes of communication in the rooms: natural acoustic propagation of sounds and sounds propagated through the sound reinforcement system. Statistically significant differences were found among data for each group of listeners in the rooms as well as between natural acoustic and reinforced sound propagation methods. In addition, statistically significant differences were found between processing functions of the reinforcement system, indicating the relative subtlety that can be addressed with that method.

These studies show the importance of the architectural features of the rooms related to the individual communication paths (source–path–receiver) that form the tectonic structure of the soundscape in an environment. Algargoosh and Soleimani (2019) conducted a study that linked the emotional perception of sounds in religious spaces with the physical environment and architectural acoustics using questionnaires, virtual reality immersion in the spaces, and monitoring heart rate and electrodermal activity. It was found that participants tended to focus on the emotional response when listening to the wet recordings of music with longer reverberation times (RTs), and participants concentrated more on the acoustics and perception when listening to the dry recordings (Algargoosh and Soleimani 2019).

### 5.4.6 Acoustical Texture

Recent advances in technology have allowed investigations in these areas that inform soundscape design. For example, Smitthakorn and Siebein (2012) conducted experiments to identify the character of acoustical texture in a virtual model of a sonic space composed only of a sound source, a receiver, and strategically located panels that directed combinations of specular and diffuse reflections to listeners at different arrival times. Orchestral music, a trumpet piece, and a piano piece were used as musical samples that were played into soundscapes with nine combinations of specular reflections and nine combinations of diffuse reflections. Smitthakorn found that the quality of texture was the primary distinguishing feature among the sound samples. Pieces that were most preferred by the listeners had specular reflections arriving in the first 40 ms after the direct sound with diffuse reflections arriving 40–80 ms, and up to 160 ms after the direct sound. Once one knows the arrival times of the reflections that are most preferred, the physical locations of diffusing panels in a sound field can be located to provide the reflections that provide preferred levels of texture for music listeners.

### 5.4.7 Interaural Cross Correlation (IACC)

Madaras (1996) used a series of studies in a 1:10 scale model of a performance hall to compare the amounts of sound diffusion from various surfaces. IACC was more strongly related with the architectural features of the room, particularly the location and amount of diffusing surfaces, when the direct sound and the first reflection were not included in the metric. This means that the sound energy reflected off the diffuse surfaces was the energy that was more heavily weighted in his metric.

## 5.5 Practical Application of Soundscape Elements in Specific Project Types

Opportunities for collaborative work to explore acoustical attributes of spaces and enhance communication abound when architects, who are interested in exploring the spatial and material definitions of space merge with soundscape designers, who are exploring the sonic flows within, between, and among spaces, to create holistic, multisensory spatial experiences in buildings and other environments. A variety of examples are considered here.

### 5.5.1 Dining Spaces

Roy and Siebein (2019) used soundscape analysis techniques in restaurants to identify specific surfaces that were responsible for directing reflected sounds across the room, making the space too loud for patrons. The owners wanted to remodel the dining spaces to include acoustical treatments. The study included taking acoustical measurements of the specific communication paths using *impulse response techniques* (Siebein and Kinzey 2010, pp. 211–229) in 44 restaurants and other dining spaces and identifying acoustical rooms where communication was desirable within the larger dining spaces. The impulse responses were analyzed to determine the paths by which sounds arrived at listener locations from people sitting across the table from each other. This communication path was called "Near." Sound reflections from people speaking across the room from the first table evaluated were called "Far." With only a few diners in the restaurant, the measured STI values for the Near and Far locations varied between 0.49 and 0.75, which means that reasonable communication could be held across the table in the Near condition and that people could also hear and understand people speaking across the room in the Far locations. As more people entered the spaces, sat down, and began to speak to each other, the measured STI values decreased to 0.21–0.31, meaning that communication became difficult to understand. The STI decreased because the conversations at the Far tables propagated across the room and become a din of background noise that interfered with communication across the Near table. People raised their voices, which is called the Lombard Effect: one tends to speak more loudly when there is noise. These sounds then propagated across the room and became noise for people seated away from them. Ultimately, this decreased the ability of people to understand conversations at their own table.

Analyses of specific impulse responses measured in existing dining spaces were presented to determine the locations of surfaces from which the reflected sounds were propagated across the room. Similar analyses could be undertaken in a 3D computer model of the room or by using graphic analyses of building sections and ray-diagramming. Tiered recommendations for average sound absorption coefficients for the restaurants were based on the amounts and locations of sound absorbing material. Those draft criteria could be used in specific spaces as targets for acoustical treatment to reduce the buildup of reflected and reverberant sounds in the room during remodeling.

The concept of sonic niches was developed by Krause et al. (2011) in the field of *eco-acoustics*. Siebein et al. (2010a) expanded the concept to apply to acoustical rooms within buildings where spaces that are smaller than the architectural boundaries of the space could be defined in space, time, pitch, level, or other acoustic parameters. This was done so that existing sounds could be preserved or enhanced, or new sounds or communication channels could be inserted and high-fidelity communication could be maintained. Sometimes architectural interventions can be used to create the niche: walls and other barriers, filters, buffers lined with sound absorbent finishes, adding textures, introducing new materials, changes in distance, floor level, or ceiling height. Acoustical principles can be applied, such as specially constructed acoustical elements, to form the niche.

Volner (2019) presents a case study of a world-renowned restaurant designed by Bjarke Ingels Group Architects and Studio David Thulrstrup that received a design award in Copenhagen. The article stated that the restaurant was noteworthy because it created intimate and varied dining experiences for patrons. Shape, materials, ceiling heights, and other environmental factors were all deliberately varied to create distinct dining experiences within one building. This concept of rooms within a larger space that are distinct environments that "engage all the senses and abound with visual and haptic delights" could be extended to include sonic delights as well. The differences among the "ever-shifting interior landscape" was unified by "ingenious overall planning" and a material palette brought together by a "Nordic flare" (Volner 2019).

### 5.5.2 Neonatal Intensive Care Unit (NICU)

A soundscape study of a neonatal intensive care unit (NICU) was performed by Siebein, Kwon and Smitthakorn (2006). They included focus group discussions, long-term sound level measurements, short-term measurements of specific acoustic events, and cataloged all the specific acoustic events using sound levels, videos, and audio recordings. A portion of the catalog is shown in Fig. 5.5. The catalog became the acoustic taxonomy of over 75 individual sources of sound identified in the study. The sound sources were then grouped into categories that included: intruding sounds from outside the NICU, building equipment noise, occupational sounds of the people, medical equipment sounds, and conversational sounds, as shown in Table 5.4. The sound level, frequency content or pitch, and times of occurrence were noted.

Sound mitigation strategies were proposed and categorized by who would be responsible for the mitigation: building design and infrastructure; hospital fixtures, furnishings, and equipment (FFE); administrative controls; and education and training. For example, a loud impact sound of a metal waste basket closing shut was grouped in the hospital FFE category because the impact could be mitigated by purchasing a waste basket with a rubber seal under the lid that reduced the impact sound. This reduction in level would occur every time that the waste basket was used.

Conversational sounds were placed in several groups based on the content of the conversation: casual conversations among staff, discussions of doctors with residents and nurses about the health and treatment of the patients, dialogue between the parents and infants, dialogue among visitors, and discussions of nonmedical staff. Methods to reduce each type of conversational sound were organized in each of the four mitigation strategies. This organization provided a framework for how

| Sound / Equipment type | Description / Purpose | Manufacturer and Model | Image | Acoustic Observations / Concerns | Sound Data $LA_{EQ}$ / $LA_{PEAK}$ |
|---|---|---|---|---|---|
| Phototherapy Lights - "bili-lights" | Bili lights use blue light to convert bilirubin that can be excreted. These lights are positioned directly above the patient. | *Draeger Photo Therapy 4000* |  | Clicking noise created from the switching of the thermal lamp unit causes concern due to the close proximity of the unit to the patient. However, the action of switching the lamp on and off is infrequent. | 72 dBA with peaks to 95 - 98 dBA as unit is switched on<br>52 - 62 dBA during operation<br>Measurements taken 1 ft. away |
| Overhead Electro Photo Light Warmer | Radiant warmers use infrared light to warm a baby, regulate a baby's temperature, and keep the patient warm during or after surgery. | *Ohmeda Ohio Infant Warmer System* | 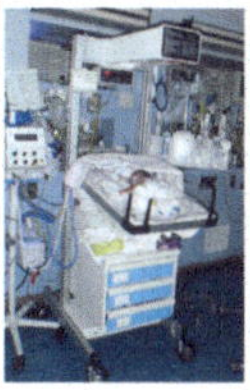 | Alarms sound in the event that equipment malfunctions, maximum and minimum temperatures need to be set, the machine becomes unplugged or the patient becomes too hot or too cold. | 78 - 80 dBA with peaks to 83 - 85 dBA when equipment malfunctions, short "beep" on 1 second interval<br>75 - 76 dBA with peaks to 79 - 81 dBA when patient temperature is out of range, 1 second beep with 1 second pause<br>Measurements taken 2 ft. away |
| Suction Pump | Suction pumps provide regulated continuous suction for tracheal and pharyngeal airway management and continuous nasogastric drainage. | *Ohmeda Medical Suction Pump* |  | The Suction Pump creates minimal motor pump noise and a constant hissing/gurgling noise made by fluids moving through the tubes. Neither noise was distinguishable beyond 2 feet from the device. | 59 - 65 dBA with peaks to 73 - 78 dBA of "hissing and "gurgling" of fluids<br>55 - 59 dBA normal motor operation<br>Measurement taken 1 ft. away |
| Ventilator | A ventilator is a machine that breathes for a patient when too weak or sick. The machine provides originated air directly to the lungs through an endotracheal tube. | *Trigger* | 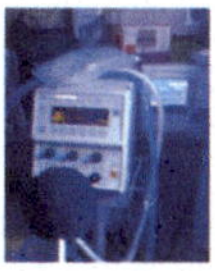 | This device alarms when overheated or in the event of equipment malfunction such as a kink in a hose. | 72 - 79 dBA alarm "beep"<br>66 - 75 dBA ambient sound level<br>Measurement taken 1 ft. away<br>Measurement taken 1 ft. away |

**Fig. 5.5** Representative section from the taxonomy table of the 75 different sounds identified in the NICU sound study

**Table 5.4** Summary of the sounds identified in the NICU and the area of responsibility for mitigation

| Types of sounds | Responsibility and mitigation |
|---|---|
| *Sounds from people talking* | |
| Parents and family members talking to infants | Strategically placed space dividers with sound absorbent facings (Design) |
| Parents and family members talking to each other, talking on cell phones, etc. while they wait for infants | Reduce propagation of sounds from any of the talking sources to other beds in unit (Design)<br>Single patient rooms— restricts flow of sound from one patient area to another (Design)<br>Sound absorbent ceiling and possibly upper walls (Design) |
| Nurses talking to each other and to parents (sometimes medically oriented discussions, sometimes gossip) | Nurses, doctors, staff, parents, and family members educated about the relative sensitivity of newborns to excessive noise (Edu) |
| Doctors and students making rounds | Distinguish between soothing conversation with parent and conversational "noise" such as gossip among staff, visitors, etc. (Edu) |
| Nurses, doctors, and staff talking during emergency events | Limit conversations in length and volume that do not have direct health and healing implications for babies (Edu) |
| *Sounds from medical equipment* | |
| Warnings, alarms, beeps, etc. of different pieces of medical equipment | Consider remote locations for continuously operating medical equipment or locate equipment in an alcove or glass enclosure within patient room (Design and Hosp)<br>Warning signals routed to PDAs used by staff set to vibrate to reduce overall noise levels (Hosp)<br>Larger size visual displays at staff locations or in patient rooms (Hosp)<br>Visual warning cues such as flashing lights to transfer emergency signals from acoustic to visual modality (Hosp) |
| *Sounds from normal occupation and use of spaces* | |
| Chairs rolling on the floor<br>Use of sinks, paper towel dispensers, etc.<br>Putting on plastic gloves<br>Emptying of trash<br>Computers, printers, etc.<br>Telephones | Look creatively at each noise source and location within the NICU and develop strategies to reduce noise created by the source (Hosp), for example, find alternate paper towel dispensers (Hosp)<br>Protect infants with private rooms (Design)<br>Limit time sources such as floor cleaning, trash removal, etc. can occur (Admin)<br>Educate staff and visitors about the sensitivity of infants to excessive noise and encourage quiet within the unit (Edu) |

(continued)

**Table 5.4** (continued)

| Types of sounds | Responsibility and mitigation |
|---|---|
| *Sounds from building equipment* | |
| Heating, ventilation and air conditioning (HVAC) system noise<br>Exhaust fans<br>Electric lights and ballasts | Design mechanical, electrical, and plumbing (MEP) systems to meet criteria for NICUs based on medical evidence (Design)<br>Careful locations of air handling units (AHU's) and terminal boxes serving NICU (Design)<br>Careful selection of HVAC units to reduce noise levels (Design)<br>Design for air velocities, use silencers to reduce duct-borne noise, vibration isolation for HVAC units, and provide adequate sound isolation for equipment room walls (Design) |
| *Sounds from outside the NICU transmitted to inside* | |
| Helicopters taking off and landing | Design exterior skin to reduce these sounds to criterion levels (Design) |
| Transportation noise such as traffic, aircraft, and trains<br>Emergency vehicles | Schedule maintenance and construction activities so that they occur to the extent possible away from areas where they can directly impact the NICU (Admin) |
| Maintenance and construction noise from the hospital | Develop alternate construction and maintenance methods to reduce noise produced (Design and Admin)<br>Limit times that construction and maintenance can occur (Admin) |

Abbreviations: *Admin* Administrative Controls by Hospital, *Design* Building Design and Infrastructure, *Edu* Education and Training, *FFE* Furniture Fixtures and Equipment, *Hosp* Hospital

sounds can be controlled practically by hospitals. This logic could be applied to other building types as well.

Ahrens and Ryherd (2019) found that exceedances of peak C-weighted sound level (LC peak) and occurrence rate measured by the maximum A-weighted sound level minus the minimum A-weighted sound level (LAmax – LAmin) used to calculate time below were related to patient outcomes in health-care occupancies. Taylor et al. (2019) interviewed medical personnel and conducted listening tests in a hospital atrium with and without natural sounds inserted in the environment. Patients sensed the insertion of the natural sounds more than nurses. Telephones ringing and babies crying were more annoying than intelligible conversations. In addition, most of these sounds were more annoying during exams than during a consult or while waiting.

### 5.5.3 Libraries

Siebein et al. (2020) discussed the evolving soundscape of a library of the twenty-first century with a historical perspective on the acoustical qualities of the earliest libraries. Public libraries in the late nineteenth and early twentieth centuries had large central reading rooms where people would come to study and read. A card catalog and central librarian's desk were also in the large central space that often had a tall domed or vaulted ceiling. The floors were made of stone. The walls and ceiling were made of masonry, wood, or plaster. The tables, desks, and chairs were not padded. The large room volume and the sound reflective materials resulted in a very reverberant room that accentuated any sounds made in the room. If one spoke to a colleague, moved a chair, or dropped a book, the sounds propagated through the room and everyone in the library knew where the sound came from and who made it. Thus, the reverberance of the room helped to maintain the quiet atmosphere because people were aware of the way that the room effectively amplified any sounds they made and carried the sounds to other locations and other patrons.

In the twentieth and twenty-first centuries, libraries began to add spaces and activities in addition to those included in the traditional library to expand its uses and diversify its users. Copy machines and other technology were added to facilitate the use of the library for modern study. Multipurpose meeting rooms were added where classes, community meetings, voting, and informal gatherings could occur. Coffee shops, book stores, dining, and open social spaces were added to encourage people to stop by the library, meet people, and possibly engage in reading or checking out digital media. Audio rooms were provided where people could listen to audio books, recordings, or watch videos. Recording studios, maker spaces, flexible use spaces, and atriums were also added to the buildings.

New acoustical analysis techniques were needed to address the reconciliation and juxtaposition of these disparate activities. Many of these activities occurred within a large, open room volume. Distance and filters or buffers rather than walls were used to divide the spaces. This allowed open views from one area to others while attempting to reduce sound transfer between spaces. The challenge for acoustical consultants and architects is to design acoustical rooms within the larger architectural rooms that have different acoustical qualities.

Sculpting sonic niches among diverse spaces for heterogeneous acoustical activities, allowing each to dwell within their own environment while other activities occur simultaneously in close proximity, is an area that would benefit from further research. How does one create a sonic transition or a buffer without the use of solid walls? Can one develop aural transitions between spaces? Can thresholds be created where one feels they are entering another, separated environment when they only move a few meters and do not close a door behind them? Can one develop architectural acoustic interventions through soundscape approaches that allow the spatial interest and excitement of these dynamic spaces to be complemented by the acoustical environment?

## 5.6 Recent Research

### *5.6.1 Disciplines Using Soundscape Ideas*

Axelsson (2020) states that Southworth was the first to use the term soundscape in a scientific journal in 1969. The article, entitled "The Sonic Environment of Cities," suggests that Southworth believed that urban design was primarily focused on the visual elements of the urban landscape but did not inherently consider all the senses (1967). Southworth's education was in architecture and urban planning. Axelsson suggests that Schafer, who is often credited with popularizing the idea of soundscapes, may have been influenced by Southworth and argues that "soundscape studies are rooted in architecture and urban design."

Indeed, a reading of Southworth's Master's thesis reveals a multifaceted study of soundscapes of 33 areas within the city of Boston that includes maps of carefully delineated acoustic attributes (1967). The study looked at the perception of sound by people with normal hearing, people who simulated those who were deaf, and people who were blind. Beautiful diagrams map the sound types in 33 "sound districts," similar to Schaffer's classification systems and Siebein's "acoustic taxonomy" (2010b). The maps show sound levels; temporal patterns similar to Schafer's "rhythms in the natural soundscape" and Siebein's "acoustic calendar" (2010b); synthesis, which includes "informativeness" and "uniqueness" as sonic qualities; activity form, which identifies separate sound events and flows; spatial form, which refers to building attributes; sound and visible form; evaluation; and the common image with localized and flowing sounds. The last component of Southworth's thesis included sonic design elements. Southworth presents three general objectives for sonic design, which include:

- To increase the diversity and informativeness of the soundscape
- To increase the number of opportunities for pure delight in sounds, particularly settings which allow individual involvement
- To increase the correlations of the sounds with visible form and activity

Southworth also states that the potential for design lies in four areas:

- Large open spaces
- Signs (similar to Schafer's soundmark)
- The sequence network (similar to Siebein's acoustic itinerary 2010b)
- Small sonically responsive spaces (similar to Siebein's acoustic room and sonic niche 2010a, b)

The soundscape was an idea defined by Schafer in 1977 as "the sonic environment" (Schafer 1977, p. 274). He explains "the term may refer to actual environments or to abstract constructions, such as musical compositions and tape montages, particularly when considered as an environment" (Schafer 1977, pp. 274–275). This definition was centered around the idea that the soundscape is composed of many players with various instruments, playing a sonic composition that varies over time.

He states, "the home territory of soundscape studies will be the middle ground between science, society, and the arts" (Schafer 1977, p. 4). Schafer imagined the soundscape approach as a cooperative study across multiple disciplines. Elizondo (2019) added that the soundscape should be evaluated in context to make design decisions that combine science and art. Elizondo looked at aspects such as perception as part of a person's cultural background, perception as a multisensorial integration and attention versus boredom, and others. His study found that soundscape design is the design of an experience that is only one part of a total reality.

Soundscape studies opened new doors for research, collaboration, and study. As more disciplines become involved, the reach of the soundscape concept expanded, folding new disciplines into the global acoustic community and increasing awareness of the various roles of the sonic environment and the soundscape approach.

Recent research includes a multitude of disciplines and multidisciplinary studies. Brown comments, "the potential outcome of adopting soundscape approaches may be that it will assist in capturing the imagination of politicians, policy makers, and a range of design professionals with respect to the management of the outdoor acoustic environment in a way that the current sole focus on environmental noise control tends not to" (2012). In fact, Kang and Aletta (2018) performed a review of recent mentions of "soundscape" in various publications and online media. They found that the number of publications and mentions have increased since 1977, with the United States and the United Kingdom contributing most of the recent soundscape research and social media mentions.

Aburawis and Yorukoglu (2018) identified links between soundscape perception and spatial experience using six soundscape perception factors and five space experience factors. They propose the use of a post-occupancy evaluation (POE) tool to construct a relationship between the soundscape perception and the spatial experience of the end users, using three phases of data collection that could be integrated into traditional POE tools during project design and construction. Paine et al. (2019) took a different approach and measured sound levels for several hours in restaurants and related architectural form, occupancy, density, and ambient sound levels in a predictive model that shows a 6 dB reduction of noise level by doubling the equivalent area of absorption developed by Rindel (2019) in three to eight locations in four restaurants. Smid (2019) used questionnaires and case studies to understand and document program, use, and acoustical issues in lobbies in performance halls. Akita et al. (2019) studied the perception of sounds that are not paid attention to in workplaces by measuring event-related responses in brain waves. He considered the sound levels and perception of different sounds from many directions heard at different levels in an office environment to understand how they are perceived.

A unique approach was used by Puyana-Romero et al. (2019). They performed surveys in which the soundscape quality and prevalent noise sources were connected to colors. Their research explored how the colors used to represent numerical data about the soundscape can influence a person's perception of the soundscape. It was found that in areas closest to the sea, the color blue was always chosen as the best representation of the individual's perception of the soundscape.

### 5.6.2 Sound and Noise

Many studies have been undertaken that involve sound and noise. Aletta et al. (2018) attempted to understand how soundscape quality affected health outcomes. They found that many soundscape studies did not include a health component. Of the studies that did, they concluded that "statistically significant associations exist between positive soundscape perceptual constructs and health benefits."

In a review of the psychophysiological effects of soundscape research on people, Erfanian et al. (2019) concluded that most research considered a maximum of two physiological responses and called for new research to consider the psychophysiology of the soundscape, similar to our understandings of noise and its health effects. Aiello et al. (2016) used social media photos of city streets to relate the apparent soundscapes present in the pictures to the human perceptions of sound. Reich (2016) looked at ecomusicology and related soundscape research to environmental issues through the study of sound. Reich explored the idea of using the latest scientific acoustic tools and technology and transforming them into an artistic expression of a musical composition to be used to raise awareness to the changing landscape and acoustic environments.

Soundscape ecology is a theoretical framework explored by Pijanowski et al. (2011), who argue that it shares parallels with landscape ecology. That field encompasses the sounds of biology or "biophony," geophysical sounds or "geophony," and human-made sounds or "anthrophony" and includes analyses of natural rhythms of various sound sources at different times of day and different months of the year. The authors assert that "society should value natural soundscapes as it does other aspects of nature."

An important aspect of soundscape design is the inclusion of other disciplines to educate those groups and to develop a broader sensitivity to sounds and noise. For example, Ruiz Arana (2019) outlined a guide for landscape architects to become more sonically aware during the design phases of projects so that they may understand how to incorporate positive sonic attributes in the landscapes they design. She also discussed "monitoring and recording how the soundscape of sites and designs evolve throughout the day and in time to help predict what future soundscapes will sound like" (Ruiz Arana 2019). Similarly, Kitapci (2019) outlined a course structure intended for interior design students that focuses on acoustical design and human perception. The study examined current course offerings in higher education institutions in Turkey and proposed a 14-week outline for an acoustics design lab to be conducted in conjunction with the interior design lab. Engel et al. (2019) conducted an interdisciplinary study on soundscape perception and urban planning using the World Health Organization's Healthy Urban Planning approach. They used questionnaires to examine urban planning concepts and acoustic metrics and determined that a correlation exists between the two indicators concepts.

Presenting the soundscape as an attribute that can be designed, controlled, and enhanced to promote the health and welfare of the user groups results in broader recognition of its importance. Serebrennikova et al. (2019) looked at differences in cultural perceptions of soundscapes to see how people might perceive the soundscape based on their cultural experiences. People of Russian and Chinese descent had similar tastes as far as preferring quieter places and places with natural sounds. Using the methods described in the ISO standards, Tsaih and Kuo (2019) conducted a soundwalk through the Longshan Temple in Taipeh and revealed that many people enjoyed the soundscape of the temple, although it was perceived as "noisy, not like a church," because the overall perception was of a calm, human-dominated, and pleasant place. This standard can be used as a tool to document how various users view the soundscape, understand which sources are the most pleasurable and which sources evoke stress, so that the unwanted sounds can be reduced and mitigated, and the wanted sounds can be preserved and enhanced. The data that result from this method, as well as other standards on acoustics, can then be transformed into the physical structures of the site or building, including acoustic interventions to reduce the unwanted sounds and to add "sonic moments," where a distinct acoustic environment is created purposefully to elicit a feeling or experience.

## 5.7 Summary

The soundscape is a term that was described by Schafer (1977) as being like a composition of sounds from many different players. If one allows the definition to remain fluid, it can begin to represent many attributes of our sonic world. By using some or all the soundscape elements outlined in this chapter, one can better understand the sonic environment and use this understanding to purposefully create a soundscape around us. The soundscape can be created by chance as the result of buildings, infrastructure, and usage sounds that are haphazardly created as a by-product of civilization. Alternately, the soundscape can be carefully crafted as a series of acoustic interventions that can enhance and preserve desired sounds, mitigate, and buffer unwanted sounds, and create new sounds to enhance architectural and acoustic environments. As architectural soundscapes become more recognized in various disciplines and among a wider range of users, a familiar and common terminology can be used to express the sonic aspects of the built world around us. There is great potential for this field in the aesthetic transformations of soundscape data to form the conceptual structure and architectural spaces in buildings of the future as dynamic, people-centered, and culturally meaningful and to form the communication channels upon which a sense of community is based.

**Acknowledgments** We express our gratitude to the editors including Dr. Brigitte Schulte-Fortkamp, Dr. Arthur Popper, Dr. André Fiebig, and Dr. Peggy Walton for insightful comments and the invitation to participate in this exciting publication. We wish to thank all the students, colleagues, practitioners, soundscape designers, and researchers whose work has contributed to this exciting field. We also wish to thank Marylin Roa for her artistic design in creating the figures and Rebecca Sutphin for many iterations of internal editing and preparation of the manuscript.

**Compliance with Ethics** Gary W. Siebein declares that he has no conflict of interest.

Keely M. Siebein declares that she has no conflict of interest.

## References

Aburawis AM, Yorukoglu PN (2018) An integrated framework on soundscape perception and spatial experience by adapting post-occupancy evaluation methodology. Build Acoust 25(1):3–16. https://doi.org/10.1177/1351010X18758478

Ahrens SA, Ryherd EE (2019) Multi-unit analysis of noise level exceedances over time in hospitals. J Acoust Soc Am 146:3022–3022. https://doi.org/10.1121/1.5137475

Aiello LM, Schifanella R, Quercia D, Aletta F (2016) Chatty maps: constructing sound maps of urban areas from social media data. R Soc Open Sci 3:150690. https://doi.org/10.1098/rsos.150690

Akita T et al (2019) Effects of sound source and its direction on subjects' impression of soundscape in workplace. J Acoust Soc Am 146:3021–3021. https://doi.org/10.1121/1.5137471

Aletta F, Oberman T, Kang J (2018) Associations between positive health-related effects and soundscapes perceptual constructs: a systematic review. Int J Environ Res Public Health 15(11):2392. https://doi.org/10.3390/ijerph15112392

Algargoosh A, Soleimani B (2019) The impact of the acoustic environments on the emotional experience of worship spaces. J Acoust Soc Am 146:3022–3023. https://doi.org/10.1121/1.5137477

Axelsson Ö (2020) Soundscape revisited. J Urban Des 25(5):551–555. https://doi.org/10.1080/13574809.2020.1810006

Bachelard G (1994) The poetics of space. Beacon Press, Boston

Banham R (1984) The architecture of the well-tempered environment, 2nd edn. University of Chicago, Chicago

Beranek L, Vér IL (eds) (2006) Noise and vibration control engineering – principles and applications. Wiley, New York

Brown AL (2012) A review of progress in soundscapes and an approach to soundscape planning. J Vib Acoust 17(2):73–81. https://doi.org/10.20855/ijav.2012.17.2302

Cervone RP (1990) Subjective and objective methods for evaluating the acoustical qualities of buildings for music. Master's thesis, University of Florida, Gainesville, FL

Chiang W (1994) Effects of architectural parameters on six acoustical measures in auditoria. Ph.D. dissertation, University of Florida, Gainesville, FL

Chiang W (2019) Performances immersed in or interacted with unique environmental contexts. J Acoust Soc Am 146:2980–2980. https://doi.org/10.1121/1.5137316

Egan DM (1988) Architectural acoustics. McGraw-Hill, New York

Elizondo FJ (2019) Contextual design of soundscapes: how to. J Acoust Soc Am 146(4):3022–3022. https://doi.org/10.1121/1.5137473

Engel MS, Caniato M, Gasparella A (2019) A study of classroom acoustics: verification of perception responses according to time-variant sound exposure. J Acoust Soc Am 146(4):2980–2980. https://doi.org/10.1121/1.5137319

Erfanian M, Mitzchell AJ, Kang J, Aletta F (2019) The psychophysiological implications of soundscape: a systematic review of empirical literature and a research agenda. Int J Environ Res Public Health 16(19):3533. https://doi.org/10.3390/ijerph16193533

Fitch JM (1975) American building: the environmental forces that shape it. Schocken, New York

Frampton K (1996) Studies in tectonic culture. MIT Press, Cambridge, MA

Gade AC (2011) Acoustics for symphony orchestras; status after three decades of experimental research. Build Acoust 18(3–4):181–206. https://doi.org/10.1260/1351-010X.18.3-4.181

Groak S (1992) The idea of building. E and FN Spon, London

Harris CM (1998) Handbook of acoustical measurements and noise control, 3rd edn. Acoustical Society of America, Woodbury

Heidegger M (1977) The question concerning technology. In: Krell D (ed) Basic writings. Harper and Row, New York

ISO (2018) Acoustics – soundscape – part 2. Data collection and reporting requirements. (12913-2), https://www.iso.org/standard/75267.html

Kang J, Aletta F (2018) The impact and outreach of soundscape research. Environments 5(5):58. https://doi.org/10.3390/environments5050058
Kitapci K (2019) Room acoustics education in interior architecture programs: a course structure proposal. In: Proceedings from Internoise 2019 noise control for a better environment
Kuhn TS (1996) The structure of scientific revolutions. The University of Chicago Press, Chicago
Kwon Y (2006) Quantitative and qualitative analyses of under-balcony acoustics with real and simulated arrays of multiple sources. Dissertation, University of Florida (Publication No. UFE0017410)
Lokki T, Patynen J (2016) Perception of music dynamics in concert hall acoustics. J Acoust Soc Am 140(5):3787. https://doi.org/10.1121/1.4967157
Madaras GS (1996) Interaural cross correlation: source of variability in concert halls. Dissertation, University of Florida (Publication No. AA00040355/00001)
Mahalingam G (1995) Acoustic sculpting: the application of object-oriented computer programming in the development of design systems for auditoria. Ph.D. Dissertation, University of Florida
Mahalingam G (2019) Acoustic sculpting revisited. J Acoust Soc Am 146(4):2979–2980. https://doi.org/10.1121/1.5137315
McCleary P (1983) An interpretation of technology. J Archit Educ 37(2):2–4
McCleary P (1988) Some characteristics of a new concept of technology. J Archit Educ 42(1):4–9
Merleau-Ponty M (1964) The primacy perception. Northwestern University Press, Evanston
Norberg-Schulz C (1980) Genius loci: towards a phenomenology of architecture. Rizzoli, New York
Odum HT (2007) Environment, power and society for the 21st century: the hierarchy of energy. Columbia University Press, New York
Paine JA, Palakapilly J, Wang LM (2019) Variations in restaurant soundscapes. J Acoust Soc Am 146(4):3022–3022. https://doi.org/10.1121/1.5137474
Park SB (2012) Soundscape of three worship spaces. Dissertation, University of Florida (Publication No. UFE0043678/00001)
Pijanowski BC, Villanueva-Rivera LJ, Dumyahn SL, Farina A et al (2011) Soundscape ecology: the science of sound in the landscape. Bioscience 61(3):203–216. https://doi.org/10.1525/bio.2011.61.3.6
Pirn R (1973) On the loudness of music rooms. J Acoust Soc Am 53:301
Puyana-Romero V, Ciaburro G, Brambilla G, Garzon C et al (2019) Representation of the soundscape quality in urban areas through colours. Noise Mapp 6(1):8–21. https://doi.org/10.1515/noise-2019-0002
Rapoport A (1969) House form and culture. Prentice-Hall, Englewood Cliffs
Raskin E (1966) Architecturally speaking. Dell, New York
Rasmussen SE (1959) Experiencing architecture. Cambridge, MIT Press
Reich MA (2016) Soundscape composition as environmental activism and awareness: an eco-musicological approach. In: Summer research. Paper 282. http://soundideas.pugetsound.edu/summer_research/282
Rindel J (2019) Restaurant acoustics. Verbal communication in eating establishments
Roy K, Siebein K (2019) Satisfying hunger, thirst, and acoustic comfort in restaurants, diners, and bars... is this an oxymoron? Acoust Today 15(2):20–28. https://doi.org/10.1121/AT.2019.15.2.20
Ruiz Arana U (2019) Thinking with my ears: guidance on sound for landscape architects. In: Proceedings from Internoise 2019 noise control for a better environment. Madrid, 2019
Sabine WC (1993) Collected papers on acoustics. Peninsula Publishing, Los Altos
Schafer RM (1977) The soundscape: our sonic environment and the tuning of the world. Destiny Books, Rochester
Schafer RM (1985) Acoustic space. In: Seamon D et al (eds) Dwelling, place and environment. Martinus Nijhoff, Dordrecht, pp 87–98

Schütz NM (2019) Urban spectralisms: relational design approaches for orchestrating and sharing space. J Acoust Soc Am 146(4):2981–2981. https://doi.org/10.1121/1.5137321

Sekler E (1965) Structure, construction and tectonics. In: Kapes G (ed) Structure in art and science. Braziller, New York

Semper G (1989) The four elements of architecture. Mallgrave and Wolfgang Hermann. Cambridge University Press, Cambridge

Serebrennikova A, Meng Q, Kang J (2019) A cross-cultural comparison in sound perception between Chinese and Russian students in urban open spaces in China. In: Proceedings from Internoise 2019 noise control for a better environment. Madrid, 2019

Shin (2012) A soundscape study in an open plan classroom in a primary school. J Acoust Soc Am 132:2071. https://doi.org/10.1121/1.4755636

Shin SB (2019) The importance of ongoing observations in educational spaces to understand and reflect changing acoustical environments. J Acoust Soc Am 146:2980–2980. https://doi.org/10.1121/1.5137318

Siebein GW (2010) Essential soundscape concepts for architects and urban planners. In: Published in the report of the designing soundscape for sustainable urban development conference. City of Stockholm Environmental and Health Administration, Stockholm, Sweden, pp 26–30

Siebein GW (2012) An exploration of the urban design possibilities offered by soundscape theory. In: Proceedings of the AESOP-ACSP joint congress, Dublin: 2013 D

Siebein GW (2013a) Creating and designing soundscapes. In: Kang et al (eds) Soundscapes of European cities and landscapes. Soundscape COST, Oxford

Siebein GW (2013b) Types of soundwalks and their applications in soundscape design. J Acoust Soc Am 134(5):4020. https://doi.org/10.1121/1.4830676

Siebein GW, Kinzey BY (2010) In: Cavanaugh et al (eds) Architectural acoustics: principles and practices. Hoboken, Wiley

Siebein GW, Siebein K (2017) Case studies that explore the soundscape of dining. J Acoust Soc Am 142:2593

Siebein, GW, Kwon, Y, Smitthakorn, P (2006) An acoustical palette for urban design. In: Proceedings of ICSV 13, Vienna, Austria, pp 2455–2460

Siebein GW, Jones C, Bum Park S, Siebein KM, et al., (2010a) The ecological basis for architectural and environmental soundscapes. In: Proceedings of Internoise 2010: noise and sustainability, INCE, Indianapolis

Siebein GW, Bettcher A, Bum Park S, Bong Shin S, et al. (2010b) Architectural and acoustical elements of soundscapes. In: EAA Euroreggio congress on sound and vibration 2010. Ljubljana, Slovenia

Siebein GW, Paek HG, Roa M, Siebein KM et al (2019) Architectural soundscapes. J Acoust Soc Am 146(4):3022–3022. https://doi.org/10.1121/1.5137476

Siebein GW, Siebein KM, Roa M, Paek HG (2020) The soundscape of 21st century libraries. Acoust Today 16(4):57–66

Smid JN (2019) The changing soundscape of the performing arts lobby. J Acoust Soc Am 146(4):2980–2980. https://doi.org/10.1121/1.5137317

Smitthakorn P, Siebein GW (2012) Architectural acoustics: effects of specular and diffuse reflections on perceived music quality. Lambert Academic Publishing, Saarbrucken

Southworth MF (1967) The sonic environment of cities. MIT Press, Cambridge, MA

Stilgoe JR (2018) What is landscape? MIT Press, Cambridge, MA

Taylor A, Moore D, Smith LSW (2019) Improving the clinical environment soundscape. J Acoust Soc Am 146(4):3023–3023. https://doi.org/10.1121/1.5137478

Tsaih LS, Kuo YT (2019) Longshan Temple soundscape. J Acoust Soc Am 146(4):2981–2981. https://doi.org/10.1121/1.5137320

Tsaih L, Shin J (2011) Soundscape of music rehearsal in band room. Dissertation, University of Florida, Gainesville, Florida (Publication No. UFE0043223)

Unwin S (2000) An architectural notebook. Routledge, London

Volner I (2019) A new home for one of the world's top restaurants creates intimate and varied dining experiences. Architect Mag 108(5):276–279. https://www.architectmagazine.com/project-gallery/noma-2-0_o

von Crawford CN (2012) Soundscapes of justice, trial courtrooms. Master's Thesis, University of Florida (Publication No. UFE0044686/00001)

Wenger Corporation (2008) Planning guide for school music facilities. https://www.wengercorp.com/Construct/docs/Wenger%20Planning%20Guide.pdf

Wilson EO (1998) Consilience: the unity of knowledge. Vintage, New York

# Chapter 6
# Psychoacoustics in Soundscape Research

**Klaus Genuit, Brigitte Schulte-Fortkamp, and André Fiebig**

**Abstract** Usually when noise effects are considered with respect to well-being and health, A-weighted sound pressure level indicators are analyzed. However, several decades ago researchers started to use measurement methods to quantify auditory sensations in more detail. Later the soundscape pioneer Murray Schafer described acoustics and psychoacoustics as the cornerstones to understanding the physical properties of sound and the way sound is perceived. This approach emphasized that all aspects of soundscape are related to perception. Psychoacoustic data are considered for a more comprehensive evaluation of acoustic environments that goes beyond the simplified use of sound level indicators. Moreover, a key consideration is that acoustic environments are perceived binaurally by humans. Thus, measurement equipment that collects spatial information about the acoustic environments is increasingly being applied in soundscape investigations and consequently is suggested in soundscape standards. Following the soundscape concept, all measurements and analyses must reflect the way soundscape is perceived by people in the appropriate context. This insight led to an increase in research and applications of psychoacoustic measurements to understand the effects of acoustic environments on humans in more detail. Although the general value of psychoacoustics is broadly acknowledged in soundscape research, several research questions remain that must be addressed to fully understand the relevance of psychoacoustic properties in different environments and contexts.

K. Genuit (✉)
HEAD acoustics GmbH, Herzogenrath, Germany
e-mail: klaus.genuit@head-acoustics.de

B. Schulte-Fortkamp
HEAD-Genuit Foundation, Herzogenrath, Germany
e-mail: bschulte_f@web.de

A. Fiebig
Department of Engineering Acoustics, Technische Universität Berlin, Berlin, Germany
e-mail: andre.fiebig@tu-berlin.de



B. Schulte-Fortkamp et al. (eds.), *Soundscapes: Humans and Their Acoustic Environment*, Springer Handbook of Auditory Research 76,
https://doi.org/10.1007/978-3-031-22779-0_6

**Keywords** Acoustic environments · Artificial head · Auditory perception · Binaural hearing · Environmental noise · Masking · Psychoacoustics · Psychoacoustic parameters · Soundscape assessment

## 6.1 Introduction

More than 40 years ago, Schultz (1978) published a dose-effect relationship that could be used for predicting community annoyance due to transportation noise of all kinds. His predictor of community reactions offered the prospect of a technical rationale for environmental noise regulations and, due to his influence, non-acoustic factors became equally important in determining a community's reaction to noise (Fidell 2003). Though the U.S. Federal Interagency Committee on Noise (FICON), the Environmental Protection Agency (EPA), and other groups had set up research projects to study annoyance caused by environmental noise, there were still no explanations for large differences in the relative rates of annoyance recorded for different communities with the same noise levels. Thus, accurate predictors of potential annoyance remained elusive.

The Directive 2002/49/EC relating to the assessment and management of environmental noise (the Environmental Noise Directive: END) since 2002 is the main instrument used in the European Union to identify critical noise pollution levels and to trigger necessary actions both at Member State and at an European-wide level. The action areas on which END focuses include: (1) the determination of exposure to environmental noise, (2) ensuring that information on environmental noise and its effects is made available to the public for preventing and reducing environmental noise where necessary, and (3) the preservation of environmental sound quality where it is good. Since environmental noise can be persistent and inescapable, a significant proportion of the population suffers from long-term exposure. Thus, the quantification of the related burden of diseases from environmental noise, including reduction in the number of years of healthy living, is an emerging challenge for policy makers (WHO 2014).

Environmental noise investigations based mostly on measurements and calculations of A-weighted sound pressure level did not allow accurate predictions of human responses to the noise. In most cases the A-weighted equivalent continuous sound level ($L_{\mathrm{Aeq}}$) or the day-evening-night average sound level ($L_{\mathrm{DEN}}$) with level adjustments for evening (+5 dB) and night (+10 dB) are considered more appropriate in the context of environmental noise exposure (e.g., ISO 1996-1 2016). However, by using those approaches, the description, classification, and assessment of environmental noise is based on a single parameter related to one property of sound, its level, and the psychoacoustic characteristics of the soundscape and the auditory sensations it elicits are completely neglected. Therefore, the need for measurements mimicking the way humans perceive sounds was increasingly acknowledged and the inclusion of more signal parameters related to psychoacoustic properties of sounds was considered. The insight that measurements and assessments must be guided by the way humans typically perceive the soundscape under scrutiny promoted the increasing use of binaural measurement systems and psychoacoustic analyses in soundscape research (Engel et al. 2021).

To understand the complexity of human responses to environmental noise, the transformation of the *sound event* (the physical event) into a *hearing event* (the perceptual event) must be considered. There are several different aspects influencing this transformation. First, the physical aspect of the sound event includes the change in the sound from a sound source to the human ears depending on the direction of incidence. Second, the psychoacoustical aspect, how the inner ear processes sound, is dependent on the time structure and frequency distribution. The third aspect, the psychological cognitive aspect, involves the context, kinds of information, individual expectations, and the attitude of the listener toward the sound that influence how the sound event is classified and interpreted (Genuit 2002). Thus, the accurate description of environmental noise requires not only the utilization of measurements capable of replicating the physical mechanisms involved in human hearing but also the application of extensive knowledge about binaural signal processing, psychoacoustics, and the cognitive aspects of human sound perception.

## 6.2 Listening to Acoustic Environments

The acoustic environment is a distinct component of the sensory experience of humans. "A Soundscape consists of events heard not objects seen" (Schafer 1994).

### *6.2.1 Spatial Hearing*

One of the most relevant aspects of the human auditory system is its ability to process the differences in information provided by the left and right ears. This binaural signal processing is essential for spatial hearing, which can provide source direction and distance, and is advantageous for pattern recognition and localization of different sources (Blauert 1996). Sound source localization is possible both horizontally and vertically, although the mechanisms involved vary along the corresponding planes (Shaw 1997; Fay and Popper 2005). In the horizontal plane, localization is based on the evaluation of interaural differences (i.e., the differences between level and phase reaching the left and right ears). Sound waves originating from sources located outside of the median plane (having a lateral offset) travel paths of different lengths toward the left and right ears of the listener, which results in different times of arrival (i.e., delayed on one side). The human brain can interpret these delays (less than one millisecond) as directional information. These differences in arrival time, typically denoted as interaural time differences (ITDs), reach their maximum when the sound source is located to the left or right of the listener. The ITD cues are considered most important for horizontal sound source localization and work especially well for low frequencies when the auditory system is able to detect pure phase differences between both ear signals. In addition to ITDs, acoustic shadowing caused by the head, shoulders, and torso of the listener introduces interaural level differences (ILDs) for signal components with small wavelengths (i.e., at higher frequencies). The combination of ITDs and ILDs are used by the human auditory system for horizontal sound source localization.

Sounds from sources located along the median plane of the human head will provide equal stimulation to the left and right ears (Blauert 1974). Nonetheless, humans are capable of localizing sound sources in the median plane, for example, distinguishing the sound source location in front from behind by moving the head to cause left and right differences. In addition, the frequency spectrum of the received sound varies with the direction of the sound source due to the direction-dependent filtering of sound caused by human anatomy: the shape of the auricles, head, shoulders, and upper body of the listener. These spectral differences can be interpreted by the brain as directional information because certain distortion patterns are associated with specific directions (Blauert 1996).

These direction-dependent modifications imposed on the received sound are often summarized as a pair of filters for each horizontal and vertical direction of incidence known as the *head-related transfer function* (HRTF). Unlike conventional sound measurement systems comprised of a microphone with an omnidirectional, flat frequency response, the sound pressure level (SPL) measured at the eardrums shows frequency-dependent variations between +15 and −30 dB related to different directions of sound incidence. These spectral modifications are introduced through interactions of the sound field with the anatomy of the listener and can be categorized as direction-dependent modifications, such as diffractions, reflections, and shadowing, as well as direction-independent alterations observed in the form of resonances. The spectral pattern induced by this directional filtering is illustrated in Fig. 6.1.

One of the major advantages of binaural signal processing is illustrated in Figs. 6.2 and 6.3. In this example, two spatially distributed loudspeakers (− 60° vs + 60° from the center) emit different signals: white noise (left loudspeaker) and diesel engine noise (right loudspeaker). Two different "receivers" were used. An omnidirectional microphone was placed at the same position (i.e., distance and height) from the sound sources as the paired microphones of an artificial head. With the omnidirectional microphone, the diesel engine noise from the right loudspeaker was masked by the white noise masker playing on the left side (Fig. 6.2). In contrast to this, the artificial head experiment shows that due to the filtering properties of human hearing, the diesel engine noise becomes identifiable in the spectrum of the right ear signal (Fig. 6.3). This means a human being could hear and distinguish the sound from the diesel engine.

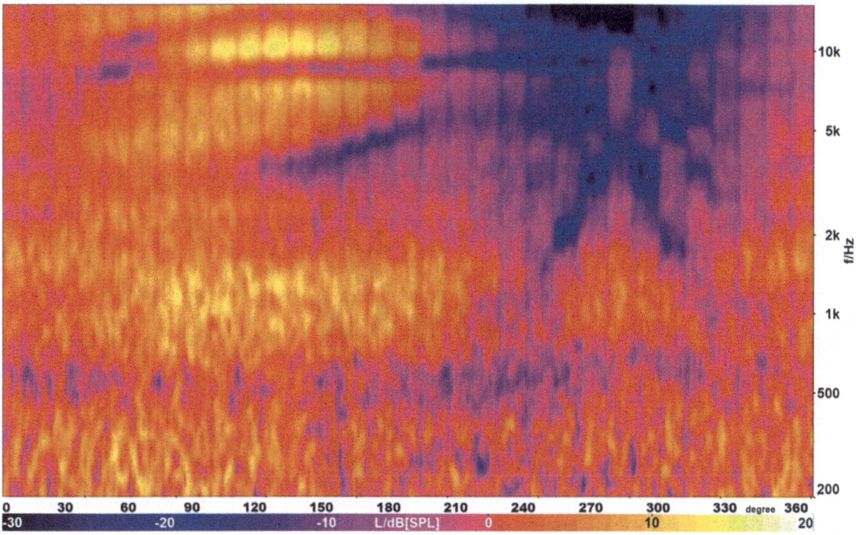


**Fig. 6.1** Spectral pattern induced by the filtering properties of the human anatomy on sound arriving at the left ear. Frequency-dependent variations between –30 and +15 dB can be observed for different directions of sound incidence. The sound source was in the horizontal plane passing through the ears, and at constant radius from the centre-point between the ears. It circled counter-clockwise as seen from above looking down, starting from the center-front direction at 0 degrees and continuing full circle back to 0 degrees. The vertical stripes in the illustration are instrumentation artifacts due to the use of fixed source positions with a resolution of 10 degrees. (Adapted from Genuit and Sottek 2010)

For the binaural condition, the diesel engine noise is effectively released from masking by white noise from the left (cf. Flanagan and Watson 1966). Comparing the panels in Figs. 6.2 and 6.3 reveals that that pre-, post-, and simultaneous masking properties will be different for the binaural condition when the masker and the masked signal have different directions of incidence. The human auditory system is able to detect single sound sources in a complex soundscape with several sound sources at different locations. It could be possible that a sound source with a lower level but with a specific, discernable pattern in the time and/or frequency domain could contribute significantly to the overall perceived annoyance.

Humans exploit the capabilities offered by binaural signal processing to enhance speech intelligibility in noisy environments (vom Hövel 1984). In the case of complex auditory environments comprised of several spatially distributed sound sources, further advantages are given by the capacity to direct auditory attention toward individual sound sources. In a noisy environment, speech intelligibility can vary by 12 dB depending on sound incidence, meaning that the level of a sound source could be decreased by up to 12 dB for specific source locations without any influence on the detectability. This specific capability of human hearing was

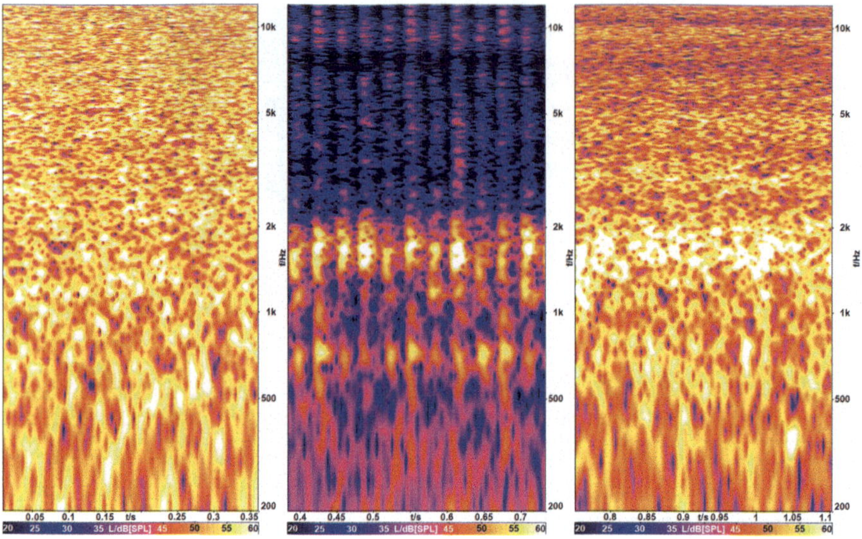


**Fig. 6.2** Sound reception by an omnidirectional microphone for two different sources at two different locations. The panels depict the spectrograms of white noise (left) and diesel engine noise (middle) placed at different positions (−60° and +60°, respectively) in azimuth. The panel on the right shows the spectrogram from the omnidirectional microphone recording. The dB scale along the bottom of each panel is color-coded with blue representing the lowest level (SPL), pink-red intermediate levels, and yellow the highest levels (up to 60 dB) at the frequencies shown on the vertical scales

observed and described long ago by Cherry (1953) but is only considered later in soundscape research.

Another relevant feature of the human auditory system is its high simultaneous resolution in the frequency and time domains, which is complemented by a high dynamic range comprising more than 120 dB. For ordinary signal analyzers using Fourier analysis (e.g., FFT), the product of spectral resolution and temporal resolution equals 1 or higher; increasing the resolution in one domain leads to a decreased resolution in the other domain. Psychoacoustic investigations have shown that the product of temporal and spectral resolutions amounts to 0.3 for the human auditory system, meaning that humans use a high-frequency resolution and, at the same time, can perceive fast temporal variations such as short amplitude or frequency changes (Genuit 1992a).

These observations illustrate the remarkable performance of human binaural sound signal processing capabilities, which are still difficult to match by current technical devices and analysis methods. They also exemplify the value of binaural measurement systems for the evaluation of complex auditory environments for which an analysis based on human perception is essential. Only binaural recording in combination with calibrated, equalized playback headphones guarantee signals for the listener that are comparable to the signals the listener would hear in the original situation.

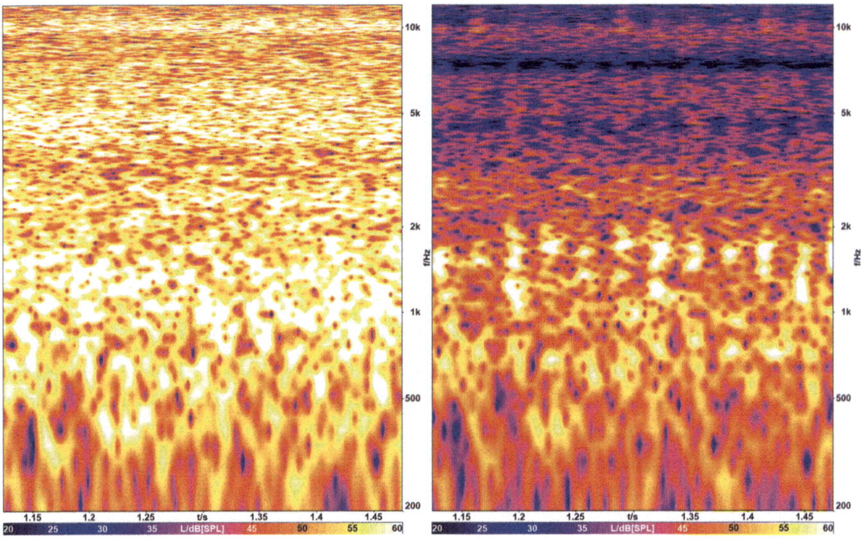


**Fig. 6.3** Example of the unmasking of spatially distributed sound sources due to the directional filtering properties of human hearing. Same sound presentation as shown in Fig. 6.2 but received by an artificial head. The figures show the spectrograms from the received sounds at the left (left panel) and right (right panel) ears. Within the right ear the diesel engine noise is clearly visible. Sound pressure level (dB SPL) and frequency scales are identical to Fig. 6.2

## *6.2.2 Aurally Accurate Measurements*

The use of binaural measurement systems is well-established in fields concerned with sound quality for which the employment of human-related sound measurements is indispensable (e.g., sound design for household appliances and automobiles). In these applications, the generation of pleasant sound experiences and the reduction of annoyances for customers is generally achieved after performing binaural measurements capable of capturing all perceptually relevant characteristics of sound-generating elements (e.g., the tires, engine, transmission, or brakes in an automobile) located at different spatial positions around the listener. Through this approach, annoying sound-emitting components can be identified and modified accordingly to elicit positive responses by the product users, leading to a higher level of customer satisfaction.

The binaural measurement system accurately simulates the acoustically relevant components of the human ear and thus is able to achieve binaural recordings of sound events that are aurally accurate. These recordings include all the features of human sound perception related to spatial hearing. Existing regulations and standards in the context of environmental noise measurements are often incompatible with the soundscape approach and the use of binaural measurement systems. Nonetheless, a few general recommendations and guidelines are available for the practical execution of soundscape measurements, for example, the height of the microphones should be chosen according to the actual or expected position and height of the receiver (Genuit and Fiebig 2014). For guidance on how to perform acoustic measurements in soundscape investigations, the technical specification ISO/TS 12913-2 (2018) provides detailed information (see Sect. 6.2.3).

#### 6.2.2.1 Binaural Recording

Binaural measurement systems are designed to mimic the directional filtering properties of human anatomy in a representative and reproducible way. The relevance of this aspect can be illustrated by considering a typical stereo microphone arrangement in which two omnidirectional microphones are placed at a distance that replicates the span between the left and right human ears. While this approach will also produce two signals with differences in time and level, these differences are applied equally for all frequencies due to the absence of natural signal filtering structures (auricles, head, shoulders). All acoustically relevant parts of the human anatomy involved in the generation of the binaural input to the brain are included in a binaural measurement system and contribute to the directional filtering. Thus, special attention must be paid to the appropriate positioning and dimensioning of the anatomical components in order to create differences in time and level between the left and right ear that vary over frequency. Therefore, a binaural measurement system and a typical stereo microphone arrangement do not record the same signals.

The exact design of a binaural measurement system is particularly relevant for the position of the artificial auricles (pinna) in relation to the head and shoulder elements, as positioning errors become apparent in the direction-dependent part of the transfer function and cannot be corrected after the measurement. Similarly, the angles of inclination and the design of the artificial cavum conchae also have a significant influence on the HRTFs.

As a result of studies on the influence of different components of the human anatomy on the sound recorded at the ears, it is possible to develop artificial heads with a simplified but mathematically accurate geometry without the loss of relevant directional information (Genuit 1984). Artificial head measurement systems can produce directional filtering patterns comparable to those generated by human anatomy and can record binaural signals with a high dynamic range. Moreover, artificial head dimensions comply with international specifications and standards as their free-field transfer functions and directional patterns are in accordance with the IEC 959 report (IEC 1990).

The fundamental principle of binaural technology includes aurally accurate recording, analysis, and reproduction. Herein, two signals, recorded by the left and right ear microphones of an artificial head are transformed into signals compatible with recordings from conventional omnidirectional measurement microphones by means of equalization. These signals can be used for analysis and parameter estimation using typical signal analyzers like level, third octave spectrum, and other parameters. These equalized signals can even be used for a loudspeaker playback system. Of course, the sound reproduction with loudspeakers cannot reproduce the same spatial impression as the playback using headphones, but at least the timbre is comparable.

Through the process of equalization, different components of the directional filtering introduced by the artificial head are reversed in accordance with the characteristics of the sound field present at the time of recording. The purpose of this equalization can be easily illustrated for a free-field (i.e., reflection-free) sound condition, when a sound source emitting a signal with a flat spectrum is placed directly in front of the artificial head at a large enough distance (i.e., in the far-field). Under this condition, the corresponding sound field equalization can be determined by performing a measurement with the artificial head followed by an equivalent measurement with a calibrated microphone. The free-field equalization is then determined by subtracting and inverting the spectra obtained from both measurements.

In addition to equalizers for the well-defined free-field and diffuse-field conditions, an equalization independent of direction (ID) was introduced. The ID equalization compensates for the direction-independent part of the artificial head's transfer function (Genuit 1992b), which is caused by resonances at the auricle cavity (cavum conchae) and the ear canal. For the case of an artificial head that is based on a mathematically describable, simplified geometry, the ID components can be determined precisely for the purpose of equalization.

#### 6.2.2.2 Binaural Reproduction

On the reproduction side, the recorded signals are corrected (i.e., equalized) once again by applying appropriate filters for playback. This is done with the intention of eliminating unwanted distortions introduced by the sound reproduction system. In addition, the signals are equalized to recreate the original sound pressure signals at the ear canals of the listener as if the listener had been present during the recording in the original sound situation. This means that an accurate reproduction of binaural recordings is only possible through the employment of a calibrated and equalized playback device. Artificial head recordings are usually reproduced through headphones as these provide better separation between the left and right ear signals and simplify control with respect to frequency and level. As no exact specifications for the transmission characteristic of headphones are available, a special hardware (or software) is used to calibrate and to equalize the individual headphones in such a way that the reproduced ear signals are comparable to the ear signals at the original sound field with respect to level and spectrum. It is highly recommended that adequate (calibrated and equalized) playback devices be used to reproduce the noise situations with a high degree of realism and to produce valid, reliable results (Genuit 2018).

A correct reproduction over loudspeakers can be achieved by employing systems capable of compensating for the unwanted crosstalk between each of the ear signals to the contralateral ear; however, to realize an adequate reproduction, a significant increase in complexity must be tackled compared to headphone playback.

Binaural measurements are of particular importance whenever a sound environment is to be reproduced accurately at a different time or in a different location, for example, in the case of further examination of the sounds in laboratory listening tests (Genuit and Fiebig 2006). Whenever evaluations under laboratory conditions are performed, the use of binaural recordings becomes indispensable. Through binaural recordings, "copies" of an acoustic environment are generated as close as possible to human perceptions, providing advantages with regard to the archiving and re-experiencing of acoustic sceneries, which also simplifies the comparability and analyses of different sound environments.

### 6.2.3 Aurally Accurate Measurements According to ISO/TS 12913-2

Given that in soundscape investigations the receiver is usually a person, the measurement height can be narrowed down to typical heights of humans. This clearly contrasts with the conventional noise measurement position according to ISO 1996-2 (ISO 2017) for which the microphone position is determined to be 0.5 m in front of an open window, and differs with the "noise maps" principle, for which the SPL calculations are related to a height of 4 m (ISO 2017). These measurement positions are obviously not typical receiver positions and constitute simple conventions for regulatory purposes. Therefore, those measurement points are not suitable for a soundscape study that aims to consider the human perception of sounds in context.

Regarding measurement time intervals, soundscape measurements should cover all variations caused by prominent sound sources, classifiable in soundscape-related terms (signals, soundmarks, or keynote sounds) as introduced by Schafer (1994). These prominent sound sources or events must not be energetically prominent; consideration must be given to whether they attract attention beyond their contribution in sound pressure level or possess a particular meaning for the local community. With respect to measurement duration, soundscape measurements must be long enough to sufficiently encompass all emission situations needed to obtain a representative, comprehensive depiction of the complete soundscape. This means that all relevant, typical sound events and sound sources must be recorded (Fiebig and Genuit 2011).

Additionally, stationary measurements are highly recommended as any movement of the measurement system and interactions of the measurement device with the person performing the measurement could potentially cause unwanted noise that does not represent the measured soundscape. For artificial head measurement systems, the use of a tripod is recommended. Outside recordings require the use of windscreens (Fig. 6.4). In general, equalization of the binaural measurement systems must be chosen with respect to the specific sound field of the investigated

**Fig. 6.4** Examples of recording sites using artificial heads in different acoustic environments. All sites are public areas with different forms of transportation noise

soundscape. The time signals of the binaural recordings must be digitally stored and sampled at a sampling rate equal to or higher than 44.1 kHz to preserve all spectral features relevant for human hearing.

San Martín et al. (2019) and Sun et al. (2018) suggested the use of multichannel recording techniques in the context of soundscape such as ambisonics. Although developed in the 1970s, ambisonics has gained weight recently since YouTube, Oculus VR, and Facebook adopted it as a standard for their 360-degree videos (see also Brambilla and Fiebig, Chap. 7). This technique can provide an alternative to binaural recordings in the context of laboratory studies of soundscapes (Davis et al. 2014) if the semantic aspects of user experience are similar in the original soundscape and its reproduction (Guastavino et al. 2005).

According to the ISO/TS 12913-2 (2018), other measurement systems like microphone arrays and surround recordings are not recommended as those systems are not yet fully standardized (Hong et al. 2017). Although those recording technologies offer some advantages, the lack of standardization makes it difficult to perform aurally accurate analyses to compute psychoacoustic parameters and indicators (cf. ISO/TS 12913-2 (2018)).

## 6.3 Psychoacoustic Analysis of Acoustic Environments

### 6.3.1 Introduction to Psychoacoustics

The discipline of psychoacoustics deals with the quantitative link between physical stimuli and their corresponding hearing sensations (Fastl and Zwicker 2007). In essence, psychoacoustic research attempts to describe sound perception mechanisms in terms of specific parameters using elaborated models. This means that mathematical descriptions are derived from measured relationships between stimulus (physics) and response (perception). By investigating different aspects of human auditory sensations, comprehensive models can be developed describing the manner of human noise perception and signal processing (Sottek 1993). Some common and established psychoacoustic parameters include *loudness*, *sharpness*, *roughness*, *fluctuation strength,* and *tonality*. Table 6.1 presents a list of basic psychoacoustic parameters and short descriptions of their meanings.

*Loudness*, a psychoacoustic parameter introduced several decades ago (e.g., Zwicker et al. 1957; Zwicker 1958), considers the basic human processing phenomenon associated with the sensation of volume. Loudness includes signal processing effects such as spectral contribution and sensitivity (i.e., frequency weighting), masking (post and simultaneous), the interactions within critical bands, and nonlinearities. The unit of loudness is the sone. The computation of this psychoacoustic parameter can be performed using the model developed by Zwicker (1982), standardized in the German standard DIN 45631/A1 (2010) and in the international standard ISO 532-1 (2017), or in the model proposed by Moore and Glasberg (Moore et al. 1997; Glasberg and Moore 2006), which is standardized in the American National Standard ANSI S3.4-2007 (ANSI 2007) and the ISO 532-2 (2017). By applying algorithms related to human auditory processes, the psychoacoustic parameter of loudness offers advantages over the A-weighted sound pressure level (Fastl and Zwicker 2007): the psychoacoustic parameter shows a much

**Table 6.1** Basic psychoacoustic parameters and their meaning in relation to human perception and availability of standard for computation of the parameter

| Psychoacoustic parameter (units) | Meaning | Standard |
|---|---|---|
| Loudness (sone) | Consideration of the distribution of critical bands and masking properties in the hearing | ISO 532-1 |
| Sharpness (acum) | Consideration of the weighted first moment of distribution of critical band rates of specific loudness, proportion of loudness of high-frequency components to total loudness | DIN 45692 |
| Roughness (asper) | Consideration of the time structure (fast modulations) of the sound signal | ECMA-418-2 |
| Fluctuation strength (vacil) | Consideration of the time structure (slow modulations) of the sound signal | ECMA-418-2 |
| Tonality (tu) | Consideration of pitch strength due to prominent tones and elevated narrow-band noise components | ECMA-418-2 |

better correspondence with loudness sensation than the $L_{Aeq}$ (Bray 2007). The parameter loudness goes far beyond simple sound-level indicators (Genuit 2006) and the advantage of loudness compared to A-weighted sound-pressure-level indicators becomes even clearer when the superposition of sounds is considered. For example, when sounds with different spectral shapes are combined, the A-weighted SPL is unable to predict the perceived loudness. Evaluating loudness becomes even more complicated when tones are added to noise. Lastly, Hellman and Zwicker (1987) showed that the A-weighted SPL can be even inversely related to loudness and annoyance.

### 6.3.2 *Psychoacoustic Analysis of Acoustic Environments*

Figure 6.6 illustrates the analysis of loudness for three simple noises with identical A-weighted SPLs ($L_{Aeq}$). Since the psychoacoustic loudness values can be interpreted as a ratio-scaled quantity, the values illustrate the great mismatch between the psychoacoustic loudness indicator and the A-weighted time-averaged SPL. The loudness values in sones can be directly compared by the intervals or differences: a loudness value twice another loudness value means that this sound is perceived as twice as loud as the other sound. Although the broadband noise is perceived as twice as loud as the narrowband noise, the sound level indicator does not indicate any difference.

Another relevant phenomenon of human perception is also illustrated in Fig. 6.5. Although the depicted European police siren sound has the same time-averaged A-weighted SPL as the synthetic (broadband and narrowband noise) sounds, the time-varying pattern of the siren sound produces loudness values that reach or surpass the representative single value of 32.8 *sone* only in very few instances. The result of this loudness analysis can be explained by the fact that the cognitive stimulus integration of humans is complex, meaning that humans do not simply average their sensation levels over time (Stemplinger 1999; Fiebig 2015). Using the statistical mean of a time-variant loudness analysis would lead to results that are too low in comparison to perceived and judged overall loudness (Fastl 1991). Thus, the percentile loudness $N_5$, indicating the loudness value reached or exceeded in 5% of the total time, expresses the perceived overall loudness more adequately and should be determined in accordance with DIN 45631/A1 (2010) and ISO 532-1 (2017). In general, the difference between the high and low loudness percentile values is an indicator for environmental noise quality (Genuit 2006). Greater loudness fluctuations indicate a strong unsteadiness with respect to loudness. Such loudness variations usually attract more attention than less-varying noise.

There is strong evidence that physiological reactions to noise correlate better with the loudness parameter than with the sound pressure level. As an example, Jansen and Rey (1962) showed that the finger pulse amplitude, an autonomous physiological reaction measured after exposure to different sounds, can vary strongly with the same sound pressure levels. The variances can be explained on the

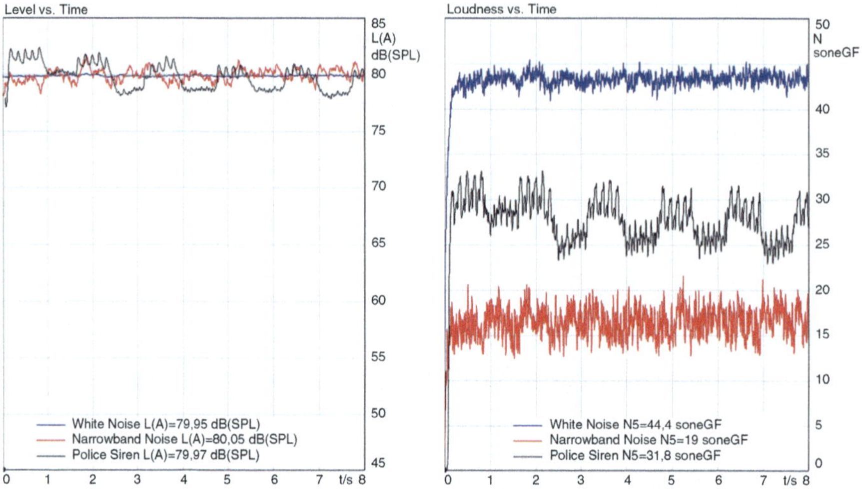


**Fig. 6.5** Comparison of the results obtained through A-weighted SPL (left) and loudness analyses according to ISO 532-1 (right) of three different sounds. Although all sounds have similar A-weighted levels, the results from the loudness analysis show significant differences. The x-axis shows the time in sec

basis of the differences in the psychoacoustic parameter loudness (Genuit and Fiebig 2007).

Similarly, the psychoacoustic parameter *sharpness*, related to the perceived spectral emphasis of a signal toward high frequencies, is a potential predictor for determining the pleasantness or annoyance of sounds, as it has been observed that sensory pleasantness decreases with increasing sharpness (Fastl and Zwicker 2007). One method for the calculation of the psychoacoustic parameter sharpness is defined in the German standard DIN 45692 (DIN 2009). However, in addition to the algorithm implemented in DIN 45692, other methods for the calculation of the parameter sharpness are also available, including the approaches introduced by Aures (1985) and von Bismarck (1974). Generally, the DIN 45692 standard and von Bismarck calculation methods produce similar sharpness results and are not dependent on the total loudness. In contrast, the sharpness computation according to Aures (1985) will increase in sharpness value for a constant spectral shape as loudness increases due to the coupling of the sharpness impression to the total loudness introduced by this method. The unit of sharpness is the *acum*.

Figure 6.6 shows the different psychoacoustic results from two simple signals: white noise and pink noise. Both have the same A-weighted sound pressure level, but the loudness of the pink noise is higher than the loudness of the white noise, and the white noise produces a higher sharpness value.

The question of which sound is less annoying or produces a higher perceived sound quality can only be answered by listening tests and statistical analyses. Usually, people prefer the louder but less sharp sound as listening tests show for synthetic signals (Fiebig 2015).

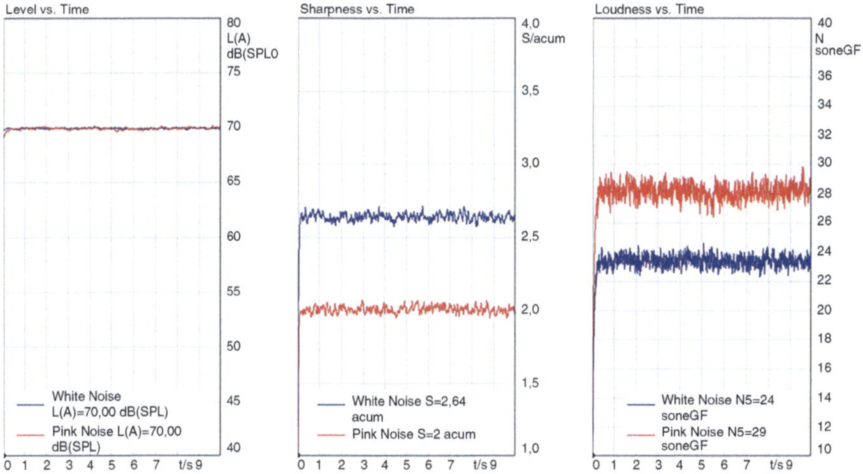


**Fig. 6.6** Analysis of white noise and pink noise with respect to sound pressure level (left), loudness (middle), and sharpness (right). White noise is random noise that has a constant power spectral density, whereas in pink noise, there is equal amplitude per octave. This means that pink noise has less energy in the higher frequency range. Both signals have the same A-weighted sound pressure level (left), but pink noise has 17% higher loudness according to the ISO 532-1 (middle) and more than 20% less sharpness (computed by means of the DIN 45692) in comparison to white noise (right). The x-axis shows the time in sec

Now the question arises as to which sound (white vs. pink noise) humans perceive as "better"? Is the signal property loudness more important with respect to the sound quality than the signal property sharpness or quite the opposite? This can only be answered by a statistical analysis of data from a jury evaluation test. A paired comparison listening test was conducted in which participants were asked to judge which sound (pink versus white noise) caused higher annoyance for the listener. The effect on the perceived annoyance of the two different signals with changing step by step the sound pressure level of the pink noise is shown in Fig. 6.7. When both signals had the same sound pressure level, but pink noise had higher loudness, only 28% of the test participants judged the pink noise as more annoying than the white noise. That means that most participants preferred the pink noise instead of the white noise although the loudness of the pink noise was higher. Obviously, the signal property of sharpness has a stronger contribution to the annoyance. Only after increasing the level of the pink noise by 10 dB did nearly all participants judge the annoyance level caused by the pink noise to be higher than the annoyance caused by the white noise.

In a real soundscape the context is very important, as acknowledged in the definition of the term soundscape in ISO 12913-1 (2014). In the laboratory with test signals, sound with higher sharpness normally has a negative correlation with the overall perceived sound quality. It is very important to distinguish between the terms *sound character* and *sound quality*. Sound character represents basic attributes (sensory properties) of sound events. Sound quality perception includes

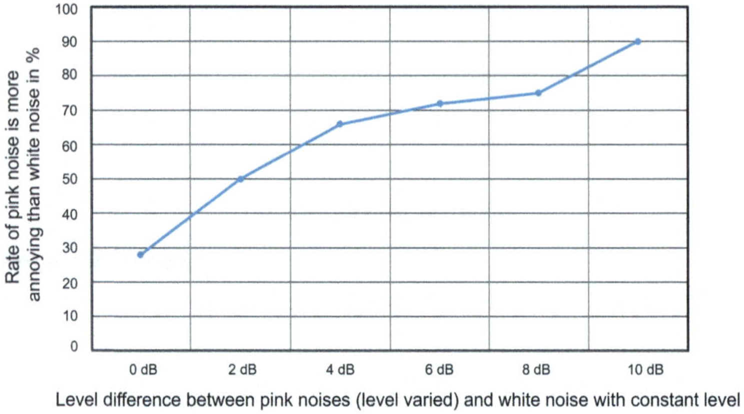


**Fig. 6.7** Percentage responses of 14 participants stating in a paired comparison test that pink noise was more annoying than white noise. The level of the pink noise varied from equalling the level of white noise until 10 dB higher level than white noise (x-axis). The level of the white noise was consistently presented at 70 dB(A) (Fiebig 2015)

non-acoustic factors that influence the interpretation of sound and is affected by context, cognition, expectations, experiences, and interactions (Blauert and Jekosch 1997).

Comparison of a relatively ugly urban place with a lot of traffic versus a beautiful park with a fountain might lead to a judgment for greater pleasantness of the park despite the higher sharpness of the fountain sound compared to the traffic noise. However, if two fountains are compared, both interpreted as pleasant sources, the less sharp fountain sound (maybe due to the fountain design) is assigned a higher sound quality (Galbrun and Ali 2012). Figure 6.8 illustrates the remarkable sharpness differences that can occur in the context of environmental noises. The fountain has a higher overall level, less loudness, but 50% more sharpness; however, most people prefer the soundscape around the fountain instead of the urban location. Fiebig (2015) gives a more detailed discussion about cognitive stimulus integration in the context of auditory sensations and sound perceptions (see Fiebig, Chap. 2).

Other psychoacoustic parameters, such as *roughness* and *fluctuation strength*, are descriptors for the human perception of temporal effects and can be indicators for the annoyance caused or the perceived "aggressiveness" of sounds, although the interpretation of the results given by these psychoacoustic analyses strongly depends on the type of sound and the source being investigated. From the physical point of view, the psychoacoustic parameters roughness and fluctuation strength are similar; they are related to modulations (both amplitude modulations and frequency modulations). However, slow modulations (e.g., modulation frequencies below 20 Hz) produce the sensation of a sound with fluctuations; in contrast, fast modulations (e.g., modulation frequencies clearly above 20 Hz) produce a sensation of an unclean, rough sound. Fig. 6.9 provides an illustration of signals with variations in roughness and fluctuation strength. A combination of a 1 kHz tone with 996 Hz and

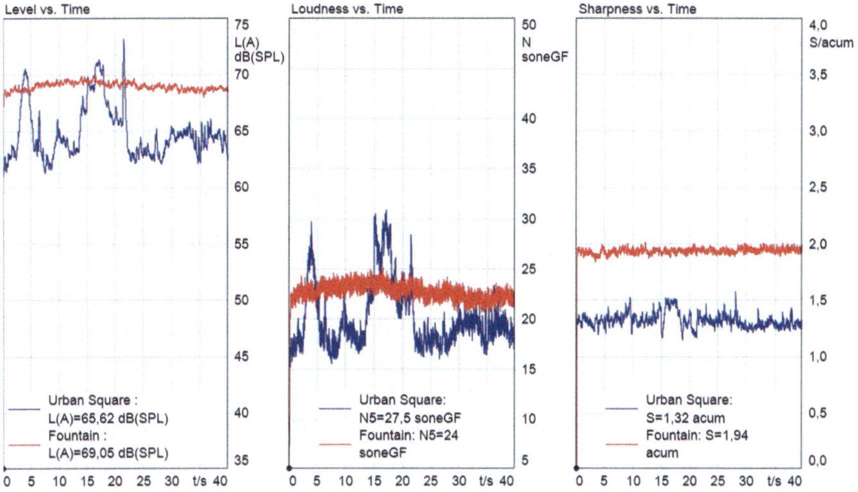


**Fig. 6.8** Analyses for two environmental sounds. Traffic at an urban square (green curves) has lower sound pressure level (left panel), but higher loudness (ISO 532-1) (middle panel) and less sharpness (DIN 45692) (right panel) than the fountain (red curves). The sound quality of the fountain was evaluated by participants in a case study as more pleasant than the urban square sounds

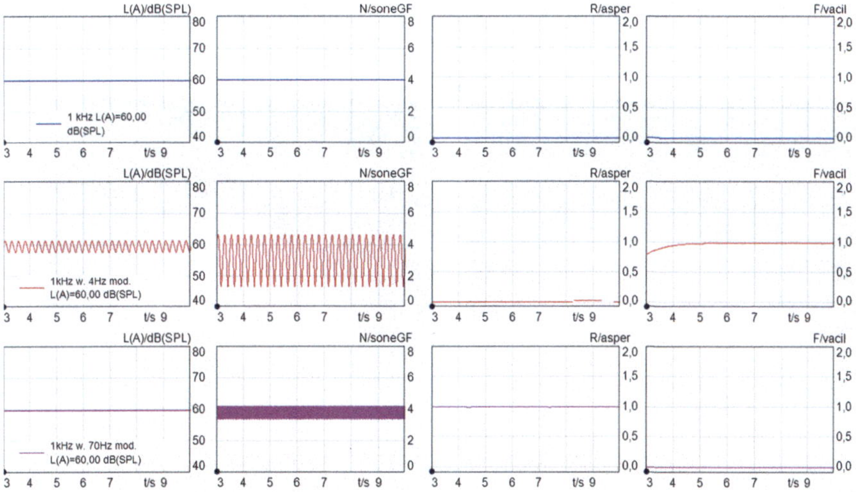


**Fig. 6.9** Comparison of three signals: 1 kHz tone (blue, top); 1 kHz tone with 996 Hz and 1004 Hz tones (red, middle); 1 kHz tone with 930 Hz and 1070 Hz tones (pink, bottom). All signals have the same sound pressure level (left panel) but have differences in loudness (second panel) and great differences in roughness (third panel) and fluctuation strength (fourth panel) according to ECMA-418-2

1004 Hz results in a slow modulation of 474 Hz, which is the fluctuation (measurement unit: *vacils*). Fluctuations are used especially for warning signals because these fluctuations create greater attention by the listener. The combination of three tones leading to a fast modulation of 70 Hz, produces a sound that is perceived as a rougher sound of 1 kHz because human hearing cannot separate the three tones and perceives only a disturbing 1 kHz tone. This is described as roughness (measurement unit: *asper*). Roughness is important for sound design and the evaluation of roughness is strongly dependent on the context, the kind of product producing the sound, and the expectations of the listener.

The *tonality* parameter (measurement unit: *tu*) describes the sensation that is related to the proportion of prominent tones or narrowband components in a signal. The newest methods for the determination of the tonality parameter are standardized in the international standard ECMA-74 (2019) or ECMA 418-2 (2020). The standards include a psychoacoustic-based tonality computation algorithm that considers relevant aspects of human auditory perception, such as hearing thresholds and masking, to determine the perceptual relevance and prominence of tonal components in a signal (Becker et al. 2019). Figure 6.10 demonstrates the importance of tonality in an example with the sound of a large widespread passenger airplane as it "takes off." In this example, the tones were synthetically removed. This effect is clearly audible even though all other parameters are unchanged (Table 6.2).

Human hearing quickly adapts to stationary signals but remains very sensitive to fluctuations and intermittent noise, as well as to prominent, salient noise events.

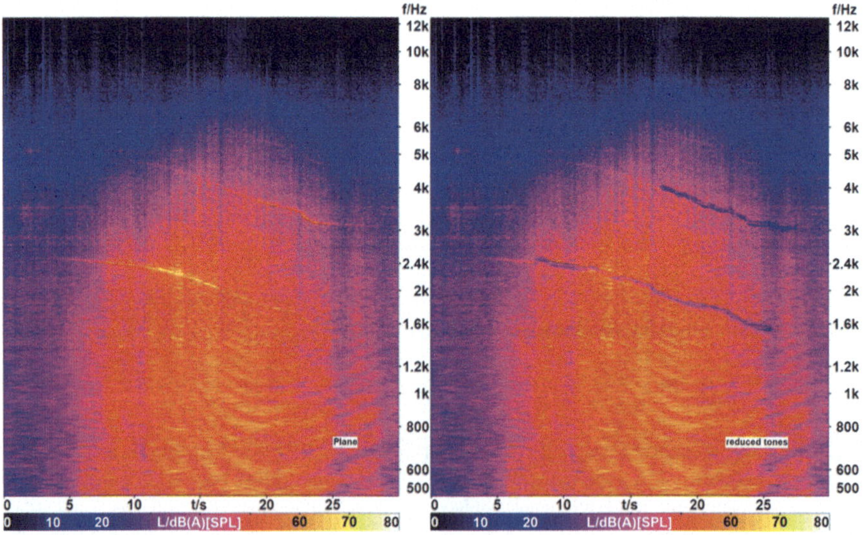


**Fig. 6.10** Comparisons of sounds from "take off" of an airplane: left panel is the original sound; the right panel is the same recording synthetically modified with reduced tonal components. FFT vs. time

**Table 6.2** Quantification of the data for aircraft takeoff illustrated in Fig. 6.10. All parameters (sound pressure level, loudness, sharpness, and roughness) are unchanged; only the tonality is significantly reduced by the removal of tonal components

| Analysis (units) | Takeoff—original | Takeoff—manipulated by reducing tones |
|---|---|---|
| Level (dB(A)) | 83.8 | 83.5 |
| Loudness (sone) | 71.8 | 71.2 |
| Sharpness (acum) | 1.04 | 1.02 |
| Roughness (asper) | 0.06 | 0.06 |
| Tonality (tu) | 0.74 | 0.38 |

Therefore, peak values and relative changes can be significant with respect to auditory perception. The use of percentile values of a parameter that has been measured over time can reveal the magnitude of fluctuations and variations (Genuit 2006). For example, if large differences are observed between values in the 5th and 95th percentiles of a measurement interval, strong fluctuations of the considered parameter are detected, which would suggest a dynamic sound situation.

In addition to the basic psychoacoustic parameters described in this section, the Relative Approach parameter developed by Genuit (1996), which is related to the detection and perception of patterns in acoustic signals, provides information about obtrusive and attention-attracting noise features. The Relative Approach analysis simulates the ability of human hearing to adapt to stationary sounds and to react to variations and patterns within the time and frequency structure of a sound. An example is shown in Fig. 6.11 in which two vehicles have pass-by events with the same sound pressure levels ($L_{\mathrm{Aeq}}$). The Relative Approach analysis shown identifies the pattern of diesel "knocking" in the second pass-by event, which is perceived as more annoying by most people.

By considering different aspects of human (binaural) signal processing through psychoacoustic analyses, pleasant and unpleasant features of sound can be identified. The first and most relevant step toward achieving meaningful results from the acoustic analysis of sound environments is to move away from indicators based on simple energy averaging (i.e., SPL values) and to adopt the usage of more detailed (psycho)acoustic parameters that consider different acoustic properties of sound events. This entails the determination of psychoacoustic parameters capable of detecting temporal and spectral patterns that are relevant to human perception. The application of well-established psychoacoustic analysis methods can advance soundscape evaluations and considerably improve perceptual assessments of environmental sound quality and the expected impacts with regard to annoyance (Genuit and Fiebig 2006). Moreover, advanced parameters, such as the Relative Approach developed by Genuit (Genuit 1996), must be further developed to improve the characterization of environmental sound conditions.

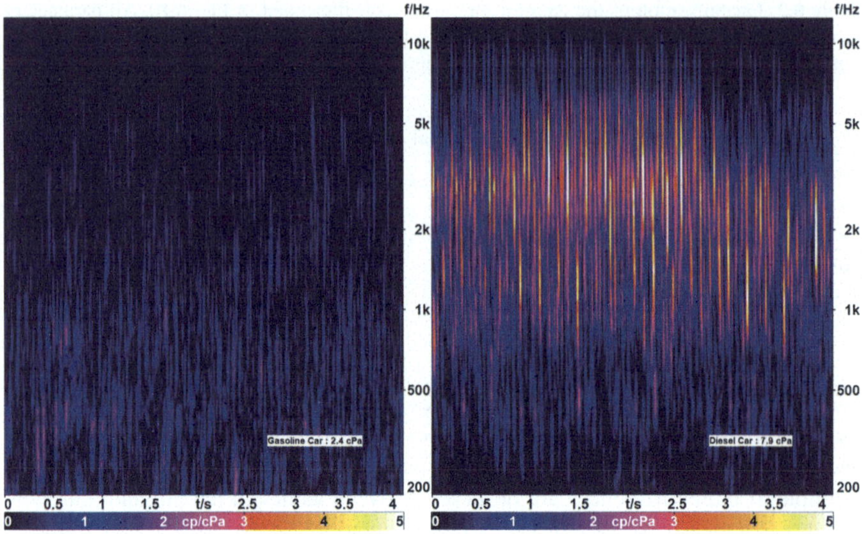


**Fig. 6.11** Comparison of Relative Approach analysis results for two vehicle engine noises. The analysis result of the gasoline engine noise is shown on the left side, and the analysis result of the diesel engine noise is displayed on the right side. Both noises possess the same A-weighted SPL. The typical "knocking" noise patterns of a diesel engine can be clearly seen in the result of the Relative Approach analysis (right)

### *6.3.3 Psychoacoustics in Soundscape*

An important step toward improving the characterization of acoustic environments and obtaining meaningful results from noise annoyance investigations is to pose specific questions about the acoustic environment under investigation. For example, in the case of complaints about noise, it might be relevant to start by asking which of the existing sound sources are causing the discomfort and are considered responsible for the noise annoyance by the person concerned. Similarly, signal attributes, such as modulations or specific patterns in the time or frequency domains, should be examined for their potential as sources of irritation. Moreover, informative features about the annoyance and the necessity (or lack thereof) of the noise should be questioned, just as the attitude and expectations of the listener should be examined (Genuit 2003). The answers to these questions will help to identify suitable acoustic analyses and appropriate measurement methods that will lead to improved and goal-oriented investigations of an acoustic environment (Berglund and Nilsson 2006). In addition, binaural recordings are often combined with psychoacoustic analysis to determine the reasons behind the annoyance. Through these processes, sound adjustments can be performed and evaluated in a perceptually relevant way before any major and potentially expensive modifications of the environment (e.g., in structural or mechanical elements) are performed.

Psychoacoustics is used frequently to develop better noise maps of certain areas and to describe acoustic properties of the area beyond the sound level distribution (Kang et al. 2016). Genuit et al. (2008) generated psychoacoustic maps of a small urban park, Nauener Platz, in Berlin. The maps of Nauener Platz showed that some psychoacoustic parameters behave differently compared to the sound pressure level. For example, in the center of the urban park the SPL dropped down due to the large distance to the roads. However, parameters like sharpness or roughness remained almost constant over the whole urban park area.

Hong and Jeon (2017) developed loudness and sharpness maps of Seoul among maps that indicated the audibility of certain sound sources and showed that consideration of a variety of parameters was advantageous for the determination of soundscape quality. Montoya-Belmonte and Navarro (2020) used a sensor network to determine loudness, sharpness, and a psychoacoustic annoyance map for a university campus in Spain. They concluded that the psychoacoustic annoyance measurement was better correlated with loudness in the locations they considered, and sharpness was only of minor importance. Those efforts reflect the increased interest in psychoacoustic parameters in soundscape investigations around the world.

Although the psychoacoustic approach for acoustic environment analysis has provided valuable information, those studies only partially cover the investigation of the sensory and mental representations of the typical sounds in urban spaces (Yang 2019). Without asking how residents feel about their surroundings and investigating the visual elements of the location under study, the results obtained from psychoacoustic analysis alone are not sufficient. Aspects such as local expectations, suitability, or acceptability of sound in their respective contexts cannot be sufficiently answered without knowledge of human responses to the locations under consideration. Psychoacoustics can analyze in detail the acoustic composition of a soundscape and the signal properties that elicit specific auditory sensations; however, a comprehensive interpretation of the results requires feedback from the listeners.

A thorough study of the acoustical properties and psychoacoustic characteristics of soundscapes is an important part of understanding the perception of the acoustic environment in context and can serve as a starting point for the classification of soundscapes. Through these analyses, acoustical properties can be identified in detail that are common across multiple locations (e.g., urban environments, urban parks, residential areas). Identification of site-specific patterns and noise features within a soundscape will continue to be necessary. The inclusion of macroscopic and microscopic analyses is required to capture the overall sound impression created by soundscape. Those combined analyses also are needed to recognize and interpret sound events that may cause strong positive or negative reactions and feelings (Schulte-Fortkamp and Nitsch 1999). The macro-level analysis is defined by descriptions embedding the noise events into the comparable soundscapes of streets, places, and urban areas. The micro-level is related to the analysis of noise events based on psychoacoustic parameters (Schulte-Fortkamp and Nitsch 1999).

In contrast to conventional environmental noise measurement regulations, the focus in soundscape investigation lies in recording and analyzing environmental sound with all relevant sound sources as perceived by individuals in context. The separation of the contributions of the different sound sources might be relevant for analytical or legal reasons and for regulatory purposes in noise policy, but the examination and assessment of the acoustic environment as a whole remains inevitable for a thorough understanding of a soundscape.

Measurement guidelines for comprehensive soundscape studies must cover both dimensions of *measurements by persons* and *measurements by instruments* (see Botteldooren, De Coensel, Aletta, and Kang, Chap. 8). In a practical sense, soundwalks, questionnaires, and explorative interviews can be used to complement the acoustical measurements and psychoacoustic analyses of soundscapes.

Psychoacoustic parameters and perceptual, visual, or contextual indicators are used to predict all kinds of soundscape-related descriptors. For example, Lionello et al. (2020) studied prediction models from the acoustic literature that measured the experience of soundscape in terms of *tranquility*, *arousal*, *valence*, *pleasantness*, or *sound quality*. They observed that a great variety of acoustic and psychoacoustic indicators were used and applied in prediction models. For example, Brambilla et al. (2013) used *sharpness* and *roughness* for the affective dimension *chaotic* versus *calm*. Aletta and Kang (2016, 2018) used *loudness*, *fluctuation strength,* and *roughness,* among other parameters, to predict *vibrancy*. Çakır Aydın and Yılmaz (2016) developed a Sound Quality Index based on *loudness*, *sharpness,* and *roughness* to predict *pleasantness* of sound. Lionello et al. (2020) pointed out that those parameters did not systematically lead to great prediction accuracy. In addition, those indicators were often combined with other parameters. Ongoing research must continue to investigate which psychoacoustic parameters are of significant value and which parameters are less closely related to the perception and assessment of acoustic environments. Although the limitations of psychoacoustics to predict human responses to soundscapes are not completely understood, the advantages of using psychoacoustic analyses beyond data from simple level indicators are indisputable.

## 6.4 Benefits and Limitations of Psychoacoustics in the Context of Soundscape

While the overall noise measured at a specific location can be analyzed in terms of several acoustical parameters, the annoyance or pleasantness level of a complex soundscape composed of several sound sources cannot be determined solely from the values obtained through such analyses. Even if the acoustic contribution of a single sound source to the overall noise does not appear significant in a physical sense, the influence of this sound source on the soundscape can be relevant perceptually. This can be explained based on how perceptual "attention" influences sound processing. Thus, to better understand the perception and evaluation of soundscapes, studies must include evaluation of typical attention processes of individuals and the possible factors that influence the (listening) focus on specific sound sources in complex environments. Selective auditory attention processes in perceiving complex (acoustic) environments continue to be very relevant research subjects (Fiebig 2015).

Auditory attention allows human beings to focus their mental resources on a particular stream of interest while ignoring others. According to de Coensel and Botteldooren (2010), most theories on attention rely on a concept that there is an interplay between bottom-up (saliency-based) and top-down (voluntary) mechanisms in a competitive selection process. For example, Knudsen (2007) discussed evidence that the perceived signal strength is influenced by bottom-up salience filters and at the same time is modulated by top-down control. In the context of soundscape, it seems crucial to understand those selection activities to be able to predict individual responses to multi-source environments more appropriately.

Assessment methods that provide different degrees of context may be applied in soundscape investigations, depending on the projected type of investigation and the resources available. These methodologies may range from evaluations on-site (e.g., by means of a soundwalk), which provide a complete range of sensory and environmental aspects, to listening tests in a laboratory environment, where there is better control over the stimuli presented and greater reproducibility (Hermida Cadena et al. 2017).

Although psychoacoustics parameters are frequently applied in soundscape investigations, a great variety of methods for psychoacoustic indicators are used with different computations and implementations. This variety of computation methods limits the comparability of the results and impedes meta-analyses of the behavior of psychoacoustic parameters in soundscape studies. Engel et al. (2021) conducted a comprehensive literature review of soundscape investigations and found that almost 30% of publications report the data and results without specific details about the computation methods and/or standards to determine certain psychoacoustic parameters. Although the ISO/TS 12913-2 (2018) states that the computation methods and/or standards used to perform the psychoacoustic analysis of the binaural measurements need to be documented, often the methods and standards for calculating the psychoacoustic parameters are not properly reported. The lack of detailed scientific reporting limits the comparability of study results and impedes progress in understanding the link between psychoacoustic properties of acoustic environments and their corresponding perception and assessment.

Another area of research that lacks consensus concerns how to represent values of time-variant noises to accurately describe overall auditory impressions and sound perceptions. Frequently different average values or percentile values are applied to quantify the psychoacoustic properties of dynamic acoustic environments. The ISO/TS 12913-3 (2019) suggests that average and percentile values can account for variation over time for certain signal properties. For example, the quotient of the loudness $N_5$ (loudness exceeded in 5% of the time interval) and loudness $N_{95}$ (loudness exceeded in 95% of the time interval) may be an indicator of the level of loudness variability (ISO/TS 12913-3 2019). However, according to Engel et al. (2021), analyzing related research outcomes cannot provide a clear direction with regard to established and acknowledged (percentile) values that correspond to an overall impression of specific auditory sensations. Fiebig (2015) concluded optimistically that predictions of overall assessments of sound perceptions are possible with fair accuracy based on indicators derived from psychoacoustic profiles that represent proxies of momentary perceptual levels. However, further research is needed to derive valid, representative, single values of time-variant noises, which are common components of soundscapes.

## 6.5 Summary

In this chapter, the need for psychoacoustic measurement in soundscapes is discussed. The mechanisms of human binaural hearing, which involve the directional filtering of sound introduced by human anatomy and the combined processing of signals to the left and right ears, are introduced with regard to the soundscape approach. Information encoded in the differences between the two ear signals (e.g., ITDs and ILDs) is used to determine the position of individual sound sources and binaural hearing provides many advantages for the identification and discrimination of individual sound sources in complex acoustic environments. These advantages include the suppression of noise and the capacity to focus on individual sound sources (due to improved signal-to-noise ratio) as well as the ability to identify spatial distribution, speed, and direction of movement of sound sources. Based on this assertion, calibrated binaural measurement systems are required in soundscape research, for which the perception and evaluation of environmental noise is a main concern. Further psychoacoustic evaluation through multi-channel systems is also desirable.

Binaural listening and the intricate signal-processing involved in human hearing provide the advantage of source focusing that in combination with spatial perception and the ability to assess the direction and speed of any movement of sound sources directly influence the perception and evaluation of environmental noise. The involvement of human perception in the evaluation of soundscapes, therefore, is particularly relevant and can only be realized with data collection methods comprising the full capabilities of the human auditory system.

Nonetheless, even if a psychoacoustic approach to the evaluation of a sound environment can aid in the interpretation of acoustic measurements and reveal critical and relevant components of a soundscape, relying solely on values obtained through psychoacoustic parameter estimation to make assertions about complex acoustic environments would disregard the significance of the emotional components of human perception. It is only through the combination of perceptual evaluation methods, which consider the context, expectations, and attitudes of the listener with psychoacoustic analyses, that the outcomes from soundscape studies can become more insightful. More relevant acoustic measurements can be performed by employing binaural measurement systems instead of single or even stereophonic microphone measurement systems (that are incapable of recreating relevant filtering properties of human anatomy). Acoustic measurement procedures described in current standards, such as ISO 1996-2, do not provide a good basis for the establishment of soundscape measurements since they do not consider the human listener as a receiver. In this chapter, the importance of suitable assessment methods that consider a broad range of sensory and environmental aspects has been stressed.

Future soundscape research must consider the need for aurally accurate measurements and psychoacoustic analyses with the distinct purpose of archiving and re-experiencing different acoustic environments. Future studies should also consider the relevance of documenting and analyzing the occurrence of a variety of sound sources since focusing on certain sound sources can change the overall assessment of soundscapes.

A common basis of measurement and data collection procedures that reflect the approach to soundscape research presented in this chapter is provided by the ISO/TS 12913-2 (2018), which has been introduced with the goal of improving the comparability and compatibility of future soundscape investigations. While this technical specification provides common ground with regard to data collection and reporting, the adoption of any soundscape standard should not limit the flexibility and interdisciplinary characteristics of the soundscape approach.

**Compliance with Ethics Requirements** Klaus Genuit declares that he has no conflict of interest.

Brigitte Schulte-Fortkamp declares that she has no conflict of interest.

André Fiebig declares that he has no conflict of interest.

## References

Aletta F, Kang J (2016) Descriptors and indicators for soundscape design: vibrancy as an example. Internoise, Hamburg

Aletta F, Kang J (2018) Towards an urban vibrancy model: a soundscape approach. Int J Environ Res Public Health 15(8):1712. https://doi.org/10.3390/ijerph15081712

ANSI S3.4 2007 Procedure for the computation of loudness of steady sounds

Aures W (1985) Berechnungsverfahren für den sensorischen Wohlklang beliebiger Schallsignale. Acta Acust Acust 59:130–141

Becker J, Sottek R, Lobato T (2019) Progress in tonality calculation. In: ICA 2019, Aachen

Berglund B, Nilsson ME (2006) On a tool for measuring soundscape quality in urban residential areas. Acta Acust Acust 92(6):938–944

Blauert J (1974) Räumliches Hören. Hirzel, Stuttgart

Blauert J (1996) Spatial hearing: the psychophysics of human sound localization. MIT Press, Cambridge

Blauert J, Jekosch U (1997) Sound-quality evaluation – a multi-layered problem. Acta Acust Acust 83(5):747–753

Brambilla G, Maffei L, Gabriele M, Gallo V (2013) Merging physical parameters and laboratory subjective ratings for the soundscape assessment of urban squares. J Acoust Soc Am 134:782–790. https://doi.org/10.1121/1.4768792

Bray WR (2007) Behavior of psychoacoustic measurements with time-varying signals. In: Noise-Con 2007, Reno
Bronkhorst AW, Plump R (1988) The effect of head-induced interaural time and level differences on speech intelligibility in noise. J Acoust Soc Am 83(4):1508–1516. https://doi.org/10.1121/1.395906
Çakır Aydın D, Yılmaz S (2016) Assessment of sound environment pleasantness by sound quality metrics in urban spaces. TU A|Z J Fac Archit 13(2):87–99
Cherry EC (1953) Some experiments on the recognition of speech, with one and with two ears. J Acoust Soc Am 25(5):975–979
Davis WJ, Bruce NS, Murphy JE (2014) Soundscape reproduction and synthesis. Acta Acust Acust 100(2):285–292
Coensel de B, Botteldooren D (2010) A model of saliency-based auditory attention to environmental sound. In: Proceedings of international congress on acoustics 2010, Sydney, Australia
DIN 45631/A1 (2010) Calculation of loudness level and loudness from the sound spectrum – Zwicker method – amendment 1: calculation of the loudness of time-variant sound. Beuth Verlag, Berlin
DIN 45692 (2009) Measurement technique for the simulation of the auditory sensation of sharpness. Beuth Verlag, Berlin
ECMA (2019) ECMA-74 – measurement of airborne noise emitted by information technology and telecommunications equipment, Geneva
ECMA-418-2 (2020) Psychoacoustic metrics for ITT-equipment, part 2: models based on human perception, Geneva
Engel MS, Fiebig A, Pfaffenbach C, Fels J (2021) A review of the use of psychoacoustic indicators on soundscape studies. Curr Pollution Rep 7:359. https://doi.org/10.1007/s40726-021-00197-1
Fastl H (1991) Beurteilung und Messung der äquivalenten Dauerlautheit, Z. f. Lärmbekämpf, vol 38. Springer, pp 98–103. Düsseldorf, Germany
Fastl H, Zwicker E (2007) Psychoacoustics. Springer, Berlin
Fay RR, Popper AN (2005) Introduction to sound source localization. In: Popper AN, Fay RR (eds) Sound source localization, Springer handbook of auditory research, vol 25. Springer, New York. https://doi.org/10.1007/0-387-28863-5_1
Fidell S (2003) The Schultz curve 25 years later: a research perspective. J Acoust Soc Am 114(6):3007–3015
Fiebig A (2015) Cognitive stimulus integration in the context of auditory sensations and sound perceptions (Dissertation), Technische Universität Berlin, epubli, Berlin
Fiebig A, Genuit K (2011) Applicability of the soundscape approach in the legal context. In: DAGA 2011, Düsseldorf, Germany
Flanagan JL, Watson BJ (1966) Binaural unmasking of complex signals. J Acoust Soc Am 40:456–468. https://doi.org/10.1121/1.1910096
Galbrun L, Ali TT (2012) Perceptual assessment of water sounds for road traffic noise masking. In: Acoustics 2012, Nantes
Genuit K (1984) Ein Modell zur Beschreibung der Außenohrübertragungseigenschaften. TH Aachen, Aachen
Genuit K (1992a) Sound quality, sound comfort, sound design—why use artificial head measurement technology? In: Daimler-Benz JRC (ed) Stuttgart
Genuit K (1992b) Standardization of binaural measurment technique. J Phys IV Proc 02:C1-405–C1-407. https://doi.org/10.1051/jp4:1992187
Genuit K (1996) Objective evaluation of acoustic quality based on a relative approach. In: Internoise 1996, Liverpool, pp 3233–3238
Genuit K (2002) Sound quality aspects for environmental noise. In: Internoise 2002, Dearborn, pp 1242–1247
Genuit K (2003) How to evaluate noise impact. In: Euronoise 2003, Naples, pp 1–4
Genuit K (2006) Beyond the a-weighted level. In: Internoise 2006. Honolulu, pp 1321–1327

Genuit K (2018) Standardization of soundscape: request of binaural recording. In: Proceedings of Euronoise, pp 2451–2458
Genuit K, Fiebig A (2006) Psychoacoustics and its benefit for the soundscape approach. Acta Acust Acust 92:1–7
Genuit K, Fiebig A (2007) Environmental noise: is there any significant influence on animals? J Acoust Soc Am 122:3082. https://doi.org/10.1121/1.2943007
Genuit K, Fiebig A (2014) The measurement of soundscapes – it it standardizable? In: Internoise 2014, pp 3502–3510
Genuit K, Sottek R (2010) Das menschliche Gehör und Grundlagen der Psychoakustik. In: Genuit K (ed) Sound-Engineering im Automobilbereich. Springer, Berlin, ISBN: 978-3-642-01415-4
Genuit K, Schulte-Fortkamp B, Fiebig A (2008) Psychoacoustic mapping within the soundscape approach. In: Internoise 2008, Proceeding, Shanghai, China
Glasberg BR, Moore BCJ (2006) Prediction of absolute thresholds and equal-loudness contours using a modified loudness model. J Acoust Soc Am 120:585–588. https://doi.org/10.1121/1.2214151
Guastavino C, Katz BFG, Polack JD, Levitin DJ, Dubois D (2005) Ecological validity of soundscape reproduction. Acta Acust Acust 91(2):333–341
Hellman R, Zwicker E (1987) Why can a decrease in dB(a) produce an increase in loudness? J Acoust Soc Am 82:1700–1705. https://doi.org/10.1121/1.395162
Hermida Cadena LF, Lobo Soares AC, Pavón I, Bento Coelho JL (2017) Assessing soundscape: comparison between in situ and laboratory methodologies. Noise Mapp 4:57–66. https://doi.org/10.1515/noise-2017-0004
Hong JY, Jeon, JY (2017) Exploring spatial relationships among soundscape variables in urban areas: a spatial statistical modelling approach. Landsc Urban Plan 157:352–353
Hong JY, Jianjun H, Bhan L, Rishabh G, Woon-Seng G (2017) Spatial audio for soundscape design:recording and reproduction. Appl Sci 2017(7):627
IEC 959 (1990) Provisional head and torso simulator for acoustic measurements on air conduction hearing aids
ISO 532-1 (2017) Acoustics—methods for calculating loudness—part 1: Zwicker method, Geneva
ISO 532-2 (2017) Acoustics—methods for calculating loudness—part 2: Moore-Glasberg method. Geneva
ISO 12913-1 (2014) Acoustics — Soundscape —Part 1: Definition and conceptual framework, Geneva
ISO 1996-1 (2016) Acoustics — Description, measurement and assessment of environmental noise — Part 1: Basic quantities and assessment procedures, Geneva
ISO 1996-2 (2017) Acoustics — Description, measurement and assessment of environmental noise — Part 2: Determination of sound pressure levels, Geneva
ISO/TS 12913-2 (2018) Acoustics — Soundscape — Part 2: Data collection and reporting requirements, Geneva
ISO/TS 12913-3 (2019) Acoustics — Soundscape — Part 3: Data analysis, Geneva
Jansen G, Rey P-Y (1962) Der Einfluss der Bandbreite eines Geräusches auf die Stärke vegetativer Reaktionen. Int Zeitschrift für Angew Physiol Einschl Arbeitsphysiologie 19:209–217. https://doi.org/10.1007/BF00697117
Kang J, Schulte-Fortkamp B, Fiebig A, Botteldooren D (2016) Mapping of soundscape, In: Kang J, Schulte-Fortkamp B (ed.) Soundscape and the built environment. Taylor & Francis incorporating Spon, London
Knudsen EI (2007) Fundamental components of attention. Annu Rev Neurosci 30:57–78. https://doi.org/10.1146/annurev.neuro.30.051606.094256
Lionello M, Aletta F, Kang J (2020) A systematic review of prediction models for the experience of urban soundscapes. Appl Acoust 170:107479
Montoya-Belmonte J, Navarro JM (2020) Long-term temporal analysis of psychoacoustic parameters of the acoustic environment in an university campus using a wireless acoustic sensor network. Sustainability 2020(12):7406. https://doi.org/10.3390/su12187406

Moore BCJ, Glasberg BR, Baer T (1997) A model for the prediction of thresholds, loudness, and partial loudness. J Audio Eng Soc 45:224–240

San Martín R, Arana M, Ezcurra A, Valencia A (2019) Influence of recording technology on the determination of binaural psychoacoustic indicators in soundscape investigations. In: Proceedings Internoise, Madrid, Spain 2019

Schafer RM (1994) The soundscape: our sonic environment and the tuning of the world. Destiny Books

Schulte-Fortkamp B, Nitsch W (1999) On soundscapes and their meaning regarding noise annoyance measurements. In: Internoise 1999, Fort Lauderdale, Florida

Schultz TJ (1978) Synthesis of social surveys on noise annoyance. J Acoust Soc Am 852–861(64):377–405

Shaw E (1997) In: Gilkey RH, Anderson TR (eds) Binaural and spatial hearing in real and virtual environments. Lawrence Erlbaum, Mahwah, pp 25–48

Sottek R (1993) Modelle zur Signalverarbeltung im menschlichen Gehör. RWTH, Aachen

Stemplinger IM (1999) Beurteilung, Messung und Prognose der Globalen Lautheit von Geräuschimmissionen. Technische Universität München

Sun K, Botteldooren D, De Coensel B (2018) Realism and immersion in the reproduction of audio-visual recordings for urban soundscape evaluation. In: Proceedings of the 47th international congress and exposition on noise control engineering, Chicago, USA

vom Hövel H (1984) Zur Bedeutung der Übertragungseigenschaften des Aussenohrs sowie des binauralen Hörsystems bei gestörter Sprachübertragung. RWTH, Aachen

von Bismarck G (1974) Sharpness as an attribute of the timbre of steady sounds. Acta Acust Acust 30:159–172

World Health Organization (2014) Burden of disease from environmental noise. In: Quantification of healthy life years lost in Europe. WHO Regional Office for Europe, Bonn, ISBN 978 92 89002295

Yang M (2019) Prediction model of soundscape assessment based on semantic and acoustic/psychoacoustic factors. J Acoust Soc Am 145:1753–1753. https://doi.org/10.1121/1.5101420

Zwicker E (1958) Über psychologische und methodische Grundlagen der Lautheit. Acustica 8:237–258

Zwicker E (1982) Psychoakustik. Springer

Zwicker E, Flottorp G, Stevens SS (1957) Critical band width in loudness summation. J Acoust Soc Am 29:548–557. https://doi.org/10.1121/1.1908963

# Chapter 7
# Measurements and Techniques in Soundscape Research

**Giovanni Brambilla and André Fiebig**

**Abstract** Soundscape research has gained importance over the last decades. Due to its emphasis on the perceptual construct of the acoustical environment in context, interdisciplinary research is conducted using a great variety of methods and tools. Inherent to the soundscape approach is the requirement that the methodology, whether focused on physical or perceptual data, reflect the way humans holistically perceive the acoustic environment in context. This general requirement is incorporated in the International Standard Organization (ISO) document ISO 12913-1: Soundscape investigations have to consider its key components, namely "people," "acoustic environment," and "context" (ISO 12913-1 2014). Currently, soundwalks, questionnaires, interviews, and recordings of sound, mimicking the way humans perceive sound, are applied to approach the subject of soundscape.

Moreover, novel sources of data, such as social networks, mobile applications, and social media combine scientific interests with broad public participation. The publication of soundscape-related international standards and technical specifications has led to a harmonization of data collection in soundscape investigations; however, at the same time, the development of new methodology in the context of the soundscape approach should be encouraged due to the multitude of unanswered research questions.

**Keywords** Binaural recordings · Interview · Observational methods · Questionnaire · Soundwalk · Triangulation

---

G. Brambilla
Department of Earth and Environmental Sciences (DISAT), University of Milano-Bicocca, Milano, Italy
e-mail: giovanni.brambilla@unimib.it

A. Fiebig (✉)
Department of Engineering Acoustics, Technische Universität Berlin, Berlin, Germany
e-mail: andre.fiebig@tu-berlin.de



B. Schulte-Fortkamp et al. (eds.), *Soundscapes: Humans and Their Acoustic Environment*, Springer Handbook of Auditory Research 76,
https://doi.org/10.1007/978-3-031-22779-0_7

## 7.1 Introduction

Murray Schafer (1977), the Canadian composer who introduced and established the concept of soundscape, argued that "soundscape" concerns any acoustic field of study, ranging from musical compositions to radio programs and acoustic environments. This statement illustrates the need for a variety of methods and tools to cope with the interdisciplinary aspects of the soundscape approach. Because studies on soundscape have evolved differently around the world and across many disciplines, there is still a diversity of opinions about its definition and aims (e.g., Brooks et al. 2014). Consequently, a multitude of methods and techniques have been and continue to be used to explore all of the sounds and their perceptions in an environment, with regard to their complexity, ambivalence, meaning, and context. Standardization efforts and rigorous scientific discourses on appropriate, effective, and reliable methods have led to a core of common soundscape tools, achieving consensus among researchers to a certain extent.

The need to consider a range of methodologies for the assessment of soundscapes is undisputed (Brown et al. 2011). However, all measurement methods, whether they collect physical or perceptual data, must be related to the way humans perceive the acoustic environment in a common context. This is the central tenet of the soundscape approach and guides the way soundscapes are measured (Kang et al. 2016b). Moreover, because the soundscape approach is becoming more popular, methods and tools now in use for audio signal processing will be developed further in the near future, along with the development of advanced analysis techniques.

Furthermore, some researchers prefer methods that produce qualitative data (e.g., Dubois et al. 2006), whereas others utilize methods that lead to quantitative data (e.g., Pheasant et al. 2010). These types of data are not inevitably incompatible; there is even a strong request to use multiple methods to collect and analyze data on different levels due to the assumed convergence and increased validity of the data. Therefore, mixed method approaches, in which different methods are systematically combined, are more frequently used to investigate soundscapes. Indeed, Steele et al. (2016) claimed that by combining qualitative and quantitative data and relying on observational or participatory methods for data collection, different aspects of public spaces can be explored behaviorally, acoustically, and perceptually. For example, Bruce and Davies (2014) combined soundwalks, semi-structured interviews, focus groups, and a soundscape simulator to study the effect of expectation on soundscape. In another approach, Bild et al. (2018) employed a mixed methods design by combining quantitative, qualitative, and spatial analyses to analyze how users of public spaces evaluated their soundscapes in relation to their activities. The authors performed parallel data gathering, which led to behavioral mapping of three urban parks.

In addition to mixed method designs, the concept of *data triangulation* and *method triangulation* is popular in the context of soundscape research (Botteldooren, De Coensel, Aletta, and Kang, see Chap. 8). In general, the idea behind triangulation is to use several sources (data, methods, theories, or analyses) to achieve higher validity and to resolve the inherent biases of a single measurement technique (Denzin 1978). Triangulation is a powerful technique that facilitates validation of data through cross-verification by applying and combining several research methodologies in the study of the same phenomenon (Schulte-Fortkamp and Fiebig 2016). Simply said, the researcher can be more confident with an experimental outcome if different methods lead to the same result. However, Kelle and Erzberger (2003) have noted that a combination of qualitative and quantitative data is still rare in empirical research and frequently lacks a methodological framework. The very same issue seems to apply for soundscape studies. Although several researchers work on the development of triangulation strategies and methods, a theoretical framework still has not been established for the systematic combination of methods and data (triangulation).

## 7.2 Methods of Early Soundscape Research

The first soundscape projects started in the late 1960s and early 1970s, laying the groundwork for future developments in the scope of soundscape methodologies. Southworth (1969) raised the question of *sonic identity* of cities and pointed out that non-visual aspects of the physical environment are among the least considered characteristics in a city. For Southworth this was surprising as a city has sounds, smells, textures, and myriad sensations of microclimate, and the perception greatly affects the interpretation of the visual information presented by the cityscape.

A few years later, Schafer (1975) and his colleagues undertook a soundscape study in five small villages of northern Europe. Despite limited time and equipment (two tape recorders, few microphones, two sound level meters, and about 100 reels of recording tape), they collected a large set of data, which included the following:

- Study of local archives for references to sound (town crier, post horns, etc.)
- Recordings and measurements of the intensity of all village signals.
- Recordings of all antique sounds in the village (blacksmith, old tools or artifacts, etc.)
- Extended recordings of characteristic ambiences in each village.
- Regular sound level recordings during the day and night, both inside and outside the village.
- Enumeration and measurement of the frequency of specific types of transportation sounds.
- Lists of sounds heard throughout the village at different times of day.
- Sound preference tests in the village school(s) in which children were asked to list their favorite and most disliked sounds in the community.
- Interviews with elderly people concerning the past soundscape of the village.

This list is rather impressive because this approach preceded the modern soundscape research by decades.

Schafer (1977), in his pioneering and far-sighted work in the World Soundscape Project, introduced the practice of soundwalking, an empirical method devised for identifying a soundscape and its components in various locations. A listening walk and a soundwalk do not represent the same process. Instead, a listening walk is simply an undirected, unassisted walk with a concentration on listening. A soundwalk, however, is an exploration of the soundscape of a given area that draws the listener's attention to sounds and ambiances detected along the way and has a functional objective. In listening to a sonic environment, the listener switches between different listening styles: from the more holistic listening, which aims to capture the "global" sound impression created by the sonic environment while waiting for familiar or important sounds to emerge (expected or not), to more specific listening in search mode, expecting particular sounds in a context, recognizing them, and interpreting them appropriately (Botteldooren et al. 2016).

Researchers have widely employed the soundwalk method to carry out research in urban environments. As summarized by Adams et al. (2008), some investigators have employed the soundwalk method as a means through which the researchers immerse themselves into the urban soundscape (e.g., Semidor 2006). But, more frequently, researchers have used this approach to engage other persons who are guided by the researcher to listen and describe the area under study (e.g., Fiebig et al. 2010).

## 7.3 Current Data Collection Methods

Due to the holistic concept of soundscape, a broad range of methods and instruments are frequently applied in soundscape studies. Although the focus of soundscape investigations and research typically lies on outdoor environments, the soundscape approach has been gaining popularity in the context of studying indoor acoustic environments as well (Torresin et al. 2019). The aim of a soundscape study is simply to assess how an acoustic environment—whether it is indoors or outdoors—is perceived, in context, by people. To study the perception of an (acoustic) environment, a variety of data collection methods are applied to address the basic elements constituting the soundscape: human perception, acoustic environment, and context. According to the ISO/TS 12913-2, the study of all of these basic elements is mandatory because in the strict sense any study that does not consider people, acoustic environment, and context cannot be seen as a full-featured soundscape study. A survey in accordance with the ISO/TS 12913-2, as Heggie et al. (2019) observed, is a reasonably onerous process as compared with traditional acoustic survey methodologies. A broad range of methods and tools for soundscape data collection are commonly used and only some of them are included in the ISO/TS 12913-2. Figure 7.1 summarizes the different data collection methods and tools with respect to the basic elements of the soundscape concept.

As indicated by Fig. 7.1, on the one hand, in most cases a comprehensive data acquisition regarding the acoustic environment is carried out, and on the other hand,

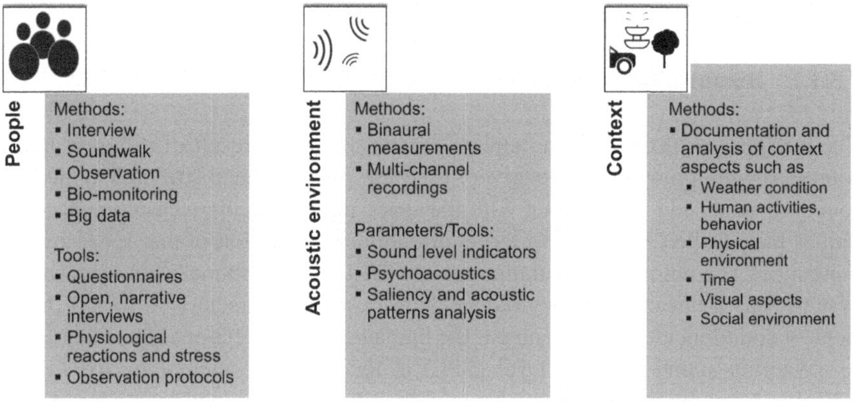


**Fig. 7.1** Methods and tools frequently applied in soundscape studies. Left: Methods and tools related to measuring human perception. Middle: Methods and tools used to measure the acoustic environment. Right: Methods to describe and document the context, where the perception of sound takes place

the mandatory measurement of the perception of sound in context is performed by means of different explicit or implicit methods. The methods range from different types of interviews and questionnaires to observation or bio-monitoring methods and finally to big data collection and analysis approaches. Data gathering via questionnaires is a way to assess the whole path from acoustic environment to soundscape, including the processes of individuals assessing and giving meaning to sound(s) and explaining their responses to the acoustic environment (ISO/TS 12913-2 2018). Questionnaires typically use questions like "To what extent do you presently hear traffic noise, sounds from human beings or natural sounds" or "How appropriate is the sound to the surrounding" or "How would you describe the quality of the surrounding sound environment" (cf. ISO/TS 12913-2 2018), which are answered in-situ by participants taking part in the data collection.

A typical questionnaire example is the Swedish Soundscape-Quality Protocol that includes scales for cross-sensory tabulation. This protocol deals with sound source identification using sounds from humans, nature, and technology, and attribute rating scales (e.g., eventful, exciting, pleasant, and calm), assessment of overall soundscape quality, and concomitant visual impressions (Axelsson et al. 2012). In 2020, a protocol for characterizing urban soundscapes has been proposed for use in the design of Soundscape Indices (SSID) and general soundscape research (Mitchell et al. 2020). The protocol to be used in designing a soundscape index consists of audio-visual recordings for use in virtual reality-based laboratory experiments and in-situ soundscape assessments via a questionnaire method that is paired with the acoustic data collection. This current approach illustrates the complexity of methods to measure perception appropriately, in context, and to permit generalization of experimental findings.

## *7.3.1 Measurements of the Physical World*

### 7.3.1.1 Binaural Measurements

The complexity of the acoustic environment is often the result of different noise sources (fixed or moving, stationary, or time varying) that are distributed in the surrounding space and that interact with the space itself. Specific measurement techniques have to be considered in terms of auditory perception in that space by the human ear. The omnidirectional microphone used in conventional sound measuring systems has a linear, frequency-independent response characteristic for all directions of sound incidence. In contrast, the human ear is a directional filter that leads to a varying sound pressure level (SPL) at the tympanic membrane by +15 to −30 dB, depending on the frequency and the direction of sound incidence. These filtering properties are due to diffractions and reflections that depend on the incidence angle of sound and are caused by the human head geometry (Genuit and Fiebig 2016). Moreover, the human hearing system introduces resonances that are independent of sound direction. The two human ears can be considered as input channels that provide interaural time and level differences between the left and right ears, which feed into dedicated brain circuits for spatial hearing. Binaural hearing yields results that are different from those that can be obtained by monaural measuring procedures. Thus, the current ISO/TS 12913-2 recommends the use of binaural measurement systems to capture all spatial information of an acoustic environment. Such binaural measurements allow the consideration of masking effects, spatial distribution, and complex phase relationships in the ways that human beings do automatically. In addition to the need for aurally accurate recordings for perception-related signal analysis, binaural recordings are essential for any (aurally accurate) reproduction of sounds (Genuit and Fiebig 2006). In other words, "copies" of the acoustic environment that are as close as possible to the human perception are needed, especially with respect to archiving and re-experiencing the acoustic scenery for comparability and analysis. Thus, aurally accurate sound-measuring technologies are fundamental for measuring the acoustic environment when perception is under consideration because this technology gathers the data necessary for comprehensive, aurally equivalent evaluations of sound.

Binaural recordings and measurements can be performed by a manikin that mimics the head, torso, and pinna of the human body and is equipped with two high-quality and calibratable microphones placed in each ear canal at the position of the eardrum (Genuit 1992). Interesting reviews on the development of manikins and artificial heads are given by Vorländer (2004) and Paul (2009). Alternatively, a human listener may be employed, wearing two miniature microphones in the ear canal or a headband supporting microphones placed at the entrance of each ear canal. The need for a calibratable and standardized system promotes the use of the manikin. The two-microphone headband is often replaced by a high-quality supra-aural headphone that is equalized for binaural recording playbacks and for which the two microphones are mounted outside each speaker. Artificial head measurement systems are often used at fixed positions, while the recording technique using some kind of binaural headset is applied when other actions are necessary for the person carrying the recording device (e.g., walking, driving a car).

For outdoor recordings, the use of windshields on the microphones is mandatory as reported in the ISO/TS 12913-2. This technical specification also requires that the binaural measurement system offers a favorable signal-to-noise ratio to be able to measure the acoustic environment without any audible distortions; the range must be chosen to avoid any overload, and any binaural measurement has to be recorded in an uncompressed format. The sampling frequency of recordings should be at least 44.1 kHz and at least a 16-bit depth is needed.

The reproduction of binaural recordings requires their equalization by adequate filters to create the same signals in the ear canal of the listener as if the listener would have been in the original sound situation. The binaural recordings are normally reproduced with headphones in order to guarantee that each channel is sent to the relevant ear. It also makes the reproduction free from reflections from the listening environment. If two loudspeakers are used, the sound from each of them would be heard with both ears (crosstalk) and this, together with reflections from the listening environment, can deteriorate the quality of the reproduction.

#### 7.3.1.2 Multi-channel Audio Techniques

Binaural recording and rendering is designed to resemble the human two-ear auditory system and reproduce sounds specifically for the two ears of a listener. However, such recorded audio is not responsive to user input, namely the audio doesn't change with movement of the head ("head-locked audio"). This limitation is settled by spatial sound field recording and reproduction technologies using a large number of microphones and loudspeakers. These technologies are aimed at replicating an acoustic scene within a region (spatial audio) in a controlled environment (typically in a laboratory) and providing an immersive sound experience (Zhang et al. 2017).

Spatial sound field recordings are commonly performed by a microphone unit composed of multiple microphone capsules and a signal processor, such as the B-Format Ambisonics microphone. In the context of soundscape research, in addition to the use of binaural measurement technologies, multi-channel recordings are required primarily for later reproduction via multi-loudspeaker set-ups in laboratories. For example, Ambisonics, the most popular audio standard for handling and delivering three-dimensional, 360-degree audio, has the advantage that it can be used for any playback setup as demanded by other surround-sound recording formats, and it allows for post-processing to vary spatial information of the recorded sound field (Hong et al. 2017). The reproduction of soundscapes that were composed in a simulator using Ambisonics data processing can yield the same soundscape dimensions as achieved in situ (Sudarsono et al. 2017). So far, Ambisonics lacks international standardization, which limits the comparability of soundscape studies based on such multi-channel recording and playback technologies. This situation is confirmed by the ISO/TS 12913-2, which acknowledges that those recording technologies striving for a future playback based on multi-loudspeaker arrays can provide a high level of immersion, but which also criticizes the lack of standardization and the difficulties of aurally accurate analyses that are required to determine psychoacoustic parameters and indicators (ISO/TS 12913-2 2018). However, there are studies available that compare the performance of different recording and playback techniques (San Martín et al. 2019) and that include video recordings of the surrounding environment at the stage of in situ data collection (Sun et al. 2018). Those studies highlight the potential of multi-channel recording and playback systems.

#### 7.3.1.3 Descriptors and Indicators

In soundscape studies, descriptors and indicators are developed to determine relevant soundscape dimensions. The ISO/TS 12913-2 defines that a *descriptor* describes the perception of any acoustic environment, whereas an *indicator* is used to predict a descriptor or a part thereof (ISO/TS 12913-2 2018). This topic is increasingly studied because of an urgent need for operational tools, like predictive models, aimed at implementing the soundscape approach in urban planning and design (Aletta et al. 2016a). Frequently, terms like *pleasantness*, *eventfulness* (Axelsson et al. 2010), *tranquillity* (Watts et al. 2013), *calmness* and *vibrancy* (Cain et al. 2013), *pleasure* and *activation*, which refer to core affects (Andringa and van den Bosch 2013), as well as *appropriateness* or *compatibility* (Schulte-Fortkamp et al. 2008), and *restorativeness* (Payne 2013) as secondary affects are potential soundscape descriptors. By means of correlation analyses, these descriptors are predicted by physical quantities, psychoacoustic metrics, and measures related to sound source occurrence. For instance, Aumond et al. (2017) claimed that pleasantness can be predicted by overall loudness and the audibility of traffic noise, bird chirping, and voices. Another study (Can and Gauvreau 2015) found that three dimensions of urban sound environments can characterize these environments in space and time: (1) the overall sound levels (sound energy), (2) the sound variations (temporal pattern), and (3) the sound spectrum (frequency distribution). This outcome confirms the results obtained in other studies (Torija et al. 2013; Di Gabriele et al. 2011).

Patterns in the sonic environment are very important. In this respect, a spectrogram can provide a useful method of measurement to visualize acoustic patterns (Genuit and Fiebig 2007). Moreover, psychoacoustic metrics like loudness, roughness, sharpness, and fluctuation strength are applied to describe the character of an acoustic environment in detail and to relate the physical phenomenon (acoustic environment) to the perceptual construct of the acoustic environment (Kang et al. 2016b). These metrics help to understand auditory sensations at least on basic, bottom-up-related sensory processing stages. Comparably, sound pressure level and psychoacoustic indicators are efficient for describing the functional aspects of soundscape perception, including preference, loudness, and communication in the context of urban shopping malls. In contrast, for playfulness and richness, a dynamic spectral center analysis (DSC) as a combined spectral-temporal analysis was found to have a significant correlation with those two perceptual factors (Yu et al. 2016). Other researchers focus on *auditory saliency* as an indicator, which refers to sound elements and events that trigger attention and contribute to the perception of the soundscape. According to Filipan et al. (2019), the human auditory cortex is sensitive to a range of spectro-temporal modulations; by evaluating those modulations, saliency can be predicted and pleasantness ratings can be estimated.

The presence of low frequencies in the sonic environment, which propagate over longer distances than mid and high frequencies, makes the soundscape description based only on frequency A-weighted SPL unsuitable (i.e., continuous equivalent A-weighted level: $L_{Aeq}$). Some investigations have been focused on figuring out more suitable indicators, such as the center of gravity of the 1/3 octave band spectrum (G), since it appears to be a good measure for the degree of pollution of the soundscape contributed by road traffic noise (De Coensel and Botteldooren 2006). Moreover, Genuit and Fiebig (2007) assumed that soundscape-related aspects (e.g., rhythm, tempo, and patterns) can be determined by focusing on relative changes of metrics and fluctuations of parameters. They reached this conclusion because human hearing adapts to steady signals and remains sensitive to fluctuations, and to prominent, salient noise events. Therefore, relative changes might be more important than absolute magnitudes (Genuit and Fiebig 2007).

Another approach has shown that the temporal structure of a soundscape can be characterized by detecting the presence of the *1/f* type feature in the spectrum of amplitude (and pitch) fluctuations on a log-log scale for the relevant time scales (0.002 Hz to 0.2 Hz) (Botteldooren et al. 2004). Suhanek et al. (2018) studied the different loudness distributions of soundscapes for listeners performing concentration-demanding tasks with respect to the evoked distraction level. In doing this, the investigators observed that not only loudness and loudness changes were mainly responsible for distraction and annoyance but also the nature of the sounds, which occurred unexpectedly and did not correspond to the general soundscape characteristics.

In the framework of acoustic ecology, many indicators have been proposed. These can be divided into two classes: indicators that estimate the amplitude, evenness, richness, and heterogeneity of an acoustic community or soundscape; and indicators that compare amplitude envelopes or, more often, frequency spectral profiles (Sueur et al. 2014).

It is important to point out that the acoustic parameters often are not sufficient to completely characterize the soundscape's features because the evaluation of these is influenced by many non-acoustic factors (see Fiebig, Chap. 2). Among these, visual aspects play an important role as has been shown frequently by laboratory studies (e.g., Seo et al. 2019). Moreover, a study carried out in some urban parks in Milan showed that vegetation was the most important aspect for the interviewees, followed by cleanliness, whereas quietness was the least important aspect and was less relevant than clean air and the sensation of personal safety (Brambilla et al. 2013a).

## 7.3.2 Measurements of Perception

### 7.3.2.1 Soundwalk

Soundwalk can be considered as a practice of the more general "sensewalking" (Henckel 2019), which is a method used by a range of disciplines to investigate and analyze how people understand, experience, and utilize spaces by focusing on sensory information. Soundwalk is a popular method that is understood as an empirical approach for identifying a soundscape and its components in various locations (Adams et al. 2008; Fiebig and Schulte-Fortkamp 2019). The ISO/TS 12913-2 (2018) defines the soundwalk method as "*a walk in an area with a focus on listening to the acoustic environment*" (page 2) and concludes that the purpose is to evaluate the soundscape in a given area by obtaining data about human sensations, responses, and outcomes by people experiencing that area (ISO/TS 12913-2 2018). Researchers have utilized soundwalk methods in various ways to investigate soundscapes (Jeon et al. 2013). For Schafer (1977), the process of conducting a soundwalk meant to explore a soundscape of a given area using a score as a guide, which consists of a map, drawing the listeners' attention to unusual sounds and ambiances to be heard along the way.

Since the early examples of soundwalks, practitioners and researchers have experimented with a huge variety of methods within the arts and humanities, social sciences, ecology studies, and engineering (Radicchi 2017). The applications of the soundwalk method differ in many aspects, such as in the way of performing acoustical measurements, the assessment methods used, the sampling and recruiting of participants, the duration of the soundwalk, the instructions given, or the incorporation of information related to other sensory modalities (visual sensations, thermal sensations, e.g., Bjerre et al. 2017). Moreover, the selection of sites to be included in a soundwalk can be the result of pre-tests, may be defined by the researchers, or can simply be chosen by the participants themselves.

Table 7.1 is a summary of the main aspects of the soundwalk method to be considered in the planning and design of data collection. However, the table cannot give a comprehensive summary of all methodical issues. For instance, when predefined discrete sites are used in a soundwalk, their order of presentation to the participants ideally should be randomized as much as possible to avoid biases due to the same sequential ordering of presentation to all participants. On this aspect, two potential caveats should be kept in mind: (1) participants may not remember the different soundscapes in the sequence equally well, and (2) the criteria/benchmark of the rating may change over time and so, for example, the evaluation of a soundscape may be dependent on the history of previous one(s). The two potential caveats may produce two types of biases: (i) ordering biases may result because the participant's rating is conditional on her/his rating order (*sequential order bias*); (ii) the evaluation of one's soundscape may directly depend on the quality of the previous one(s) (*sequential history bias*). Thus, if the participants are divided into paired groups,

**Table 7.1** Main aspects of the soundwalk method

| Aspects | General options | Recommended by ISO/TS 12913–2 |
|---|---|---|
| Acoustic recording technology | Monaural<br>Binaural<br>Multi-channel | Binaural |
| Acoustic recording conditions | Stationary<br>Mobile | Stationary |
| | Short-term recordings<br>Long-term recordings | At least a few minutes (not shorter than 3 minutes) |
| Sampling of participants | Ad hoc sample<br>Random sample<br>Systematic sample | *Issue not systematically addressed so far* |
| | Visitors<br>Locals<br>Noise experts | In favor of local experts allowing to collect ecologically valid data |
| Sample size | Small scale (single)<br>Large scale (groups) | At least 20 in groups up to 5 participants |
| Duration of soundwalk | Short-term visit<br>Long-term visit | *Issue not addressed* |
| Sites | Walking route<br>Discrete sites<br>  Predefined<br>  Chosen by participants | In favor of discrete sites (predefined locations or individually chosen places the participants prefer to listen to) |
| Instruction | Level of attention directed toward sounds<br>Emphasis on multi-modal experience<br>No explicit focus on listening | Focus on listening (silent mode) |
| Visual data collected | Description<br>Videos | Documentation by means of photographs or videos |

they can be asked to walk along the planned path in two opposite directions, clockwise and counter-clockwise (Brambilla et al. 2017).

As shown by Fiebig (2016), general measurement requirements for the soundwalk method to guarantee a high level of reliability cannot be defined because certain locations require clearly longer measurement intervals than other locations due to their larger acoustical variability. This variability can occur across time, such as seasons, but also within 24 h due to a time-dependent different use of the same place: a square can be an outdoor market during the day and a meeting place during the night for dancing activities.

This means that the appropriate soundwalk design depends on the object of investigation and the environment under scrutiny taking into account its special features. Usually, measurements of a few minutes are considered to be necessary at minimum to allow participants to immerse into the soundscape and to be able to provide reliable ratings. However, as Payne and Guastavino (2018) suggest, longer periods of exposure to a soundscape (around 40 min) can influence soundscape assessments, in particular for the components *fascination* and *extent* as essential parts of restoration according to the *Attention Restoration Theory* (Kaplan 1995).

Frequently, sound recordings are taken during the soundwalk to determine the acoustic and psychoacoustic parameters characterizing the acoustic environment (Genuit and Fiebig 2010) and to reproduce the acoustic environment later in laboratory trials. The experimental protocol of these laboratory trials should require ecological validity in order to allow participants to react, to some extent, as if they were in a natural situation (Guastavino et al. 2005). In other studies, the soundwalk method has been integrated with lightwalk, having in mind the presence of other sensory factors, such as light and smell, that also affect the environment appraisal in addition to sound (Henckel 2019).

#### 7.3.2.2 Questionnaire

Next to interview techniques, questionnaires are the most commonly used tool in soundscape investigations (Botteldooren et al. 2016). Rating and semantic scales are very often applied in soundscape studies because rating scales are convenient, easy to explain, and produce straightforward data (Rohrmann 2007). The style of rating scales varies strongly, differing in attributes, (verbal) qualifiers, and number of categories, for example. According to Engel et al. (2018), question types and answers adopted in soundscape investigations are frequently open-ended questions when dealing with sound source identification or when collecting demographic and socioeconomic information. In contrast, semantic differential scales, Stapel scales, and rating order scales are commonly considered as closed-ended questions dealing with soundscape quality, sound source evaluation, dominance, background, satisfaction, and visual quality. A *Stapel scale* is a special unipolar rating scale comprising categories ranging from negative to positive values and is only verbally labeled in the middle of the scale. *Semantic differential scales* are bipolar scales that use opposites at the ends of the scale, such as loud vs. quiet or sharp vs. dull. Table 7.2 illustrates the great variety of applied rating scales in soundscape investigations, which substantially impedes the comparability of results over different studies. In the informative annex C, referring to data collection methods, the ISO/TS 12913-2 proposes several rating scales for collecting overt responses, ranging from ordinal category scales to unipolar continuous rating scales.

In soundscape studies, it is popular to provide soundwalk participants with a paper questionnaire rather than to use face-to-face interviews. Before starting, the experimenter gives instructions on how the participants have to fill out the printed questionnaire during the walking route or at discrete sites (predefined or chosen by participants). Usually, the participants are asked to answer all the questions in the order they are presented in the questionnaire. To have a more efficient data collection, it might be reasonable to use a web-based questionnaire that can be filled out by the participants on their electronic devices (e.g., smartphone). Many software solutions, known as CAWI (Computer-Assisted Web Interviewing), are available on

**Table 7.2** Different aspects of rating scales used in soundscape investigations (not exhaustive)

| Aspects | Rating scales typically used |
|---|---|
| Polarization | Unipolar<br>Bipolar |
| Discretization | Discrete<br>Analog |
| Qualifiers (labels) | Intensity related (e.g., not at all – slightly – moderately – very – extremely)<br>Agreement related (e.g., strongly agree – agree – neither agree nor disagree – disagree – strongly disagree)<br>Quality related (e.g., very good – good – neither good nor bad – bad – very bad)<br>Frequency related (e.g., never – rarely – sometimes – often – very often) |
| Judgment dimension(s) (attributes) | Unpleasantness (pleasantness)<br>Annoying<br>Loud<br>Calm<br>Vibrant<br>Eventful/uneventful<br>Monotonous |
| Number of categories | 5 categories<br>7 categories<br>9 categories<br>11 categories |

the web and these tools are now popular. The way of submitting the questionnaire to survey participants is strictly related to the modality of their interview.

#### 7.3.2.3 Interview

Interviews are common evaluation methods to assess acoustic environments (Engel et al. 2018) and have a long tradition in the context of environmental noise annoyance research. The perceptual reality of humans is verbally investigated by different types of interviews, such as *narrative interviews,* mainly working with open questions, or *guideline interviews,* using a set of open and closed questions. By means of interviews, associations, feelings, interpretations, and emotions concerning the acoustic environment are explored in depth. These additional data implicitly contain location-specific (e.g., identification and classification of sources at a certain area) and person-specific aspects (general preferences, noise sensitivity, expectation, personal coping, and restoration strategies). As Kang et al. (2016a) claimed, open interviewing provides valid soundscape-related data, leading to *"a detailed picture of the soundscape as perceived by the people concerned."*

Qualitative data gained by the use of guideline, narrative, or in-depth interviews using open questions are frequently analyzed with Grounded Theory. The Grounded Theory approach is becoming an increasingly important methodological approach in soundscape studies (Aletta and Kang 2018). According to Glaser (1992), the Grounded Theory approach is a general methodology of analysis linked with data collection that uses a systematically applied set of methods to generate an inductive theory about a substantive area. Based on such a text-analysis approach, concepts recurring in the interview data are coded using more general concepts, which finally reach the status of categories. This iterative coding process aims at a stepwise enhancement of the level of abstraction and at an increase of the generalizability of outcomes, leading to a new theory (Fiebig and Schulte-Fortkamp 2004). Thus, the Grounded Theory is frequently applied as a text analysis method for soundscape-related qualitative interview data (e.g., Liu and Kang 2016; Schulte-Fortkamp and Fiebig 2006). Because the performance and analysis of qualitative interviews is time-consuming, typically only small sample sizes are considered. According to Holloway (1997), it must be understood that although the sample size in qualitative research is relatively small, ideally each case provides a lot of information.

In selecting the people to be interviewed, *local experts* are important to include. These are people generally living in or enjoying the area under investigation (i.e., residents in a housing area or tourists in a tourist area) (see Schulte-Fortkamp and Jordan, Chap. 3). Local expert participation sharply focuses the subsequent analysis of acoustical and perceptual data as the information collected from local experts often enhances the investigator's sensitivity to the subtle particularities of the examined areas (Schulte-Fortkamp 2016).

During survey interview sessions, acoustic data is often collected next to each respondent or group of respondents to be representative of the perceived sonic environment and to link the acoustical data to the corresponding perceptual data. According to the ISO/TS 12913-2, considering detailed guidelines for performing narrative interviews is important to guarantee compatible data collection that is related to individual perceptions. Therefore, the technical specification provides a guideline interview example. However, despite this, the form, style, and extent of the interview methods currently in use for the exploration of soundscapes vary strongly over soundscape investigations.

#### 7.3.2.4 Observational Methods

Observational methods are fundamentally different from experimental methods for which a participant actively and consciously takes part in the data collection. The active participants have ample time to think about the evaluation and they can exert control over it (Fiebig 2015); thus, they are particularly prone to demand characteristics effects. Demand characteristics effects describe the phenomenon that subjects tend to change their behavior, when they know that they are part of an experimental study, and they conjecture the experimental objectives and behave accordingly (Zizzo 2008). Those effects are expected to particularly occur in controlled experimental studies. Observation is the true foundation of the scientific method and many commonly held beliefs can often be overturned by simple observation according to Sussman (2015). In the context of soundscape investigations, asking visitors of a place to describe their listening experience automatically triggers an attentive and descriptive listening mode and is probably not the most representative listening style in relation to an everyday soundscape experience (Botteldooren et al. 2016). Therefore, combining surveys with other non-participatory methods like observation can yield a more holistic understanding of a participant's experience with respect to activity and the evaluation of acoustic environments (Steele et al. 2016).

In *non-participatory observation methods,* the observed participants are usually not aware that they are part of a study; therefore they might behave more naturally (Lavia et al. 2018). In certain cases, the use of non-participatory observations appears reasonable and can include quantitative measurements regarding walking speed, proximity to others, head movements, or occupation time (ISO/TS 12913-2). Moreover, Witchel et al. (2013) studied body language indicators for assessing the effects of soundscape quality and of specific interventions on individuals. They found evidence that changes in soundscape can be associated with subsequent objective and statistically significant changes in body language. Frequently, the effect of background music on movement and non-movement behaviors is studied by means of non-participatory observation methods (Aletta et al. 2016b; Meng et al. 2018). The advantage of observational, non-participatory methods is obvious. If humans are not aware of being observed and studied, they feel and behave more naturally, probably resulting in higher validity of the experimental outcomes. Moreover, as studies can focus on specific crowd movement and non-movement behaviors (e.g., path, speed, location of stop points), data can be collected that would be impossible to obtain or at least difficult to assess by questionnaires or interviews (Meng et al. 2018).

Observational measures of behavior increase a study's validity and generalizability; they address robustness, representativeness, and relevance particularly important for person–environment research in which the applicability is paramount and external validity is important to establish (Sussman 2015). Typically, such methods make use of video recordings from which the needed information is extracted, and there is no active interaction between the experimenter and the participants (e.g., Aletta et al. 2016b).

Although the benefit of non-participatory methods is evident, robust protocols for behavioral observation methods must be developed to make non-participatory soundscape studies and their results comparable for particular use cases (Lavia et al. 2018). Aletta et al. (2016a) suggested that controlling the experimental conditions is hard, except for one variable at a time, and drawing far-reaching conclusions with regard to cause and effect is risky. Moreover, research using non-participatory observational methods must always be aware of ethical rules, principles, and conventions that all research and investigations are bound to uphold.

#### 7.3.2.5 Bio-Monitoring

There is accumulating research that indicates how various forms of sounds affect the activity and functionality of the central and peripheral nervous systems (Erfanian et al. 2019). Consequently, Botteldooren et al. (2016) pointed out that bio-monitoring techniques could be used to assess different responses "more objectively than questionnaires." As Hume and Ahtamad (2013) claimed, a more complete objective assessment of soundscapes can be achieved by means of recorded objective physiological responses in association with assessments of individual sounds. This approach might even become more popular.

On one hand, rapid developments of wearable devices that allow monitoring of physiological reactions are making it increasingly simple to record such biological data. On the other hand, it is widely acknowledged that the magnitude of physiological reactions due to stress, activation, and arousal does not solely depend on the SPL or loudness. Those responses are also significantly related to the emotional reactions triggered by sounds, their character, and meaning to the listener. Physiological experiments have demonstrated that the body and brain respond to emotional content, as well as to sound pressure levels (Davies et al. 2013).

A study by Bradley and Lang (2000) supported this assumption with their observations that specific patterns of physiological reactions were elicited when participants were listening to affective sounds such as screams, babies crying, women sighing, a jet taking off, or a whirring fan. Moreover, Hume and Ahtamad (2013) observed significant changes in physiological measures (e.g., heart rate, respiratory rate, and forehead electromyography) in response to the presentation of sound-clips considered to be unpleasant. In contrast, Erfanian et al. (2019) conducted an extensive literature review and concluded that there is not sufficient evidence to make conclusions about physiological manifestations related to soundscapes. Physiological expressions evoked by the soundscape are not always aligned with the perpetual attributes of the soundscape. Thus, Erfanian et al. (2019) asked for more research to explain inconsistencies in results regarding fundamental physiological alterations evoked by acoustic environments, taking into account the temporal variation and duration of environmental sounds in particular. Although it remains questionable whether bio-monitoring techniques provide a more "objective" view of perception, as Botteldooren et al. (2016) assumed because researchers are still struggling with large inter-individual differences, the wider application of bio-monitoring due to the availability of easy-to-use wearable sensors for physiologic monitoring can be expected.

#### 7.3.2.6 Big Data

Big data is a very popular term associated with the idea that by collecting very large data sets new information is gained that exceeds the potential of conservative database systems. In particular, if there is a need to involve the public with respect to their quality of life and well-being, today's smart technological solutions allow for collecting environment-related information directly from the people concerned. Information and communication technology can provide the data for the development of urban policies that respond to community demands (Kang et al. 2018). According to Radicchi et al. (2017) and Radicchi (2019), smart technologies that lead to large amounts of data are expected to play an important role for acousticians, city planners, and policy makers. The authors proposed a mobile application that allows for the bottom-up production of informative and descriptive data sets of the way people experience quietness in cities as a part of everyday life by using the perception-oriented, context-sensitive concept of soundscape (Radicchi et al. 2017). There are many other applications available on the web, some of which are limited to measure sound pressure levels and others that allow feedback by users in terms of the type of perceived sound sources and attributes of the acoustic environment (for an overview of available apps see Radicchi 2019). A disadvantage of this approach is that the performance of these applications, often using the microphone on smartphones, cannot be compared with traditional instrumentation (i.e., lower sensitivity and accuracy).

The general idea of involving the public in the collection process of environmental data is in the same spirit as "Citizen Science" (Dickinson and Bonney 2015). If the applications are broadly applied, the data collected by this form of "crowd sourcing" and "participatory sensing" can be used to draw noise maps more representative of the real acoustic environment to which the citizens are exposed. In contrast, traditional noise maps, such as those required by the European Directive 2002/49/EC on environmental noise (European Directive 2002/49/EC 2002), rely solely on sound pressure levels. An interesting example of the approach mentioned above has been proposed by Picaut et al. (2019) and applied in the city center of Lyon, France. This approach to noise mapping is a paradigm shift from the traditional one as it can additionally represent the appraisal of the acoustic environment by the people exposed. The noise maps representing the spatial distribution of psychoacoustics parameters, like loudness and sharpness, are more related to the actual hearing process, which can also lead back to people's perceptions of the acoustic environment (Kang et al. 2016a).

Noise mapping could also be used as a tool to monitor the spatial distribution and audibility of preferred sounds instead of focusing only on the assessment of the exposures to unwanted sounds (Aletta and Kang 2015). Moreover, Kang et al. (2018) developed visual maps that depicted the spatial variation of the selected perceptual attributes across the study area and thereby provided information about the soundscapes. The consideration of multi-dimensional aspects in soundscape maps, according to Kogan et al. (2017), can overcome the limits of conventional noise maps and, as centered directly in the population's perception and preferences, can represent a key instrument for policy makers and urban planners who deal with the design and management of acoustic environments.

Another source of big data is related to the widespread use of on-line social networks that allow people to share their opinions and feelings on the internet: they write about their personal interests, opinions, but also about their feelings about noisy activities and sounds they hear during their daily life. A methodology to analyze these opinions automatically has been proposed by using machine learning and natural language processing technologies (Gascó et al. 2019). Another approach for analyzing data shared on social networks relies on tagging information of geo-referenced pictures of cities (Aiello et al. 2016) to assess the relationship between soundscapes and emotions and to map which areas are chaotic, monotonous, calm, or exciting.

### 7.3.3 Data Collection According to ISO/TS 12913-2

The technical specification ISO/TS 12913-2 (2018) deals with data collection and reporting requirements on soundscape and is intended to harmonize the collection of data by which information on the key components (people, acoustic environment, and context) is obtained, measured, and reported. The technical specification also proposes data collection tools and methods such as soundwalks, questionnaires, interviews, binaural recording, and other measurement technologies. Furthermore, a full-featured soundscape study requires investigation of a soundscape from several viewpoints and the use of multiple measurement methods that promote convergent validity. The ISO/TS 12913-2 also defines minimum reporting requirements in soundscape studies that should be adopted to ensure a sufficient level of clarity regarding the data collection process and to support meta-analyses. In particular, the ISO/TS 12913-2 proposes two different questionnaires: a guided interview and sound source taxonomy to collect soundscape data. The questionnaire *method A* contains a section to determine the identified sound sources and their audibility. *Method A* also includes eight 5-point ordinal category scales to determine the perceived affective quality of soundscapes by means of a two-dimensional model (Axelsson et al. 2010). Moreover, participants should assess the *appropriateness* and the *quality* of the overall sound environment on a scale ranging from "not at all" to "perfectly" and from "very good" to "very bad," respectively.

*Method B* described in ISO/TS 12913-2 provides 5-point unipolar continuous category scales referring to *loudness*, *unpleasantness*, *appropriateness,* and *desire to revisit the site again*. Moreover, this method asks for the recognized sound sources and their ranking with regard to their dominance level without providing any predefined options. In contrast to method A, the soundwalk participants are additionally encouraged to report on their feelings in their own words in written form. According to Aletta et al. (2019), the two questionnaire methods of the ISO/TS 12913-2 resulted in similar soundscape assessment outcomes with a statistically significant level of association, showing that the two methods discriminate similarly between "positive" and "negative" soundscapes. However, they recommended the use of both methods complementarily. Extensive work by Tarlao et al. (2019) showed that respondents across sites, including indoor and outdoor settings, assigned similar meaning to the same scales, supporting the attempt to develop a standardized soundscape questionnaire portable to multiple environments.

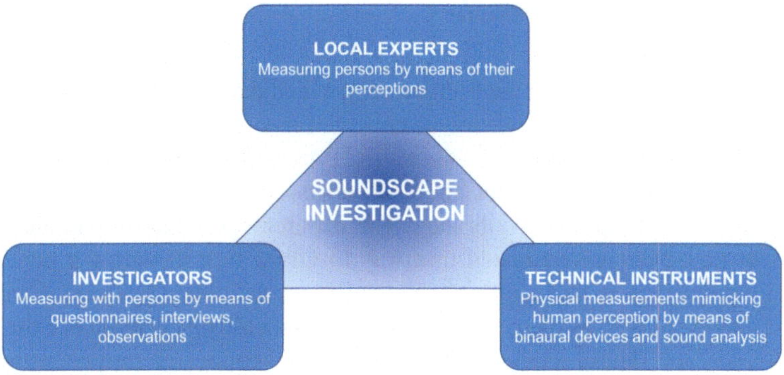


**Fig. 7.2** Methodological triangulation in soundscape analysis. (Adapted from Lercher and Schulte-Fortkamp 2013)

The translation of questionnaires with the respective instructions, items, and qualifiers to be used internationally poses a major challenge that impedes the comparability of soundscape studies. Initial attempts to make cross-national comparisons with standardized data collection procedures, using the same set of stimuli, conditions, and equipment, illustrated the great efforts needed to validate the different linguistic versions across different languages (Jeon et al. 2018).

The general approach of investigating a soundscape from several points of view using multiple measurements is known as triangulation. Triangulation is an established research strategy in the social sciences. Within this approach, different investigation techniques are applied to study a subject or phenomenon in order to improve the validity of the research outcome and to avoid systematic errors that can occur when relying only on one technique (Lercher and Schulte-Fortkamp 2013). A schematic view of the elements of triangulation to be involved in soundscape studies is given in Fig. 7.2, which is adapted from Lercher and Schulte-Fortkamp (2013). Botteldooren, De Coensel, Aletta, and Kang, Chapter 8 provides further information about how to combine multiple measurements systematically (Botteldooren, De Coensel, Aletta, and Kang, see Chap. 8).

## 7.4 Analysis Approaches

The acoustic and perceptual data collected either in situ or in laboratory environments must be analyzed for many purposes: description, classification, modeling, and so forth. Multiple methods can be applied, depending on the major aim of the data analysis, and further progress can be expected due to the increasing popularity of neural networks and machine learning methods. The ISO/TS 12913-3 (2019) provides guidance on how to analyze data collected in agreement with ISO/TS 12913-2.

### 7.4.1 Correlation Approaches

Many studies can be found in the literature for which data, mostly related to acoustics and ratings of (sonic) environments collected in-situ, are examined in terms of correlation to identify potential relationships. Typical perceptual metrics predicted by different independent variables are pleasantness, acoustic comfort, sound quality, tranquility, restoration, or vibrancy. The different explanatory parameters, as dependent variables, are estimated in a way to give a best fit with the perceptual data. The input data is frequently not limited to acoustic data because visual and thermal factors, as well as the general satisfaction with the place, influence the quality of soundscape. These relationships, even if useful for design purposes, must be applied with caution because their reliability is limited to the specific scenarios under scrutiny and often cannot be generalized.

A diverse assortment of classical regression analyses is used in soundscape studies to determine relationships between soundscape descriptors and (acoustic) indicators (e.g., multiple linear, logistic, probit, or multivariate regression models), which take into account the data type of the dependent variables. Although these correlation analyses are very common in empirical research and can be applied to detect factors that significantly affect soundscape descriptors from larger sets of potential factors (Yang 2019), critical discussions of the regressions are frequently missing. Strong correlations are often carelessly interpreted as indicating causality; however, correlations only show associations, which are *potentially* related to causal relationships.

Among the studies reported, Liu and Kang (2015) performed Pearson correlation and stepwise regression analyses between physical parameters and perceived loudness (occurrence) of individual sound categories. They observed limitations in the general predictability of soundscape composition parameters using physical and psychoacoustic parameters. In addition, Pheasant et al. (2010) used correlation analyses to identify auditory and visual factors in "restorative" or "tranquil" environments. They developed a *Tranquility Rating Prediction Tool* (TRAPT) by which a relationship is proposed between acoustical features ($L_{Aeq}$), percentage of natural contextual features, and taking into account the presence of litter and water (Watts et al. 2013; Watts and Marafa 2017). This tranquility index is an exemplary illustration of the multi-sensory soundscape notion, which in this case incorporated acoustical and visual variables in one metric.

Another example is the experimental study carried out by Brambilla et al. (2013b) on 20 urban squares as large open areas in the center of Rome. By means of multivariate stepwise regression, a model was developed to predict the ratings on the soundscape attribute "chaotic" vs. "calm" on the basis of the $L_{Aeq}$, and the psychoacoustic parameters roughness and sharpness. In general, the widespread and established sound pressure level indicators averaged over time, such as $L_{Aeq}$ or $L_{DEN}$, could only explain a small amount of the perceptual data. The metric $L_{DEN}$ is the day-evening-night level and is used in the area of environmental noise assessment as a descriptor based on energy equivalent noise level ($L_{Aeq}$) over a whole day with different penalties for nighttime noise and for evening noise (see Lercher and Dzhambov, Chap. 9).

Furthermore, not only the perceived soundscape quality influenced the assessment of the overall environment, as this depended on other non-acoustical factors, too (see regressions in Fig. 7.3 for five urban parks in Milan, Italy). The regression line of the data (solid black line) is shifted down from the line (dotted black line) corresponding to the equal percentage of respondents for both the overall environment and the soundscape. With the exception of 5% of the observations, the percentage of respondents rating the perceived quality of the overall environment as "good" was greater than that observed for the soundscape, inferring that other factors influenced the assessment in addition to the soundscape. Over the years, researchers have sought to establish correlations between physical and perceptual data, but as Kang et al. (2016b) have pointed out, such correlations are not necessarily useful as they often neglect the involvement of all related stakeholders. Moreover, neural networks and machine learning methods have been applied to determine relationships between multiple independent variables and a few dependent variables (e.g., Fen et al. 2018; Verma et al. 2019). Rapid assessment and large-scale quantification of environmental attributes are also possible through deep learning (Verma et al. 2019).

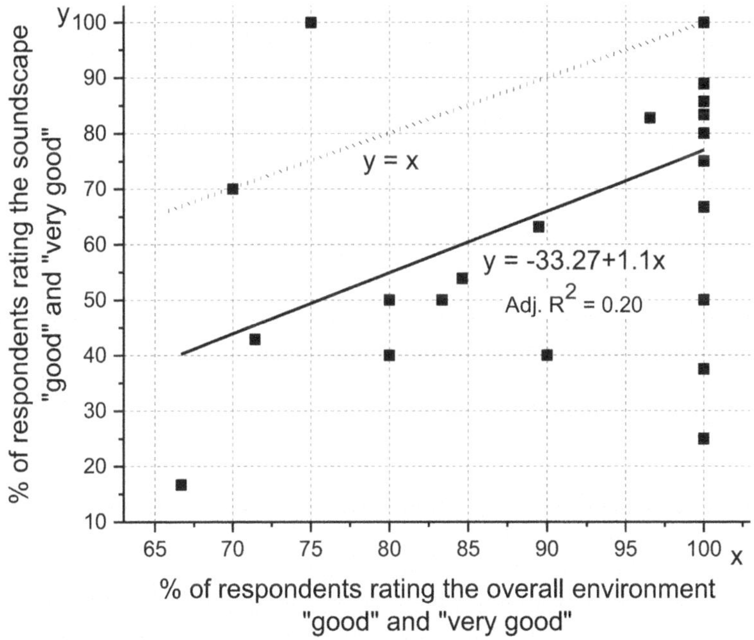


**Fig. 7.3** Perceived quality of soundscape and overall environment in five urban parks in Milan (Brambilla et al. 2013a). Solid black line corresponds to the regression line of the experimental data and dotted black line to the equal percentage of respondents for both the overall environment and the soundscape

Large data sets, which might become increasingly available due to crowd-sourcing and big data (social networks) (To and Chung 2019), promise to yield excellent predictions by intelligently connecting input variables in complex models. Often these approaches are also used to automatically recognize sound sources and classify soundscapes. According to Llorca (2018), the use of artificial intelligence to solve some persisting problems in the field of urban sound design seems plausible. However, artificial intelligence and machine learning algorithms are often countered with great skepticism by soundscape researchers who prefer focusing on open and narrative interviews.

### *7.4.2 Classification Approaches*

The classification of soundscapes is a common goal of various investigations and surveys with the aim of providing guidelines for designing and planning urban soundscapes. For this purpose, multiple input data are typically processed by clustering methods to identify relevant features that allow categorization and classification of soundscapes. Those methods are intended to identify similar features of soundscapes and to use those to develop classification tables and to allow a reliable assignment of soundscapes to clusters or categories. An example is the analysis carried out by Rychtáriková and Vermeir (2013) on 370 binaural recordings in urban public places. These were divided into 20 proposed categories and defined by a set of acoustical parameters related to $L_A$, the temporal changes of the sound were evaluated through roughness R and fluctuation strength F, the frequency spectrum through the sharpness parameter S, and the spaciousness via the "urban interaural level difference" (the sound level difference between left and right ear).

Another example is the experimental work by Brambilla et al. (2017) in which the authors developed a classification model to predict the soundscape category on the basis of selected features of the soundscape and the perceived sound sources. The data set was divided into two subsets, one for training the model and the other to test it and evaluate its classification performance. A multinomial logistic regression was applied to develop the model because the soundscape grouping resulted in three categories by the hierarchical cluster analysis (Fig. 7.4). The results were satisfactory but could not be generalized to other contexts, and only the data processing methodology was of potential interest for further applications.

Another approach has been proposed by De Coensel et al. (2008) based on the behavior of ant clustering, which is governed by fuzzy rules. These rules are optimized by a genetic algorithm, specially designed to achieve the optimal set of homogeneous clusters. Soundscape similarity is expressed as the fuzzy resemblance to the shapes of the SPL histogram, the frequency spectrum, and the spectrum of temporal fluctuations. These represent the loudness, the spectral content, and the temporal fluctuations and patterns of the soundscapes. Sun et al. (2019) proposed a coarse hierarchical classification scheme that offers an alternative to the core affect model from Axelsson et al. (2010). The hierarchical classification could probably be used to automatically classify soundscapes without involving participants. The

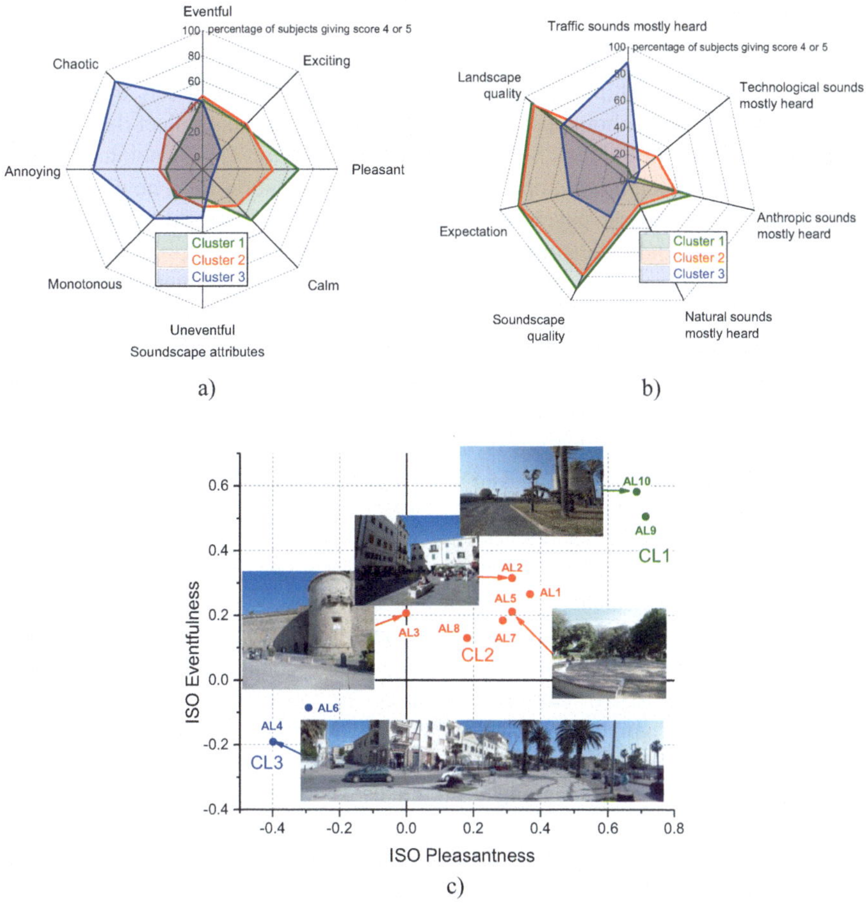


**Fig. 7.4** Features and statistical classification of the sites in terms of: (**a**) 8 different attributes of the soundscape; (**b**) appraisals on the type of sounds perceived, soundscape and landscape quality, and individual expectation; (**c**) plot according to ISO/TS 12913-3:2019. (Adapted from Brambilla et al. 2017)

authors classify soundscapes first with regard to whether they were perceived as a background or contained foregrounded sound elements; for those attracting attention, distinction was made between disruptive or supportive sonic environments and, finally, between calming and stimulating supportive environments (Sun et al. 2019).

The research on this topic is lively and further approaches are being applied, such as Neural Networks (e.g., Puyana-Romero et al. 2016; Verma et al. 2019), Structural Equation Modelling (e.g., Hong and Jeon 2015) and Principal Component Analysis (Aumond and Lavandier 2019; Axelsson et al. 2010). For instance, the Principal Component Analysis by Axelsson et al. (2010) provided a two-dimensional soundscape model of perceived affective quality (e.g., pleasantness–eventfulness, or calmness–vibrancy).

## 7.5 Summary

Although there is a tradition of a few decades of soundscape research, there is still an ongoing discussion about appropriate methods and tools that reflect the inherent holistic concept of the subject. Several decades ago, Schafer (1977) pointed out the need for a multi-disciplinary approach, laying the foundation for a new discipline that combines acoustics, psychoacoustics, sociology, and art. Although several trends on popular and widespread methods can be observed, the process of laying the groundwork for a new discipline and providing the necessary methods and tools is not yet complete.

In this chapter, a variety of state-of-the-art methods and techniques for investigating soundscapes from different points of view have been summarized. The different methods vary in many aspects: the focus on perception or acoustics, the degree of standardization (qualitative vs. quantitative), and the degree of participant involvement in a study (participatory vs. non-participatory). Also, there are multiple differences in structured or unstructured questionnaires regarding layout, qualifiers, scale format, etc. Soundscape methods are based on the principle of measuring with people as when participants are asked to report on their perceptions, are physiologically monitored by (medical) technologies, or their overt behavior is observed. Yet, great efforts will be made to use computer programs that measure aspects of soundscape in a human-mimicking way, exploiting computational intelligence (Botteldooren et al. 2016). Currently, a variety of soundscape methods are even discussed from a pragmatic point of view with regard to the practical and logistical constraints for consulting firms of varying sizes, addressing the technical demands of each project, while still achieving minimum data collection requirements (Heggie et al. 2019).

These observations clearly suggest that it is still necessary for the international community of soundscape researchers to collaborate to identify and agree on relevant soundscape descriptors and methods in order to move this area of research forward (Aletta et al. 2016a). The publication of soundscape-related international standards and technical specifications that recommend definitions, data collection techniques, and analysis methods provide a reference for soundscape investigators and researchers to draw on and can potentially lead to a harmonization of the data collection process in soundscape investigations. This, however, will not lead to a stagnancy of method development in the context of the soundscape research, as there is no perfect method for measuring anything; an interplay of errors and bias afflicts all measurements (Stevens 1975).

Research on methods and techniques capable of taking into account the demands defined in the ISO/TS 12913-2 for a holistic approach in the analysis of environments will remain essential. Mixed-methods and triangulation are recognized as adequate means to explore the complexity of human perception of acoustic environments. However, reconciling data from multiple methods remains a challenge for soundscape research (Steele et al. 2016). Consequently, research strategies based on the application of multi-methods or mixed methods must be elaborated to guide research and to approach the overarching goal of being able to systematically apply the soundscape approach for city planning and urban design.

**Compliance with Ethics Requirements** Giovanni Brambilla declares that he has no conflict of interest.

André Fiebig declares that he has no conflict of interest.

## References

Adams MD, Bruce N, Cain R, Jennings P et al (2008) Soundwalking as a methodology for understanding soundscapes. Proc ICA 30(2)

Aiello LM, Schifanella R, Quercia D, Aletta F (2016) Chatty maps: constructing sound maps of urban areas from social media data. Roy Soc Open Sci 3:150690. https://doi.org/10.1098/rsos.150690

Aletta F, Kang J (2015) Soundscape approach integrating noise mapping techniques: a case study in Brighton, UK. Noise Mapp 2:1–12. https://doi.org/10.1515/noise-2015-0001

Aletta F, Kang J, Axelsson Ö (2016a) Soundscape descriptors and a conceptual framework for developing predictive soundscape models. Landsc Urban Plan 149:65–74. https://doi.org/10.1016/j.landurbplan.2016.02.001

Aletta F, Lepore F, Kostara-Konstantinou E, Kang J, Astolfi A (2016b) An experimental study on the influence of soundscapes on people's behavior in an open public space. Appl Sci 6:276. https://doi.org/10.3390/app6100276

Aletta F, Kang J (2018) Towards an urban vibrancy model: a soundscape approach. Int J Environ Res Public Health 15:1712. https://doi.org/10.3390/ijerph15081712

Aletta F, Guattari C, Evangelisti L, Asdrubali F et al (2019) Exploring the compatibility of "Method A" and "Method B" data collection protocols reported in the ISO/TS 12913-2:2018 for urban soundscape via a soundwalk. Appl Acoust 155:190–203. https://doi.org/10.1016/j.apacoust.2019.05.024

Andringa TC, van den Bosch KA (2013) Core affect and soundscape assessment: fore- and background soundscape design for quality of life. Proceedings of InterNoise 2013, Innsbruck, September 2013

Aumond P, Can A, Coensel BD, Botteldooren D et al (2017) Modeling soundscape pleasantness using perceptual assessments and acoustic measurements along paths in urban context. Acta Acust United Acust 103(3):430–443. https://doi.org/10.3813/AAA.919073

Aumond P, Lavandier C (2019) Links between representation of outdoor soundscape, noise annoyance at home and outdoor acoustic measurement. Proceedings of InterNoise 2019, Madrid, June 2019

Axelsson Ö, Nilsson ME, Berglund B (2010) A principal components model of soundscape perception. J Acoust Soc Am 128(5):2836–2846. https://doi.org/10.1121/1.3493436

Axelsson Ö, Nilsson ME, Berglund B (2012) The Swedish soundscape-quality protocol. J Acoust Soc Am 131(4):3476. https://doi.org/10.1121/1.4709112

Bild E, Pfeffer K, Coler M, Rubin O et al (2018) Public space users' soundscape evaluations in relation to their activities. An Amsterdam-based study. Front Psychol 9:1593. https://doi.org/10.3389/fpsyg.2018.01593

Bjerre LC, Larsen TM, Sørensen AJ, Santurette S et al (2017) On-site and laboratory evaluations of soundscape quality in recreational urban spaces. Noise Health 19(89):183–192. https://doi.org/10.4103/nah.NAH_109_16

Botteldooren D, De Coensel B, De Muer T (2004) The temporal structure of the urban soundscape. Proceedings of CFA/DAGA 2004, Strasbourg, pp 361–362

Botteldooren D, Andringa T, Aspuru I, Brown AL et al (2016) From sonic environment to soundscape. In: Kang J, Schulte-Fortkamp B (eds) Soundscape and the built environment. CRC Press Taylor & Francis Group, Boca Raton, pp 17–41
Bradley MM, Lang PJ (2000) Affective reactions to acoustic stimuli. Psychophysiology 37:204–215
Brambilla G, Gallo V, Zambon G (2013a) The soundscape quality in some urban parks in Milan, Italy. Int J Environ Res Public Health 10:2348–2369. https://doi.org/10.3390/ijerph10062348
Brambilla G, Maffei L, Di Gabriele M, Gallo V (2013b) Merging physical parameters and laboratory subjective ratings for the soundscape assessment of urban squares. J Acoust Soc Am 134(1):782–790. https://doi.org/10.1121/1.4768792
Brambilla G, Pedrielli F, Masullo M (2017) Soundscape characterization and classification: a case study. Proceedings of ICSV 2017, London, July 2017
Brooks BM, Schulte-Fortkamp B, Voigt KS, Case AU (2014) Exploring our sonic environment through soundscape research & theory, vol 10. Acoustics Today, Winter, pp 30–40
Brown AL, Kang J, Gjestland T (2011) Towards standardization in soundscape preference assessment. Appl Acoust 72:387–392. https://doi.org/10.1016/j.apacoust.2011.01.001
Bruce NS, Davies WJ (2014) The effects of expectation on the perception of soundscapes. Appl Acoust 85:1–11. https://doi.org/10.1016/j.apacoust.2014.03.016
Cain R, Jenning P, Poxon J (2013) The development and application of the emotional dimensions of a soundscape. Appl Acoust 74(2):232–239. https://doi.org/10.1016/j.apacoust.2011.11.006
Can A, Gauvreau B (2015) Describing and classifying urban sound environments with a relevant set of physical indicators. J Acoust Soc Am 137(1):208–218. https://doi.org/10.1121/1.4904555
Davies WJ, Adams MD, Bruce NS, Cain R et al (2013) Perception of soundscapes: an interdisciplinary approach. Appl Acoust 74:224–231. https://doi.org/10.1016/j.apacoust.2012.05.010
De Coensel B, Botteldooren D (2006) The quiet rural soundscape and how to characterize it. Acta Acust United Acust 92:887–897
De Coensel B, Botteldooren D, Debacq K, Nilsson ME, et al (2008) Clustering outdoor soundscapes using fuzzy ants. IEEE Congress on Evolutionary Computation, Hong Kong, pp 1556–1562, June 2008
Denzin NK (1978) The research act. In: A theoretical introduction to sociological methods. Mc Graw-Hill, New York
Di Gabriele M, Maffei L, Brambilla G, Gallo V, et al (2011) Towards a soundscape classification of urban spaces. Proceedings of Forum Acusticum 2011, Aalborg
Dickinson JL, Bonney R (eds) (2015) Citizen science: public participation in environmental research. Cornell University, Ithaca
Dubois D, Guastavino C, Raimbault M (2006) A cognitive approach to urban soundscapes: using verbal data to access everyday life auditory categories. Acta Acust United Acust 92:865–874
Engel MS, Fiebig A, Pfaffenbach C, Fels J (2018) A review of socio-acoustic surveys for soundscape studies. Curr Pollut Rep 4:220–239. https://doi.org/10.1007/s40726-018-0094-8
Erfanian M, Mitchell AJ, Kang J, Aletta F (2019) The psychophysiological implications of soundscape: a systematic review of empirical literature and a research agenda. Int J Environ Res Public Health 16:3533. https://doi.org/10.3390/ijerph16193533
European Directive 2002/49/EC 2002 of the European Parliament and of the Council of 25 June 2002 (2002) relating to the assessment and management of environmental noise. Official Journal of the European Communities L 189/12
Fen J, Tung F, Li W, Pasquier P (2018) Soundscape emotion recognition via deep learning. Proceedings of Sound and Music Computing Conference, Limassol, July 2018
Fiebig A, Schulte-Fortkamp B (2004) The importance of the Grounded Theory with respect to soundscape evaluation. Proceedings of DAGA 2004, Strasburg, March 2004
Fiebig A, Acloque V, Basturk S, Di Gabriele M, et al (2010). Education in soundscape – a seminar with young scientists in the COST short term scientific mission "soundscape – measurement, analysis, evaluation." Proceedings of ICA 2010, Sydney, August 2010
Fiebig A (2015) Cognitive stimulus integration in the context of auditory sensations and sound perceptions. epubli, Berlin

Fiebig A (2016) Reliability of in-situ measurements of acoustic environments. Proceedings of DAGA 2016, Aachen, March 2016
Fiebig A, Schulte-Fortkamp B (2019) Methods to study soundscapes. In: Bull M, Cobussen M (eds) The Bloomsbury handbook of sonic methodologies. Bloomsbury, London
Filipan K, De Coensel B, Aumond P, Arnaud C et al (2019) Auditory sensory saliency as a better predictor of change than sound amplitude in pleasantness assessment of reproduced urban soundscapes. Build Environ 148:730–741. https://doi.org/10.1016/j.buildenv.2018.10.054
Gascó L, Asensio C, De Arcas G, Clavel C (2019) Evaluating noise perception through online social networks: a text mining approach to designing a noise-event alarm system based on social media content. Proceedings of InterNoise 2019, Madrid, June 2019
Genuit K (1992) Sound quality, sound comfort, sound design – why use artificial head measurement technology? JRC, Daimler-Benz, Stuttgart
Genuit K, Fiebig A (2006) Psychoacoustics and its benefit for the soundscape approach. Acta Acust United Acust 92(6):952–958
Genuit K, Fiebig A (2007) The acoustic description of patterns in soundscapes. Proceedings of InterNoise 2007, Istanbul, August 2007
Genuit K, Fiebig A (2010) Psychoacoustics for the creation of acoustically green city areas. Proceedings of ICA 2010, Sydney, August 2010
Genuit K, Fiebig A (2016) Human hearing–related measurement and analysis of acoustic environments. In: Kang J, Schulte-Fortkamp B (eds) Soundscape and the built environment. CRC Press Taylor & Francis Group, Boca Raton, pp 133–160
Glaser BG (1992) Basics of grounded theory analysis. Sociology Press, Mill Valley
Guastavino C, Katz BFG, Polack J-D, Levitin DJ et al (2005) Ecological validity of soundscape reproduction. Acta Acust United Acust 91:333–341
Heggie C, Smyrnova J, Smith B, Allen M, et al (2019) The practicalities of soundscape data collection by systematic approach according to ISO 12913-2. Proceedings of InterNoise 2019, Madrid, June 2019
Henckel D (2019) Soundwalks as sensewalks: the case for integrated sensewalks. Proceedings of InterNoise 2019, Madrid, June 2019
Hong JY, Jeon JY (2015) Influence of urban contexts on soundscape perceptions: a structural equation modeling approach. Landsc Urban Plan 141:78–87. https://doi.org/10.1016/j.landurbplan.2015.05.004
Hong JY, He J, Lam B, Gupta R et al (2017) Spatial audio for soundscape design: recording and reproduction. Appl Sci 7:627. https://doi.org/10.3390/app7060627
Holloway I (1997) Basic concepts for qualitative research. Blackwell Science Ltd, Oxford
Hume K, Ahtamad M (2013) Physiological responses to and subjective estimates of soundscape elements. Appl Acoust 74:275–281. https://doi.org/10.1016/j.apacoust.2011.10.009
International Organization for Standardization (2014) ISO 12913-1:2014 acoustics-soundscape-part 1: definition and conceptual framework. Geneva
International Organization for Standardization (2018) ISO/TS 12913–2:2018 acoustics-soundscape-part 2: data collection and reporting requirements. Geneva
International Organization for Standardization (2019) ISO/TS 12913–3:2019 acoustics-soundscape-part 3: data analysis. Geneva
Jeon JY, Hong JY, Lee PJ (2013) Soundwalk approach to identify urban soundscape individually. J Acoust Soc Am 134(1):803–812. https://doi.org/10.1121/1.4807801
Jeon JY, Hong JY, Lavandier C, Lafon J et al (2018) A cross-national comparison in assessment of urban park soundscapes in France, Korea, and Sweden through laboratory experiments. Appl Acoust 133:107–117. https://doi.org/10.1016/j.apacoust.2017.12.016
Kang J, Schulte-Fortkamp B, Fiebig A, Botteldooren D (2016a) Mapping of soundscape. In: Kang J, Schulte-Fortkamp B (eds) Soundscape and the built environment. CRC Press, Taylor & Francis Group, Boca Raton, pp 161–195
Kang J, Aletta F, Gjestland TT, Brown LA et al (2016b) Ten questions on the soundscapes of the built environment. Build Environ 108:284–294. https://doi.org/10.1016/j.buildenv.2016.08.011

Kang J, Aletta F, Margaritis E, Yang M (2018) A model for implementing soundscape maps in smart cities. Noise Mapp 5:46–59. https://doi.org/10.1515/noise-2018-0004

Kaplan S (1995) The restorative benefits of nature: toward an integrative framework. J Environ Psychol 15:169–182. https://doi.org/10.1016/0272-4944(95)90001-2

Kelle U, Erzberger C (2003) Qualitative und quantitative Methoden: kein Gegensatz. In: Flick U, von Kardorff E, Steinke I (eds) Qualitative Forschung. Ein Handbuch. Rowohlt Taschenbuch Verlag, Reinbek bei Hamburg

Kogan P, Turra B, Arenas JP, Hinalaf M (2017) A comprehensive methodology for the multidimensional and synchronic data collecting in soundscape. Sci Total Environ 580:1068–1077. https://doi.org/10.1016/j.scitotenv.2016.12.061

Lavia L, Witchel HJ, Aletta F, Steffens J et al (2018) Non-participant observation methods for soundscape design and urban planning. In: Aletta F, Xiao J (eds) Handbook of research on perception-driven approaches to urban assessment and design. IGI global

Lercher P, Schulte-Fortkamp B (2013) Soundscape of European cities and landscapes – harmonising. In: Kang J et al (eds) Soundscape of European cities and landscapes, 1st edn. Oxford by Soundscape-COSTTD0804

Liu J, Kang J (2015) Soundscape design in city parks: exploring the relationship between soundscape composition parameters and physical and psychoacoustic parameters. J Environ Eng Landsc Manag 23(2):102–112. https://doi.org/10.3846/16486897.2014.998676

Liu F, Kang J (2016) A grounded theory approach to the subjective understanding of urban soundscape in Sheffield. Cities 50:28–39. https://doi.org/10.1016/j.cities.2015.08.002

Llorca J (2018) Virtual reality for urban sound design: a tool for architects and urban planners. In: Aceves-Fernandez MA (ed) Artificial intelligence: emerging trends and applications. IntechOpen, London

Meng Q, Zhao T, Kang J (2018) Influence of music on the behaviors of crowd in urban open public spaces. Front Psychol 9:596. https://doi.org/10.3389/fpsyg.2018.00596

Mitchell A, Oberman T, Aletta F, Erfanian M et al (2020) The soundscape indices (SSID) protocol: a method for urban soundscape surveys—questionnaires with acoustical and contextual information. Appl Sci 10:2397. https://doi.org/10.3390/app10072397

Payne SR (2013) The production of a perceived restorativeness soundscape scale. Appl Acoust 74(2):255–263. https://doi.org/10.1016/j.apacoust.2011.11.005

Payne SR, Guastavino C (2018) Exploring the validity of the perceived restorativeness soundscape scale: a psycholinguistic approach. Front Psychol 9:2224. https://doi.org/10.3389/fpsyg.2018.02224

Paul S (2009) Binaural recording technology: a historical review and possible future developments. Acta Acust United Acust 95(5):767–788

Pheasant RJ, Horoshenkov KV, Watts GR (2010) Tranquillity rating prediction tool (TRAPT). Acoust Bull:18–24

Picaut J, Fortin N, Bocher E, Petit G et al (2019) An open-science crowdsourcing approach for producing community noise maps using smartphones. Build Environ 148:20–33. https://doi.org/10.1016/j.buildenv.2018.10.049

Puyana-Romero V, Maffei L, Brambilla G, Ciaburro G (2016) Modelling the soundscape quality of urban waterfronts by artificial neural networks. Appl Acoust 111:121–128. https://doi.org/10.1016/j.apacoust.2016.04.019

Radicchi A (2017) A pocket guide to soundwalking. Some introductory notes on its origin, established methods and four experimental variations. In: Besecke A (ed) Perspectives on urban economics, Berlin

Radicchi A, Henckel D, Memmel M (2017) Citizens as smart, active sensors for a quiet and just city. The case of the "open source soundscapes" approach to identify, assess and plan "everyday quiet areas" in cities. Noise Mapp 4:104–123. https://doi.org/10.1515/noise-2018-0001

Radicchi A (2019) Mobile applications for environmental noise and soundscape evaluation. Proceedings of InterNoise 2019, Madrid, June 2019

Rychtáriková M, Vermeir G (2013) Soundscape categorization on the basis of objective acoustical parameters. Appl Acoust 74:240–247. https://doi.org/10.1016/j.apacoust.2011.01.004

Rohrmann R (2007) Verbal qualifiers for rating scales: sociolinguistic considerations and psychometric data. http://www.rohrmannresearch.net/pdfs/rohrmann-vqs-report.pdf. Accessed Jan 2021

San Martín R, Arana M, Ezcurra A, Valencia A (2019) Influence of recording technology on the determination of binaural psychoacoustic indicators in soundscape investigations. Proceedings of InterNoise 2019, Madrid, June 2019

Schafer RM (1975) Five village soundscapes. World Soundscape Project. https://www.sfu.ca/sonic-studio-webdav/WSP/FVS/fvs.html. Accessed Jan 2021

Schafer RM (1977) The tuning of the world. Alfred a. Knopf, Inc, New York

Schulte-Fortkamp B, Fiebig A (2006) Soundscape analysis in a residential area. An evaluation combining noise and people's mind. Acta Acust United Acust 96:875–880

Schulte-Fortkamp B, Genuit K, Fiebig A (2008) Developing a public space in Berlin based on the knowledge of new experts. Proceedings of InterNoise 2008, Shanghai, October 2008

Schulte-Fortkamp B, Fiebig A (2016) Impact of soundscape in terms of perception. In: Kang J, Schulte-Fortkamp B (eds) Soundscape and the built environment. CRC Press, Taylor & Francis Group, Boca Raton, pp 161–195

Schulte-Fortkamp B (2016) When soundscape meets architecture. Noise Mapp 3:216–231. https://doi.org/10.1515/noise-2016-0015

Semidor C (2006) Listening to a city with the soundwalk method. Acta Acust United Acust 92(6):959–964

Seo R, Jo HI, Jeon JY (2019) VR environment-based evaluation of impact factors on the urban soundscape recognition. Proceedings of ICA 2019, Aachen, September 2019

Southworth MF (1969) The sonic environment of cities. Environ Behav 1:49–70. https://doi.org/10.1177/001391656900100104

Steele D, Bild E, Tarlao C, Martín IL, et al (2016) A comparison of soundscape evaluation methods in a large urban park in Montreal. Proceedings of ICA 2016, Buenos Aires, September 2016

Stevens SS (1975) Psychophysics. Introduction to its perceptual, neural, and social prospects. Wiley, New York

Sueur J, Farina A, Gasc A, Pieretti N et al (2014) Acoustic indices for biodiversity assessment and landscape investigation. Acta Acust United Acust 100:772–781

Suhanek M, Djurek I, Grubeša S, Petošić A (2018) Psychoacoustical approach in soundscape characterization. Adv Sci Technol Eng Syst J 3:211–221. https://doi.org/10.25046/aj030419

Sudarsono AS, Lam YW, Davies WJ (2017) The validation of acoustic environment simulator to determine the relationship between sound objects and soundscape. Acta Acust United Acust 103:657–667

Sun K, Botteldooren D, De Coensel D (2018) Realism and immersion in the reproduction of audio-visual recordings for urban soundscape evaluation. Proceedings of InterNoise 2018, Chicago, August 2018

Sun K, De Coensel B, Filipan K, Aletta F et al (2019) Classification of soundscapes of urban public open spaces. Landsc Urban Plan 189:139–155. https://doi.org/10.1016/j.landurbplan.2019.04.016

Sussman R (2015) Observational methods. The first step in science. In: Gifford R (ed) Research methods for environmental psychology. Wiley, pp 9–27. https://doi.org/10.1002/9781119162124.ch2

Tarlao C, Steele D, Guastavino C (2019) Investigating factors influencing soundscape evaluations. Proceedings of InterNoise 2019, Madrid, June 2019

To WM, Chung AWL (2019) Exploring the roles of artificial intelligence and next-generation virtual technologies in soundscape studies. Proceedings of InterNoise 2019, Madrid, June 2019

Torija AJ, Ruiz DP, Ramos-Ridao AF (2013) Application of a methodology for categorizing and differentiating urban soundscapes using acoustical descriptors and semantic-differential attributes. J Acoust Soc Am 134(1):Pt. 2791–802. https://doi.org/10.1121/1.4807804

Torresin S, Albatici R, Aletta F, Babich F et al (2019) Assessment methods and factors determining positive indoor soundscapes in residential buildings: a systematic review. Sustainability 11:5290. https://doi.org/10.3390/su11195290

Verma D, Jana A, Ramamritham K (2019) Classification and mapping of sound sources in local urban streets through AudioSet data and Bayesian optimized neural networks. Noise Mapp 6:52–71. https://doi.org/10.1515/noise-2019-0005

Vorländer M (2004) Past, present and future of dummy heads. Proceedings of Tecniacústica 2004, Guimarães, September 2004

Watts GR, Miah A, Pheasant R (2013) Tranquillity and soundscapes in urban green spaces – predicted and actual assessments from a questionnaire survey. Environ Plann B Plann Des 40:170–181. https://doi.org/10.1068/b38061

Watts G, Marafa L (2017) Validation of the Tranquillity Rating Prediction Tool (TRAPT): comparative studies in UK and Hong Kong. Noise Mapp 4:67–74. https://doi.org/10.1515/noise-2017-0005

Witchel HJ, Lavia L, Westling CEI, Healy A, et al (2013) Using body language indicators for assessing the effects of soundscape quality on individuals. Proceedings of AIA-DAGA 2013, Merano, March 2013

Yang M (2019) A review of regression analysis methods: establishing the quantitative relationships between subjective soundscape assessment and multiple factors. Proceedings of ICA 2019, Aachen, September 2019

Yu B, Kang J, Ma H (2016) Development of indicators for the soundscape in urban shopping streets. Acta Acust United Acust 102:462–473. https://doi.org/10.3813/AAA.918965

Zhang W, Samarasinghe PN, Chen H, Abhayapala TD (2017) Surround by sound: a review of spatial audio recording and reproduction. Appl Sci 7:532. https://doi.org/10.3390/app7050532

Zizzo DJ (2008) Experimenter demand effects in economic experiments, Social science research network. http://dx.doi.org/10.2139/ssrn.1163863

# Chapter 8
# Triangulation as a Tool in Soundscape Research

**Dick Botteldooren, Bert De Coensel*, Francesco Aletta, and Jian Kang**

**Abstract** Triangulating information is an essential practice in soundscape studies, and the application of this analysis tool has important implications for soundscape data collection and theory development. In this chapter, we introduce the concept of triangulation, how it is defined in the current ISO 12913 series on soundscape, and why it is relevant for the soundscape approach more broadly. Different types of triangulation analysis in soundscape studies are presented and some examples from the existing scientific literature are discussed. Different ways of "measuring" soundscapes and approaches for linking qualitative and quantitative data are presented, paving the way for soundscape prediction, shifting the focus from mere soundscape characterization to soundscape design. A brief outlook on research trends and research agenda concludes the chapter, highlighting how triangulation should play a fundamental role in any theoretical and methodological development in future soundscape studies.

**Keywords** Soundscape analysis · Soundscape theory · ISO 12913-3 · Soundscape modeling · Prediction tools · Soundscape design · Soundscape research · Environmental sounds perception · Soundscape mixed methods · Soundwalk · Narrative interviews

*Bert De Coensel passed away before publication of this work was completed.

D. Botteldooren · B. De Coensel (Deceased)
Department of Information Technology, Ghent University, Ghent, Belgium
e-mail: dick.botteldooren@ugent.be

F. Aletta · J. Kang (✉)
Institute for Environmental Design and Engineering, The Bartlett, University College London, London, UK
e-mail: f.aletta@ucl.ac.uk; j.kang@ucl.ac.uk



B. Schulte-Fortkamp et al. (eds.), *Soundscapes: Humans and Their Acoustic Environment*, Springer Handbook of Auditory Research 76,
https://doi.org/10.1007/978-3-031-22779-0_8

## 8.1 Introduction

Collecting reliable data on how people perceive(d) and experience(d) acoustic environments is a difficult process per se. Making sense of it is probably an even more challenging task. Each measurement method has its own advantages, but each also has flaws and limitations. Each comes with a specific domain of applicability, and each brings its corresponding uncertainties to the dataset. The main elements within the scope of the soundscape approach are the *acoustic environment*, *people*, and *context*. Focusing only on one of those when gathering data may result in a characterization of the soundscape that does not adequately reflect the social and physical reality of the place. Combining different perspectives and "inquiries of the reality" to learn about the soundscapes is commonly seen as a desirable approach in soundscape studies.

Such a research approach is often referred to as "triangulation." Webb et al. (1966) provided one of the first discussions of this concept within the broader panorama of the social sciences. The idea underpinning triangulation is that one can achieve a more robust representation of observed phenomena when different surveying approaches are complementary and mutually confirmatory, leading to the same (or consistent) results. In the context of soundscape data collection, the triangulation technique can facilitate the understanding and interpretation of data through cross-examination of the three-fold soundscape framework (i.e., acoustic environment/people/context) (Schulte-Fortkamp and Fiebig 2006; Zannin et al. 2013).

Triangulation is explicitly mentioned in the ISO 12913 series on soundscape *Part 3: Data analysis* (International Organization for Standardization 2019). In particular, Section 8 (Triangulation) introduces the technique and *Annex E* provides more extensive descriptions. Annex E of Part 3, refers to sociological literature (Denzin 1989) and reports that four different types of triangulations are typical and applicable, in principle:

- *Data triangulat*ion: use of multiple datasets sourced in different spatiotemporal settings with either a single or different protocols/instruments for data collection.
- *Methodological triangulation*: "within-method" and "between-method" where the former consists of applying the same method on different occasions (effectively data triangulation under a longitudinal design) or using multiple techniques within a given method (e.g., different protocols), and the latter refers to different methods within a single research design (i.e., observation).
- *Investigator triangulation*: referring to the observation of the same context by multiple researchers/evaluators to ensure that different viewpoints (e.g., cultural attitude, discipline background, technical knowledge, and expertise) are represented.
- *Theory triangulation*: alternative or competing theories are applied in examining the same datasets.

Data triangulation and methodological triangulation are the types that occur most frequently in soundscape studies (Aletta et al. 2016; Engel et al. 2018). While the techniques seem to be well-rooted in the soundscape research community, it is interesting to note that very few scholars refer explicitly to the term "triangulation" in the soundscape literature. An October 2020 search performed in the Scopus scientific database for urban soundscape studies showed that only a few items (less than ten) plainly refer to a triangulation framework.

When dealing with both physical and perceptual phenomena, triangulation can often address both quantitative and qualitative data. Qualitative research intrinsically promotes a multi-method attitude. Indeed, scholars have noted that in its original usage, triangulation synthesized different qualitative research methods and the combination of qualitative and quantitative approaches was outside of its scope (Denzin 2012). Each qualitative methodology (e.g., interviews, behavioral observations, etc.) is based on certain epistemological notions, emerging from a history of disciplinary development; therefore, combining these interpretive methods could already be a barrier itself. Soundscape requires an additional layer of complexity with regard to physical measurements (e.g., sound levels, psychoacoustic indicators, etc.) to characterize the acoustic environments with objective parameters. Triangulation that uses both qualitative and quantitative methods as a strategy adds rigor, richness, and depth to any soundscape investigation and achieves a better understanding of the investigated phenomena.

While the framework for implementing triangulation in soundscape studies is technically already in place, and supported by Part 3 of the ISO 12913, the research community is still working to facilitate its widespread adoption by both researchers and practitioners. In a field study, Aletta et al. (2019) carried out a soundscape investigation on a university campus in Rome (Italy) in accordance with two parts of the ISO/TS 12913-2 technical specifications (Aletta et al. 2019). The aim of the study was to triangulate results from Method A and Method B of the Technical Specifications in which each method was used by two groups of participants exposed to the same acoustic environments in the same soundwalk. Results showed that the two methods resulted in similar soundscape categorizations/assessments in a statistically significant way, but some differences between groups were still observed. Consequently, the authors recommended further investigations of aspects of triangulation and revision or integration of the two methods into a new data collection instrument.

Similarly, Jo et al. (2020) attempted to extend the scope of the previous study by triangulating data from all three of the methods proposed in Part 2 of ISO 12913 in a laboratory experiment (i.e., questionnaires for Method A and B; interviews for Method C). They found good comparability among the three protocols (Jo et al. 2020). Findings from these studies seem to suggest that triangulation may contribute to soundscape theory development insofar as it promotes the comparison and juxtaposition of different methods. This process may enable the synthesis of these methods, ultimately leading to newer and more comprehensive data collection instruments and methodologies.

In the three-fold relationship between the acoustic environment/people/context, one should always keep in mind that the main driver for investigating soundscapes, in fact, is people. Individuals are central to the whole data collection and analysis exercise, either as *agents* or *subjects*. This chapter will begin with defining the processes of *measuring people* as opposed to *measuring with people* and measuring in a way that mimics the human experience. Subsequently, these three approaches are discussed with regard to application to examine how they inform soundscape mapping and impact assessment and how their triangulation can eventually pave the way for soundscape prediction. In the last section of the chapter, open questions and possible challenges will be discussed.

## 8.2 Multiple Approaches for Analyzing Soundscape

### *8.2.1 Measuring People*

In a strict interpretation, soundscape does not refer to the physical sound environment but to a mental image, a cognitive map (Tversky 1993), in the mind of those experiencing the soundscape (Botteldooren et al. 2016). In this view, the challenge in assessing soundscape lies in getting a glimpse of this mental image. Although brain imaging allows researchers to identify some basic responses of people to soundscape (Irwin et al. 2011; Li and Kang 2019), these techniques are still too elaborate to be deployed in an ecologically valid setting. Hence, one must rely on behavioral responses, even today.

Experiencing the sonic environment is seldom the purpose for people to be in a public place. Hence sound in the space often goes unnoticed and does not contribute to the overall experience of the place (Sun et al. 2019). Interacting with users of the space during their normal activities may place an unnatural focus on the sonic environment and thus change the way their brains process environmental sounds. Retrospective, in-depth interviews (Schulte-Fortkamp and Fiebig 2006) allow investigators to explore the perception, understanding, and appraisal of the sonic environment without affecting perception. In these narratives, participants should be encouraged to focus not only on the sounds and the activities and sources that caused them, but also on the effect the sounds had on the listener. Likewise, mind map drawing could be used to describe the soundscape (Marry 2011). As a drawing may be less appropriate to represent a sonic environment, participants may need stronger convincing to engage in that type of evaluation. Whatever method of expression is used, a significant effort should be put into getting information beyond the obvious, and formal techniques are available for that purpose (Tenbrink 2015). This form of assessment is henceforth called *measuring people*.

*Measuring people* for soundscape assessment has several advantages:

- It gives a complete and holistic picture of how the community feels about the soundscape at a given location and time.
- A thorough assessment of the cognitive map of the place is also a sound basis for setting the goals of soundscape design: it reveals expectations of the users of the

space, their envisaged use, their understanding within a socio-cultural context, etc.
- It may disclose non-acoustic concerns, raising awareness, and strategic considerations of the local population.

*Measuring people* also has some significant drawbacks:

- It is very labor-intensive and often only a handful of participants can be included.
- Future scenarios can hardly be imagined by people without focusing on specific sounds; hence, the approach does not permit the study of future scenarios.

### 8.2.2 Measuring with People

Perception of the sonic environment is a key element in soundscape formation. The development of psychoacoustic knowledge driven by applications in the acoustic design of commercial products and appliances has produced a variety of indicators and tools that analyze a sound in a way that resembles human perception as closely as possible. In contrast to this application of psychoacoustics, the perception of an urban soundscape involves a diverse and sometimes subtle mixture of sounds to which a person is exposed. Therefore, auditory scene analysis (ASA) (Bregman 2015) is a key element in the perception of a complex sonic environment. Moreover, ASA does not rely on a single sense. The brain regions responsible for attention and gating, which primarily steer how auditory streams and objects are formed, perform these tasks in a multisensory way (Musacchia and Schroeder 2009). Using a person as a measurement instrument automatically includes these complex listening phenomena. The process of setting up a test situation with very specific listening instructions combined with a structured question-and-answer scheme will be referred to as *measuring with people*.

Soundwalking, as introduced in detail in Chap. 7 (Brambilla and Fiebig), has become a very popular method for measuring soundscape with people; however, the methodology creates a challenge when it comes to assessing future scenarios or comparing different places in the world. If the intervention will only slightly modify an existing environment, augmented reality (AR) may be used to add or even suppress sonic elements. Although this technology has been widely used in recreating historical events, archaeological sites (Sikora et al. 2018), and as art (Oberman et al. 2020), its application in urban soundscape design remains limited. Virtual reality (VR) is now widely used for evaluating urban development scenarios, but quite often the acoustic and sonic components are underdeveloped. Nevertheless, a strong interaction has been demonstrated between the appreciation of the visual and auditory design, even in a virtual environment (Echevarria Sanchez et al. 2017). Soundwalking in VR often limits the freedom to move by assigning a predefined path (Oberman et al. 2018), but VR, or a more simplified version using a 2D visual display, can bring an urban public place to the lab and allow movement between places without the need for traveling (Aumond et al. 2017; Sun et al. 2019).

*Measuring with people* requires clear instructions and often a well-defined set of standardized questions (ISO/TS 12913-2:2018). A set of questions referring to the emotions the sonic environment evokes, as described by the classical circumplex model of affect (Russell 1980), could be used (Method A of the standard). Experimenters invite the participants to transfer the effect to the sounds that created it. Arguably the arousal component is more universal between persons and situations or even sensory modalities, but the valence component largely depends on context and expectations.

A more direct assessment of the sonic environment relates to psychoacoustic descriptors (Method B of the standard). The associated questions require the person to distance him/herself from the sources of the sound.

*Measuring with people* for soundscape assessment has several advantages:

- Comparisons across populations, regions, and time can be made due to standardization and focused questioning.
- A detailed analytical image of the soundscape can be constructed as it requires attentive listening, thereby identifying sounds and their frequency of occurrence while actively looking for meaning and emotion.
- The training and effort needed by the investigator remains moderate.
- Effects of interventions can be assessed since the response-variance between persons is expected to be low due to the structured questioning and clear instructions.
- Future scenarios can be analyzed using artificially created audio-visual environments.

*Measuring with people* also has some significant drawbacks:

- In-depth impact of the sonic environment when experienced within context and use of the space are not assessed because the unbiased relationship between the study participants and the environment is changed by the instructions given prior to the observation.
- Monitoring changes and the effects of interventions still require a huge effort.

### *8.2.3 Measuring in a Human-Mimicking Way*

The remarkable capabilities of the human hearing system in analyzing the sonic environment are hard to match with electronic equipment. The further task of assigning meaning and creating a mental representation of the sonic environment is an even harder challenge.

Today's commercial sound level meters are stuck in the technology and knowledge of the previous century. A-weighted equivalent sound levels are the 1970's effort to capture the loudness that a human ear would assign to a spectro-temporally fluctuating sound. Newer standards on electronic loudness evaluation (International Organization for Standardization 2017) have advanced the field and are able to approximate the human auditory periphery much better. Such loudness evaluation is readily available in equipment for binaural sound quality evaluation that could be deployed for mobile soundscape assessment (see Genuit, Schulte-Fortkamp, Fiebig, Chap. 6). Also, the equipment has been stripped down for applications in smart city sensor networks (Segura et al. 2015). Additional psychoacoustic indicators, such as sharpness and fluctuation strength, have also been related to soundscape; nevertheless, whether these indicators can work as a proxy for sound recognition (and can assign meaning) or if these indicators are intrinsically important for the perception of the sonic environment remains unclear.

Soundscape goes beyond perception to include meaning and understanding of the sonic environment by persons or society (e.g., the soundscape definition given in ISO 12913-1 2014). Meaning and understanding are formed within a certain context, and either may be influenced by expectations about the place or common beliefs (Filipan et al. 2017). Artificial intelligence (AI) is still a long way from assigning meaning and understanding to a sensory perception within a context emerging from situational awareness. Yet, if meaning could be reduced to associations evoked by the sound (Kohonen 2012), then the problem is simply reduced to finding a suitable verbal label that describes the concept. Convolutional deep neural networks have made progress in the recognition of environmental sounds (Salamon and Bello 2017), although the technology seems far less advanced than its visual counterpart or even applications in speech recognition.

In complex sonic environments, AI systems for sound recognition may give confusing outcomes. The human brain has the useful ability to steer attention and gate out irrelevant or useless information. As a side effect, this also implies that during many uses of a public place, environmental sounds can remain subliminal. Attention modeling is only now being introduced in AI with the objective to select the most informative soundbites for the task at hand (Bahdanau et al. 2015). In contrast, for the task of analyzing urban sonic environments, the activity or use of the space often does not involve active listening.

Our main interest is the identification of sounds that attract the attention of the user. Saliency is a key element in this analysis. Sensory saliency is determined by the ability of the periphery and the auditory cortex to detect specific elements in the sound, which may steer attention in a bottom-up way (Elhilali et al. 2009; Kayser et al. 2005). Semantic saliency or incongruence (Gygi and Shafiro 2011) refers to deviant detection at a higher level of cognition. Predictive coding theory postulates that perception results from a combination of sensory input and prior prediction (Sedley et al. 2016). Mismatch between prediction and sensory input then results in surprise and may trigger attention. Because this variant of saliency relies on situational awareness and expectations, it is much harder to include in AI systems for assessing the sonic environment.

People understand the environment not only from its auditive content but from a combination of sensory information of which the visual input is the most important, although the particular weight of each modality in multi-sensory perception is person dependent (Sun et al. 2018). Artificial intelligence is only starting to fuse data from multiple sensors for scene analysis (Essid et al. 2017). The main purpose is to create AI that has situational awareness and can conduct dialogue that refers to objects in the scene.

Measuring with (human-mimicking) AI refers to the ensemble of technologies from loudness measurement to AI-based machine listening. The main advantages as a soundscape measuring technique are:

- Machine-based measurement allows practitioners to monitor the evolution of a soundscape over longer periods of time, facilitating the discovery of diurnal or seasonal patterns.
- In a mobile variant, the soundscape in large areas can be mapped more continuously than with the techniques discussed previously.
- Predicting the effects of interventions through modeling is relatively straightforward.
- Standardization of the basic concepts, such as loudness, allows for comparison amongst studies and regions.

But there are also significant drawbacks:

- AI may be able to identify the sounds that people may recognize but replicating understanding and meaning within a context remains a difficult problem.
- Machine listening in a multisensory context continues to be very challenging.
- The lack of standardization for advanced machine listening techniques prohibits widespread implementation.

## 8.3 Purpose of the Analyses

Each of the methods for analyzing soundscape that was introduced in Sect. 8.2 has its disadvantages, but triangulation enables the combination of these methods and circumvents their flaws (Turner et al. 2015). The implementation of triangulation that is most appropriate depends on the purpose of the analyses. In soundscape, two purposes are common: theory development and testing on the one hand; mapping, trend analysis, and impact prediction on the other. Theory development and testing refers to the more academic endeavor to fully understand processes in people's minds. Mapping, trend analysis, and impact prediction are essential for bringing soundscape analysis and design into urban planning and design practice.

### *8.3.1 Theory Development and Theory/Hypothesis Testing*

The theory of urban soundscape aims to relate the sonic environment at a (public) place within its context and typical use, in other words, how it is perceived and understood by people and society. Such a theory could take an analytical, stepwise approach: identify noticeable and salient auditory objects; assign meaning and create a mental image or cognitive map; and, finally, appraisal (Botteldooren et al. 2016). A considerable portion of the literature has been influenced by environmental psychology theory, which has led to models to characterize soundscapes (Axelsson et al. 2010; Andringa and van den Bosch 2013).

Yet, the development of a soundscape theory would benefit from a *holistic triangulation* approach. *Measuring people*, as described above, could give the most complete view on the elements that affect the relationship between the sonic environment and its soundscape. Early work using verbal descriptions identified the importance of auditory objects and meaning in the appraisal of sonic environments (Dubois et al. 2006). Object formation comes before qualitative assessment; however, we advocate that soundscape theory development would benefit from reaching beyond the direct assessment of soundscape to include neuroscience, psychology, evolutionary theory, and even artificial intelligence.

Neuroscience research is providing increasing insight in the working of the brain, often through new imaging techniques. These insights may contribute to the understanding of how complex sonic environments are processed, assigned meaning, and remembered by the human brain. To this end, relevant neuroscientific findings should be identified and translated at an algorithmic level to the field of soundscape research (Hassabis et al. 2017). A few examples illustrate the development of soundscape theory on this basis. Topographically separated regions of specialization in the auditory cortex may allow researchers to assess and identify specific sensitivities of human hearing system. Using functional MRI, Schönwiesner and Zatorre (2009) showed that there are specific areas tuned to detecting modulation ripples. These findings provide inspiration for the development of sensory saliency models. Another example of useful insight from the realm of neuroscience refers to individual differences with regard to sensitivity to the living environment. Neurological circuits related to this sensitivity have been identified (Kliuchko et al. 2016). They are related to the circuits that control gating out, leading to the hypothesis that noise-sensitive people lack part of the ability to gate out uninformative sound and, therefore, would appreciate complex sonic environments less than others with that ability. These neural insights could lead to a better understanding of the relationship between the sonic environment and human responses to the soundscape.

The field of psychology could contribute to the development of soundscape theory through the vast amount of knowledge on psychoacoustics (Fastl and Zwicker 2007); however, linking psychological experiments with optimized stimuli and an ecologically valid context may not be trivial. For example, it has been argued that the ability to track stimulus statistics and generate predictions supports the choice of what to attend to and what to ignore. Rogalla et al. (2020) showed in mice that the ability to predict the scene decreases saliency and surprise increases saliency. Although the authors suggest that this might indicate a difference in the perception of saliency compared to humans, who would recognize the most informative signal as more salient, one could also argue that this difference is simply caused by a difference in listening style. This would lead to the hypothesis that in an inattentive listening condition, which is the case in an everyday human experience in an urban sonic environment, saliency and directing attention to a sound would arise from unexpected stimuli. In contrast, in a focused listening condition, which occurs during soundwalks, the most informative sounds would attract the most attention, as suggested for the human listening style by Rogalla et al. (2020).

Modern experimental psychology relies strongly on brain monitoring to evaluate how sensory stimuli and sound are processed. One of the strongest signals in EEG, the alpha oscillation, is recognized as informative for many cognitive brain functions (Sadaghiani and Kleinschmidt 2016). Observation of these oscillations may inform soundscape theory with regard to how the brain switches between different sound streams. For example, such observations revealed that general inhibition of sensory inputs is selectively released for specific sounds and free brain wandering may be replaced by focused attention, thus putting forward inhibition as the default state.

Psychological studies may shed light on yet another hot topic in soundscape theory: audio-visual interaction. Using simple stimuli, Giard and Peronnet (1999) showed that there may be clear differences between persons on how they process congruent and contradicting audio-visual stimuli, which is consistent with the hypothesis that there are significant differences among people in how they experience a sonic environment in a visual context. Visually dominant people may be exclusively led by the visual environment when evaluating the sonic one (Sun et al. 2018). The importance of the visual component in soundscape perception was described long ago (Viollon et al. 2002), yet soundscape theory may gain additional benefit from recognizing the strong personal differences in audio-visual integration. Techniques for measuring people could put a stronger focus on these differences.

Soundscape theory can benefit from consideration of the evolution of hearing in general (van den Bosch et al. 2018). The evolution of the auditory system and hearing in animals provided critical advantages: first and foremost, hearing can detect danger that is associated with sounds, such as an approaching predator; secondly, sound localization can produce a cognitive map of the environment; thirdly hearing can be used by the listener to find specific sites (e.g., a babbling stream), prey, or potential mating partners. In humans, the latter capability has been expanded to complex verbal communication. Indeed, human hearing has been tuned evolutionarily to the spectral content and typical modulations found in human speech.

More interesting insights can be gained from evolution theory when it comes to danger and its counterpart, safety. In a more naïve view, danger would solely be attributed to loud sudden sound events. Today, these no longer imply danger, but could simply be attributed to activities of other persons that may not be indicative of immediate danger (e.g., a plane, a car passing). However, when considering the absence of safety rather than the presence of danger, another view is more appropriate. The generalized unsafety theory of stress (Brosschot et al. 2018) predicts that safety constantly needs reconfirming. In this view, the complete absence of sound could lead to a state of stress, which implies that hearing typical environmental sounds confirms that the environment is safe.

As soundscape theory becomes more mature, testing benefits from *convergent triangulation*. Structured questionnaires like those proposed in the ISO standard (International Organization for Standardization 2018) could be a typical tool for this testing. Such questionnaire surveys are a format that can focus the investigation on one hypothesis. For example, the circumplex model of valence-arousal proposed in the environmental psychology literature (Russell 1980) inspired the hypothesis that the effect of the sonic environment on people could be described in a similar way with a matching classification of soundscapes in a pleasantness-eventfulness plane (Axelsson et al. 2010). In the context of theory testing, widely applying this classification could lead to a confirmation of the hypothesis that this is indeed an appropriate way to classify sound environments. But that does not exclude that there might be other classifications that could turn out to be more appropriate.

### *8.3.2 Mapping, Trend Analysis, and Impact Prediction*

Holistic triangulation also allows a more complete, holistic, and contextual portrayal of the unit(s) under study. Here, one single method is not uniquely capable of mapping and analyzing the complex phenomena of how people *perceive* and *understand* the sonic environment. There is no unique or generally applied method to combine several techniques for impact prediction, trend analysis, and mapping of soundscape, yet some general suggestions can be made, and good practices can be illustrated as valuable examples.

*Impact assessment* naturally starts with setting goals and identifying where the impact will be the largest. This is naturally done by *measuring people*, assessing how they feel about their living environment as a whole. More specifically, the assessment must determine where they expect the sonic environment to contribute to the livability of their neighborhood or where unwanted sound would prohibit the possible use of a space. This process could include a narrative analysis but should evaluate causal relationships between sound and the general appreciation of the living environment. In addition, inhabitants may have general feelings of stress or feel a lack of restoration that may not be related directly to high background sound levels. This knowledge allows the investigator to refine the questionnaires used to continue *measuring with people* (De Coensel et al. 2010) by focusing on: particular needs for the area; disturbing sounds or sounds that contribute to the identity of the place; specific times of the day or periods of the year; specific locations where sounds can be heard that should be removed or conserved, etc. One could argue that this focused knowledge is sufficient to start an urban redevelopment process or environmental restoration process and then assess its impact.

A co-creation process (Van Renterghem et al. 2020a, b) may rely heavily on knowledge gained from the stakeholders involved. Some interventions, such as adding water features (Jeon et al. 2010) or natural sound playbacks (Schulte-Fortkamp 2010; Van Renterghem et al. 2020a, b), may be conducted based on knowledge of the local experts: the population at large. Quite often, however, measuring in a *human mimicking way* is required to link the perception and understanding of the sonic environment to the physics of sound emission and propagation by different sources. The latter can be used to assess the impact of noise reduction methods such as screening by noise barriers, buildings, or berms. Other considerations include noise emission limits, pavement state, volume of traffic, and the location of flight paths.

*Mapping* of soundscape and *trend analysis* also benefit from triangulation. In mapping and trend analysis, reproducibility and reliability of measurements are crucial to compare over space and time respectively, while often attempting to detect subtle differences. Therefore, physical measurements are often a first choice. For well-known sound sources that can be described by known statistics (e.g., traffic), average exposure can be modeled quite accurately. Yet, a direct link between such measurements and soundscape is often missing (see below), and a complementary direct assessment with people is needed.

Well before soundscape became a popular approach, noise annoyance surveys using standardized questions (Fields et al. 2001) were promoted for mapping and trend analysis. Noise annoyance surveys conducted in many countries focus on a specific context (in and around your home) and one particular effect (annoyance), yet they have been interpreted as a form of measuring with people. Even with such relatively narrow scope (i.e., only the domestic context, and only the annoyance-related outcome), there could be considerable methodological challenges. Trend analysis based on longitudinal annoyance surveys has been conducted for many years; however, the approach is very sensitive to methodological details such as placement of questions, response alternatives, and season (Brink et al. 2016). Thus, a very strict protocol must be followed. Mapping and trend analysis in more general soundscape studies are likely to also require such methodological rigor. Standardization in soundscape assessment started in 2009 (International Organization for Standardization 2018) and, therefore, trend analysis over periods of several years is lacking. Questionnaire-based soundscape studies mostly focus on specific areas such as parks, boulevards, waterfronts, or commercial areas. Most often, interviews are taken at the site either as part of a soundwalk or with arbitrary visitors to the place. When asked to report on sounds that are audible, it is difficult for participants to reflect on the whole visit or on a typical day rather than to listen to the current sonic environment.

An alternative way of obtaining a wholistic impression on the soundscape and its effect on humans is to ask participants to point at places in their daily living environment that evoke a certain state or understanding, such as pointing at tranquil or vibrant areas or areas where sound disturbs intended activities (De Coensel et al. 2017). In summary, mapping and trend analysis would also benefit from triangulating several measurement techniques.

## 8.4 Linking Mixed Methods

Section 8.3 identified the need for and uses of triangulation. This section explores the different methods that have been used to combine different measurement techniques for these purposes.

Figure 8.1 shows the relative amount of scientific effort that has been put into linking the different measurement techniques, both for developing soundscape theory and for reducing dimensionality in mapping, trend analysis, and impact prediction. The numbered links refer to:

1. Linking measurements in a human-mimicking way, including the use of simple A-weighted levels, psychoacoustic quantities, and sound recognition, to measurements with persons, including closed questionnaires administered during soundwalks and surveys with arbitrary users of the space. As the fields of psychoacoustics, general psychology, and even artificial intelligence try to bring measurements closer to the perceptions of humans, it can be expected that a lot of effort is put into connecting these two. Moreover, the ab initio models that are available to calculate acoustic and psychoacoustic quantities from source characteristics and propagation is steadily growing giving this modeling path the potential for eventually becoming the main prediction tool.
2. Using people to quantify how a sonic environment is perceived must be linked to the actual impact of the environment by its day-to-day users. This difference may sound subtle, but it could be attributed to several factors. Firstly, at the level of perception itself, day-to-day users of a public place may be less aware of the sonic environment and pay less attention to it than the participants in a soundwalk or the soundscape professional who is visiting the place. Secondly, for local inhabitants, the context, and the way the environment is understood may be significantly different than for casual visitors. Finally, the overall effect of a public place on mental restoration and quality of life depends on the local situation.

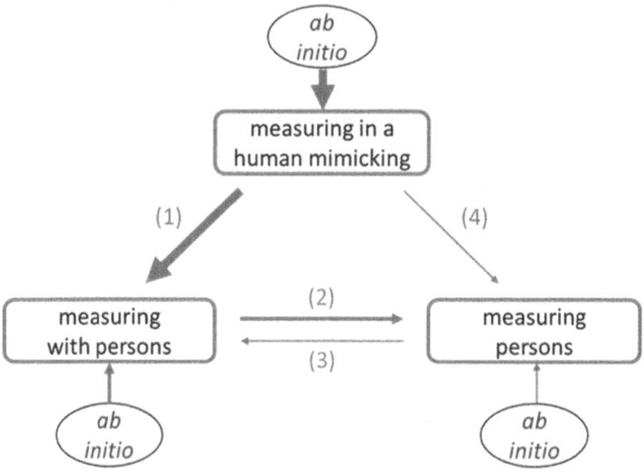


**Fig. 8.1** Linkage between the groups of soundscape measurement methods; thickness of the arrow indicates the amount of work conducted in these areas; the arrow between the measurement and the "ab initio" label indicates the opportunities for ab initio modeling and prediction of these measurements based on activities, topography, etc. Examples are connections which are represented by the numbers: (1) linking measurements in a human-mimicking way; (2) using people to quantify how a sonic environment is perceived; (3) identifying a soundscape structured response from persons brought to a place; (4) linking a physical measurement and an overall state for individuals

Very often today, the link between what focused listeners hear in a park and the effect on everyday visitors is made for what is called "the average person," which is a construct that does not exist. Nevertheless, there is a growing body of literature on the linkage between the responses on structured questionnaires (International Organization for Standardization 2018) and the overall effect of a particular sonic environment.

3. The reverse link that tries to identify the structured response from persons brought to a place based on the general understanding of and expectations for that place is of less practical use and is less often studied. From a conceptual point of view, investigating this link is nevertheless quite interesting. It can show how representative and relevant the common interview methodologies are. For example, models predicting the expected answers on a structured questionnaire from a narrative would highlight the relevance of the questionnaire as an easier and more reproducible alternative.
4. The direct link between a physical measurement and an overall state for individuals has been often studied at an epidemiological level for exploring things like health effects. This often results in explained variances that are rather low. This could be at least partly attributed to poor representations of the sonic environment and how it is perceived and understood by persons. In soundscape research, exploring this link is much less common. It is indeed easier to follow the path (1) followed by (2), discussed previously.

Ab initio modeling is used in Fig. 8.1 to refer to models that predict the specific type of measurement based on general knowledge on drivers such as traffic, population density, and topography. Today, most of such models used in environmental sound research and in practice depend on the calculation of long-term averaged $L_{Aeq}$ (A-weighted equivalent sound level). Such energy exposure models may not be representative, but they are much easier to calculate as the distribution over time and frequency of the individual contributions to the sonic environment are completely ignored. Improved ab initio modeling for more advanced psychoacoustic parameters remains a challenge (Genuit 2018); hence, the alternative approach of directly predicting perception by measuring with people became a valid alternative (Hong and Jeon 2017).

### 8.4.1 Imprecision and Vagueness

Perception and understanding of the sonic environment are conceptually well-defined, but mathematically remain inherently imprecise and vague. To handle the imprecision, an additional degree of freedom is typically introduced. Thus, rather than putting a crisp label on a soundscape (e.g., this is a *tranquil* environment) or to describe it by a crisp attribute, the degree to which the environment belongs in a certain category is evaluated or a vague range of values for that attribute is identified.

*Fuzzy set theory* is well suited to handle the categorization vagueness, such as when considering these three sets of soundscapes: those that contribute to the liveliness of the place, those that contribute to calmness, and those that are disruptive. In crisp logic, each soundscape would be classified into one of these classes. Fuzzy set theory on the contrary allows a soundscape to be 0.9 lively and 0.4 disruptive at the same time (Sun et al. 2019).

The vagueness of concepts can also be embedded in the logic used to model relationships. For example, if many bird sounds are heard, the soundscape will be perceived as rather tranquil. In this example, "many" and "rather" could be vaguely represented in fuzzy rules and fuzzy inference could be used to derive the overall result of multiples of these rules. This method has been applied for annoyance prediction (Botteldooren et al. 2022) and, after that, for soundscape quality assessment (Maristany et al. 2016) or for deriving the preservation value of a soundscape (Jia et al. 2020).

## 8.4.2 Causality Links

Linking different observations inherently raises the question of causality, and this is not different when considering soundscapes. In particular, this is apparent when stepping away from the more direct effects and exploring the links labeled (2) and (3) in Fig. 8.1. Proof of causality could be obtained from longitudinal, cohort studies (e.g., to assess the impact on mental health) or from intervention studies where only soundscape characteristics are changed. Very few of these studies have been reported today. For the case of cohort studies, the differences in spatiotemporal granularity of soundscape assessment and the long-term benefits for health and well-being introduce important methodological issues. An interesting alternative for intervention studies could be a co-creation approach for which participants are allowed to instantaneously change the soundscape (Steele et al. 2019; Van Renterghem et al. 2020a, b). Such studies could assess the perceived soundscape immediately after it has been modified by the user, but knowledge can also be extracted from the sounds added directly by the users.

On a shorter timeframe, a change in perception or appraisal immediately following a change in sonic environment may indicate a causal relationship. Such changes could occur naturally or could be induced at a place, but they could also occur when people move around in a city. Granger causality (Bressler and Seth 2011) can be used to assess the direction of induction for such short time frames. A time series (e.g., change in the pleasantness rating of the soundscape) is caused by another time series (e.g., a salient sound event) if the former can be predicted on the basis of previous values of the latter (Filipan et al. 2019).

Fitting a regression model is the most common approach for linking different observations. Such models range from regressions (simple linear regression, polynomial or logistic regression) to shallow artificial neural networks that can incorporate strong nonlinearities. In soundscapes, people play a central role, specifically when paths (1) or (4) in Fig. 8.1 are considered. Mostly, the differences between persons in perceiving and understanding the audio-visual environment cannot be fully included in the models. Such models allow to include the person as a random factor introducing an offset or when interaction is allowed, a different dependence on one of the independent factors. These could reflect, for example, the sensitivity of people or simply the way they use rating scales. Alternative ways of accounting for unknown personal factors are the use of a z-score or any other function to normalize the data.

Regression models can also be used for spatial interpolation between measurements with humans. These models could include the distance to relevant sound sources, the type of area or land use, or even calculated noise maps for different sources. Kriging methods (Rathbun and Stein 2000) have been very popular for linear interpolation between spatial data and are readily available in software packages. Using these methods without careful consideration of the physics behind sound propagation often ignores the important effects produced by such things as screening by buildings in an urban context. As a result, investigators may miss tranquil spots and micro-parks. Adding knowledge through calculated maps may be a better approach (Wei et al. 2016). When using regression models with few degrees of freedom, a careful choice should be made with regard to whether logarithmic exposure indicators should be used. Indeed, addition on logarithmic values naturally implies an amplification or attenuation of a sound and is very suited to fit a distance-dependent model. Yet avoiding logarithmic (dB) values for weighted contributions from different sources may be preferred. Using strong nonlinear regression models with more degrees of freedom, such as (shallow) artificial neural networks, circumvents this choice.

In Boes et al. (2018) and Sun et al. (2019), logistic regression was used to relate the sounds that are heard and perceived calmness, liveliness, and disruption to physical indicators and sound recognition model outcomes. Linear regression and artificial neural network models were compared for predicting soundscape quality in Puyana Romero et al. (2016). For mapping and interpolation between measurements with persons, artificial neural networks were used in Yu and Kang (2009), while Hong and Jeon (2017) compared kriging and inverse distance weighting methods.

## 8.5 Toward Soundscape Prediction

So far, triangulation has been treated as an analysis technique aimed at confirming observed soundscape phenomena or theories. The ISO/TS 12913-3:2019, which covers soundscape data analysis, supports this approach and focuses on the idea that associations should be sought between acoustic data sets (i.e., measured with instruments in a *human-mimicking way*) and perceptual data sets (i.e., measured *with people*). In the Technical Specifications, Section A.4 suggests that potential relationships linking Method A questionnaire results to acoustic data should be investigated via statistical analyses, such as correlation analyses or analysis of variance (ANOVA). In the specific case of ordinal data (as per the scales of Method A), the Spearman's rank correlation coefficient should be computed. Likewise, *Annex B,* which addresses the link between Method B results and acoustic data, similarly suggests a correlation analysis, but since Method B includes interval data, the Pearson correlation coefficient should be calculated.

### 8.5.1 From Characterization to Design

The proposed studies in the ISO/TS 12913-3:2019 represent a confirmatory approach, but the associations between acoustic and perceptual variables that are revealed do not establish causality. The confirmatory approaches are useful for characterizing the *status quo*, but they offer little insight into hypothetical soundscape scenarios. The natural evolution of linking methods and observations is probably moving toward "modeling," which could provide the ability to predict or anticipate how physical acoustic environments would be assessed by people or communities. The social, cultural, and physical phenomena interacting in soundscapes are hugely complex, making soundscape modeling/prediction an incredibly demanding task. Not surprisingly, soundscape "design" comes with limitations. For some scholars, the overarching goal is to be able to make such predictions without surveying people.

Currently, prediction tools are scarce (Kang 2017; Kang et al. 2019). Aletta et al. (2016) argue that the rationale for modeling the causal relationships between the physical (i.e., acoustic) and perceptual properties of a sound environment is two-fold: on one hand, a soundscape prediction model could be used to anticipate how people would experience the sound environment while avoiding the (often) lengthy task of actually *asking* people; on the other hand, an accurate predictive model can disclose the underlying reasons of the perceived features, effectively becoming an investigation and design tool. A typical example in soundscape literature is the Tranquility Rating Prediction Tool (TRAPT) by Watts and colleagues (2011; Pheasant et al. 2009), where the perceived tranquility of a place is a function of sound levels and visible green features. Therefore, when designing a tranquil soundscape, one could modulate the visual and auditory factors. This example highlights that for soundscape modeling to be meaningful, it is essential that the right soundscape *descriptor* is selected, so that corresponding soundscape *indicators* can be identified to predict the former. Soundscape indicators and descriptors are generally defined as "measures used to predict the value of a soundscape descriptor" and "measures of how people perceive the acoustic environment" (Aletta et al. 2016).

### 8.5.2 Overview of Available Models

A predictive model for a soundscape descriptor could relate to either a single dimension (as in the previous tranquility example) or to a more general valence-related construct of "soundscape quality" (is the soundscape overall "good" or "bad"). Considering the highly contextual nature of soundscape experience, however, accurate models would probably focus on one dimension at a time. The next challenge then becomes identifying the appropriate dimension around which to build the prediction strategy. Depending on context, models for different dimensions may be needed as tranquility (or calmness, quietness, and similar constructs) is not the desirable criterion used to assess the quality of any urban space. The soundscape circumplex model proposed by Axelsson et al. (2010) for soundscape characterization suggests that vibrant/exciting/lively soundscapes are just as desirable as calm/tranquil/quiet soundscapes because both are related to "positive" (i.e., pleasant) constructs. For instance, Aletta and Kang (2018) proposed a prediction model for vibrancy that was based on a set of psychoacoustic parameters (roughness, fluctuation strength, loudness), and on the presence of people in the visual scene and music in the auditory scene as soundscape indicators. These objectively measurable parameters accounted for approximately 76% of the variance in the mean individual vibrancy scores. The social presence (visible people within the place) was later confirmed to be an important predictor for vibrancy-related constructs (Sun et al. 2019).

As previously shown, researchers traditionally use regression analysis to triangulate acoustic and perceptual data. This implies that both the acoustic and perceptual data should be numerical, meaning in turn that the most viable option for gathering perceptual data is semantic scales (Engel et al. 2018). Therefore, future research efforts should further develop the semantic scales currently used in soundscape studies and make more use of the already standardized ones (as in the ISO/TS 12913-2:2018) for comparative purposes (International Organization for Standardization 2018). For this to happen, it is crucial that broad international consensus is sought on which soundscape dimensions are to be prioritized. Harmonization has resulted in the establishment of scientific networks with the goal of international consensus (COST TUD Action TD-0804 2013). When working with semantic scales, verification of the reliability of the data collection instrument is also important. In a literature example where Method A was used, researchers showed that some "scaling bias" emerges when people are asked to assess the vibrant-monotonous, calm-chaotic, and eventful-uneventful soundscape constructs on a five-point semantic scale, as specified by the Technical Specifications (Lionello et al. 2019).

A literature review by Lionello et al. (2020) suggested that soundscape modeling can be abstracted to a three-component framework; namely: (1) soundscape indicators, (2) soundscape descriptors, and (3) the set of rules mapping the former to the latter. The mapping can rely on either linear (e.g., multiple regressions) or on non-linear prediction (e.g., fuzzy-logic, Support Vector Regression Machine, Artificial

Neural Networks, etc.). The systematic review highlights that soundscape indicators are generally related to the following elements: acoustic and psychoacoustic metrics, environmental and contextual information, visual information, and person-related factors. With regard to the modeling strategy in the reviewed studies, the main approach was using linear regression, while fewer studies implemented non-linear models. The target values (i.e., the soundscape descriptor variables) are generally considered as the average values of individual scores across the whole sample of participants/assessors. A crucial aspect to contemplate is whether the model performs adequately in terms of its actual predictive capability. This should be considered in terms of correlations between the predicted soundscape outcomes and a specific validation dataset. More importantly, accuracy will be subject to several factors and sources of uncertainty. These may include, among others: data collection methods, sample size, socio-cultural context, physical measurements, and experimental methods.

When reviewing published applications of linear and non-linear models, linear methods appear to be easier to apply and are generally interpreted more easily, which explains their relative popularity. On the other hand, non-linear approaches may lead to more accurate results, but they carry additional layers of complexity in terms of the definition of the model, extraction of the features, and analysis of the data. If only acoustic data are fed into such models, the benefits may be limited, and visual and contextual data would be a natural extension of the scope. In general, using predictors that represent a higher level of abstraction, such as embedding time dynamics or other perceptual data (e.g., perceived greenness), results in better model performance and accuracy. In summary, choices made during data collection and the size of the datasets fed into the models should be carefully considered. Including subjective information (i.e., perceptual data) in the indicators could positively contribute to the performance of the soundscape prediction model, but this raises a practical question: If perceptual data are needed to predict perceptual outcomes, why not actually gather the target perceptual data? From an engineering perspective, this is indeed less suitable, as the goal for some research is to reduce dependence (in design terms) on data sourced directly from people and to utilize objective parameters instead (Mitchell et al. 2020). In this regard, non-linear methods do seem to provide more accurate results than linear ones, but the challenges related to their implementation sometimes lead researchers to other choices.

## 8.6 Summary: Soundscape and Triangulation

Soundscape research over the past couple of decades has been changing how stakeholders look at environmental acoustics and urban sounds in the built environment. As an emerging discipline, soundscape research has had to triangulate its way into scientific and public discourse, and place itself at the intersection of academia, practice, and the attitudes of the general public. It is equally important that soundscape research is well-positioned in public health science and environmental science so that the societal benefits of adopting a soundscape approach become evident and, therefore, public investment and allocation of community resources are acceptable (Andringa et al. 2013). The standardization process supported by international networks and agencies clearly played a pivotal role in raising awareness about soundscape as a viable approach to characterize, manage, and design urban sound environments. Nevertheless, more efforts are definitely needed to attract the attention of stakeholders to additional challenges ahead (Aletta and Xiao 2018a, b).

Part of the answer will come from the education domains: researchers and practitioners with a broad understanding of this multi-faceted issue who can use the triangulation philosophy are needed to find new solutions. For this purpose, training programs that focus on soundscape in universities and other institutions of higher education along with continuous professional development opportunities would be very beneficial. This may lead to a new professional role for soundscape researchers (or urban sound planners) who will lead a paradigm shift toward new transformation processes for the built environment (Alves et al. 2015; Echevarria Sanchez et al. 2018).

### *8.6.1 Triangulation as a Theoretical Approach*

This chapter has described how soundscape, as a discipline, is rooted in the concept of triangulation around its main components: *acoustic environment*, *people*, and *context*. However, while this is a vital framework, triangulation does not necessarily offer the advancement of knowledge per se. Based on the trends in previous research, triangulation theory development will move in three main directions:

1. *Scale*—the level and/or spatial scale at which the soundscape approach should be implemented is debated. While many scholars insist that soundscape analysis is only effective at very local scales (impact prediction, "right here, right now," causal effects, etc.), others advocate for much larger-scale applications with soundscapes considered at city and regional scales (e.g., mapping, zoning, planning, etc.). From the community perspective, soundscapes could be considered in a "hierarchical" way, considering soundscapes assessments derived from individual experiences (labeled Type I soundscapes), soundscapes derived from groups of individuals; i.e., "collective perceptions" (Type II), and soundscapes derived from higher level concerns (Type III) (COST TUD Action TD-0804 2013).
2. *Aim*—there seems to be consensus that triangulation should deal to some extent with modeling and anticipation of future soundscapes. From a design perspective, the people/context/acoustic environment framework could easily be translated into a program/context/idea system (Bauer 2016). Developing this theoretical strand will boost the ambition to inform design and cultivate innovative concepts for public spaces and community life.
3. *Scope*—different disciplines, such as environmental psychology, physiology, acoustics, and many more, inform the soundscape approach. However, triangula-

tion should still be looking at the science of perception while gathering information and developing theoretical models across different sensory domains, such as vision and smell (Aletta and Xiao 2018a, b). Theoretical triangulation will be needed to synthesize a consistent system of analysis for the urban realm from a human-centered perspective.

### *8.6.2 Triangulation as an Analysis Method*

Being able to link ever-growing datasets with different data architectures will be the challenge for soundscape in the future. The pervasiveness of sensor networks in smart cities will generate constant and massive streams of data that can be used for soundscape analysis. This will require more and more sophisticated triangulation techniques to make sense of such data and convert it into useful information. Advanced machine learning will likely offer the tools for this goal. Deep neural networks (DNN) and convolutional neural networks (CNN) (Lecun et al. 2015) have become the dominant methodology for visual and sound data classifications. Thus, a brief discussion of their applicability for linking soundscape assessment methodologies is warranted.

The success of DNN and CNN largely relied on a vast amount of image and sound data that have become available online. Similar labeled data are not available for training perception and understanding of sonic environments in the proper contexts. Moreover, the available labeled data have not been centralized in a way remotely, like the large image databases we can query today. This should not imply that these advanced techniques cannot have any benefits in soundscape. A first promising approach could be to use sound recognition as a front-end training to identify sound sources as part of the measurement system. To this end, transfer learning, which is duplicating an already trained network from another environment to the problem at hand, could be used. Promising examples are the trained CNN used for detecting bird vocalizations in the context of biodiversity monitoring (Grill and Schluter 2017) and the 2019 DCASE challenge (Detection and Classification of Acoustic Scenes and Events), which mainly focuses on indoor or voice activity detection networks (Zhang and Wu 2013). The audibility of voices, natural sounds, and mechanical sounds is strongly related to ratings for soundscape quality and adding a few layers to the network that already learned how to identify these sounds could be beneficial. Specific initiatives have studied CNN for direct categorization of very short (a few seconds) environmental sounds on an arousal-valence scale (Fan et al. 2018). Although the authors conclude that a trained CNN performs well in evaluating arousal and a transfer learning approach gives good estimates for valence, their excerpts remain very short, and their ecological validity is limited.

Apart from these techniques, support vector machines, Bayesian inference networks, and many similar techniques that require less training data are available today. Using a biologically inspired front-end that identifies the features that are primarily detected by the human auditory system may be beneficial (Thoret et al. 2020).

### *8.6.3 Triangulation as a Research Agenda*

Clearly, the triangulation concept has been and will continue to be central to soundscape studies. While the starting point was an analyzing technique, the literature shows that triangulation is, in fact, a more general approach to soundscape as a whole that contributes to theory development and articulation. Triangulation also provides a useful lens to identify future research trends and lines of investigation. Very few scientific works in soundscape studies explicitly refer to triangulation as a reference framework, but the reality is that the concept underlies the vast majority of soundscape research and practice. The research community should make more effort to raise awareness about this important topic and invest resources into its definition. Triangulation should therefore be put at the forefront of the soundscape conversation and used as a compass for this field and as a metric to assess the rigor and relevance of research outcomes.

**Acknowledgments** The authors would like to dedicate this chapter to our late colleague and friend, Dr. Bert De Coensel, he will be deeply missed.

The authors are also grateful to Andrew Mitchell and Xiaochao Chen for their help in editing some parts of this manuscript.

**Compliance with Ethics Requirements** Dick Botteldooren declares that he has no conflict of interest.

Bert De Coensel† declares that he has no conflict of interest.

Francesco Aletta declares that he has no conflict of interest.

Jian Kang declares that he has no conflict of interest.

## References

Aletta F, Kang J (2018) Towards an urban vibrancy model: a soundscape approach. Int J Environ Res Public Health 15(8):1712. https://doi.org/10.3390/ijerph150

Aletta F, Xiao J (2018a) Handbook of research on perception-driven approaches to urban assessment and design. IGI Global, Hershey. https://doi.org/10.4018/978-1-5225-3637-6

Aletta F, Xiao J (2018b) What are the current priorities and challenges for (urban) soundscape research? Challenges 9(1):16. https://doi.org/10.3390/chal

Aletta F, Kang J, Axelsson Ö (2016) Soundscape descriptors and a conceptual framework for developing predictive soundscape models. Landsc Urban Plan 149:65–74. https://doi.org/10.1016/J.landurbplan.2016.02.001

Aletta F, Guattari C, Evangelisti L et al (2019) Exploring the compatibility of "Method A" and "Method B" data collection protocols reported in the ISO/TS 12913-2:2018 for urban soundscape via a soundwalk. Appl Acoust 155(C):190–203. https://doi.org/10.1016/J.apacoust.2019.05.024

Alves S, Estévez-Mauriz L, Aletta F et al (2015) Towards the integration of urban sound planning in urban development processes: the study of four test sites within the SONORUS project. Noise Mapp 2(1):57–85. https://doi.org/10.1515/noise-2015-0005

†Deceased

Andringa TC, van den Bosch K (2013) Core affect and soundscape assessment: fore-and background design for quality of life. In: Proceedings of the Internoise 2013 Conference. Innsbruck
Andringa TC, Weber M, Payne SR (2013) Positioning soundscape research and management. J Acoust Soc Am 134:2739–2747. https://doi.org/10.1121/1.4819248
Aumond P, Can A, De Coensel B et al (2017) Global and continuous pleasantness estimation of the soundscape perceived during walking trips through urban environments. Appl Sci 7(2). https://doi.org/10.3390/app7020144
Axelsson Ö, Nilsson ME, Berglund B (2010) A principal components model of soundscape perception. J Acoust Soc Am 128(5):2836–2846. https://doi.org/10.1121/1.3493436
Bahdanau D, Cho K, Bengio Y. Neural machine translation by jointly learning to align and translate. In: 3rd International Conference on Learning Representations, ICLR 2015 – Conference Track Proceedings
Bauer J (2016) The new ISO-standard on "Soundscape" - Maximizing the benefit for the Architectural design process. In: Proceedings of the Internoise 2013 Conference. Innsbruck
Boes M, Filipan K, De Coensel B, Botteldooren D (2018) Machine listening for park soundscape quality assessment. Acta Acust united with Acust 104(1):121. https://doi.org/10.3813/AAA.919152
Botteldooren D, Andringa T, Aspuru I et al (2016) From sonic environment to soundscape. In: Soundscape and the built environment, 1st edn. CRC Press, Boca Raton, p 26
Botteldooren D, Verkeyn A, Lercher P (2022) Noise annoyance modelling using fuzzy rule based systems. Noise Health 4(15):27
Bregman AS (2015) Progress in understanding auditory scene analysis. Music Percept An Interdiscip J 33(1):12–19. https://doi.org/10.1525/MP.2015.33.1.12
Bressler SL, Seth AK (2011) Wiener–Granger Causality: a well established methodology. NeuroImage 58(2):323–329. https://doi.org/10.1016/J.NEUROIMAGE.2010.02.059
Brink M, Schreckenberg D, Vienneau D et al (2016) Effects of scale, question location, order of response alternatives, and season on self-reported noise annoyance using ICBEN scales: a field experiment. Int J Environ Res Public Health 13(11). https://doi.org/10.3390/IJERPH13111163
Brosschot JF, Verkuil B, Thayer JF (2018) Generalized unsafety theory of stress: unsafe environments and conditions, and the default stress response. Int J Environ Res Public Health 15(3). https://doi.org/10.3390/IJERPH15030464
COST TUD Action TD-0804 (2013) In: Kang J, Chourmouziadou K, Sakantamis K et al (eds) Soundscape of European cities and landscapes. Soundscape-COST, Oxford
De Coensel B, Bockstael A, Dekoninck L et al (2010) The soundscape approach for early stage urban planning: a case study. 39th Int Congr Noise Control Eng 2010, INTER-NOISE 2010 3:2333–2342
De Coensel B, Sun K, Botteldooren D (2017) Urban soundscapes of the world: selection and reproduction of urban acoustic environments with soundscape in mind. INTER-NOISE 2017 - 46th Int Congr Expo Noise Control Eng Taming Noise Mov Quiet 2017-Janua
Denzin NK (1989) The research act: a theoretical introduction to sociological methods, 3rd edn. Prentice-Hall, Englewood Cliffs
Denzin NK (2012) Triangulation 2.0. J Mix Methods Res 6(2):80–88. https://doi.org/10.1177/1558689812437186
Dubois D, Guastavino C, Raimbault M (2006) A cognitive approach to urban soundscapes: using verbal data to access everyday life auditory categories. Acta Acust united with Acust 92:865–874
Echevarria Sanchez GM, Van Renterghem T, Sun K et al (2017) Using virtual reality for assessing the role of noise in the audio-visual design of an urban public space. Landsc Urban Plan 167:98. https://doi.org/10.1016/j.landurbplan.2017.05.018
Echevarria Sanchez GM, Alves S, Botteldooren D (2018) Urban sound planning: an essential component in urbanism and landscape architecture. In: Aletta F, Xiao J (eds) Handbook of research on perception-driven approaches to urban assessment and design. IGI Global: Hershey, pp 1–22

Elhilali M, Xiang J, Shamma SA, Simon JZ (2009) Interaction between attention and bottom-up saliency mediates the representation of foreground and background in an auditory scene. PLoS Biol 7:e1000129. https://doi.org/10.1371/journal.pbio.1000129
Engel MS, Fiebig A, Pfaffenbach C, Fels J (2018) A review of socio-acoustic surveys for soundscape studies. Curr Pollut Rep 4:220–239. https://doi.org/10.1007/S40726-018-0094-8/FIGURES/8
Essid S, Parekh S, Duong NQK et al (2017) Multiview approaches to event detection and scene analysis. In: Virtanen T, Plumbley M, Ellis D (eds) Computational analysis of sound scenes and events. Springer, Cham. https://doi.org/10.1007/978-3-319-63450-0_9
Fan J, Tung F, Li W, Pasquier P (2018) Soundscape emotion recognition via deep learning. In: Proc 15th Sound Music Comput Conf Sonic Crossings, SMC 2018 289–294
Fastl H, Zwicker E (2007) Psychoacoustics: facts and models. In: Psychoacoustics: facts and models. Springer Berlin Heidelberg, pp 1–463
Fields JM, De Jong RG, Gjestland T et al (2001) Standardized general-purpose noise reaction questions for community noise surveys: research and a recommendation. J Sound Vib 242(4):641–679. https://doi.org/10.1006/jsvi.2000.3384
Filipan K, Boes M, De Coensel B et al (2017) The personal viewpoint on the meaning of tranquility affects the appraisal of the urban park soundscape. Appl Sci 7(1):1–16. https://doi.org/10.3390/app7010091
Filipan K, De Coensel B, Aumond P et al (2019) Auditory sensory saliency as a better predictor of change than sound amplitude in pleasantness assessment of reproduced urban soundscapes. Build Environ 148:730–741. https://doi.org/10.1016/J.BUILDENV.2018.10.054
Genuit K (2018) Application of psychoacoustic within soundscape, the new challenge for acoustic consultants. In: INTER-NOISE 2018 - 47th international congress and exposition on noise control engineering: impact of noise control engineering
Giard MH, Peronnet F (1999) Auditory-visual integration during multimodal object recognition in humans: a behavioral and electrophysiological study. J Cogn Neurosci 11:473–490. https://doi.org/10.1162/089892999563544
Grill T, Schluter J (2017) Two convolutional neural networks for bird detection in audio signals. In: 25th Eur Signal Process Conf EUSIPCO 2017 2017-Janua:1764–1768. https://doi.org/10.23919/EUSIPCO.2017.8081512
Gygi B, Shafiro V (2011) The Incongruency advantage for environmental sounds presented in natural auditory scenes. J Exp Psychol Hum Percept Perform 37:551–565. https://doi.org/10.1037/A0020671
Hassabis D, Kumaran D, Summerfield C, Botvinick M (2017) Neuroscience-inspired artificial intelligence. Neuron 95:245–258. https://doi.org/10.1016/J.NEURON.2017.06.011
Hong JY, Jeon JY (2017) Exploring spatial relationships among soundscape variables in urban areas: a spatial statistical modelling approach. Landsc Urban Plan 157:352–364. https://doi.org/10.1016/J.LANDURBPLAN.2016.08.006
ISO 12913-1 (2014) Acoustics-soundscape-part 1: definition and conceptual framework. Geneva, Switzerland
International Organization for Standardization (2017) ISO 532-2:2017(en), Acoustics — Methods for calculating loudness — Part 2: Moore-Glasberg method. https://www.iso.org/obp/ui/#iso:std:iso:532:-2:ed-1:v1:en. Accessed 16 May 2022
International Organization for Standardization (2018) ISO/TS 12913–2:2018(en), Acoustics — Soundscape — Part 2: data collection and reporting requirements. https://www.iso.org/obp/ui/#iso:std:iso:ts:12913:-2:ed-1:v1:en. Accessed 16 May 2022
International Organization for Standardization (2019) ISO/TS 12913–3:2019(en), Acoustics — Soundscape — Part 3: data analysis. https://www.iso.org/obp/ui/#iso:std:iso:ts:12913:-3:ed-1:v1:en. Accessed 16 May 2022
Irwin A, Hall DA, Peters A, Plack CJ (2011) Listening to urban soundscapes: physiological validity of perceptual dimensions. Psychophysiology 48:258. https://doi.org/10.1111/j.1469-8986.2010.01051.x

Jeon JY, Lee PJ, You J, Kang J (2010) Perceptual assessment of quality of urban soundscapes with combined noise sources and water sounds. J Acoust Soc Am 127(3):1357. https://doi.org/10.1121/1.3298437

Jia Y, Ma H, Kang J, Wang C (2020) The preservation value of urban soundscape and its determinant factors. Appl Acoust 168:107430. https://doi.org/10.1016/j.apacoust.2020.107430

Jo HI, Seo R, Jeon JY (2020) Soundscape assessment methods: compatibility of questionnaires and narrative interview based on ISO 12913-2. In: INTER-NOISE and NOISE-CON Congress and Conference Proceedings. Institute of Noise Control Engineering, pp 3509–3518

Kang J (2017) From dBA to soundscape indices: managing our sound environment. Front Eng Manag 4(2):184–192. https://doi.org/10.15302/J-FEM-2017026

Kang J, Aletta F, Kachlicka M, et al (2019) Towards soundscape indices. In: Towards soundscape indices. Proceedings of the International Conference on Acoustics ICA 2019. Aachen

Kayser C, Petkov CI, Lippert M, Logothetis NK (2005) Mechanisms for allocating auditory attention: an auditory saliency map. Curr Biol 15:1943. https://doi.org/10.1016/j.cub.2005.09.040

Kliuchko M, Heinonen-Guzejev M, Vuust P et al (2016) A window into the brain mechanisms associated with noise sensitivity. Sci Rep 6. https://doi.org/10.1038/srep39236

Kohonen T (2012) Self-organization and associative memory. Springer Science & Business Media, Berlin

Lecun Y, Bengio Y, Hinton G (2015) Deep learning. Nature 521(7553):436–444. https://doi.org/10.1038/nature14539

Li Z, Kang J (2019) Sensitivity analysis of changes in human physiological indicators observed in soundscapes. Landsc Urban Plan 190:103593. https://doi.org/10.1016/j.landurbplan.2019.103593

Lionello M, Aletta F, Kang J (2019) On the dimension and scaling analysis of soundscape assessment tools: a case study about the "Method A" of ISO/TS 12913-2:2018. In: Proceedings of the International Congress on Acoustics ICA 2019. Aachen

Lionello M, Aletta F, Kang J (2020) A systematic review of prediction models for the experience of urban soundscapes. Appl Acoust 170:107479. https://doi.org/10.1016/J.APACOUST.2020.107479

Maristany A, López MR, Rivera CA (2016) Soundscape quality analysis by fuzzy logic: a field study in Cordoba, Argentina. Appl Acoust 111:106. https://doi.org/10.1016/j.apacoust.2016.04.013

Marry S (2011) Assessment of urban soundscapes. Organised Sound 16:245–255. https://doi.org/10.1017/S1355771811000252

Mitchell A, Oberman T, Aletta F, Erfanian M, Kachlicka M, Lionello M, Kang J (2020) The soundscape indices (SSID) protocol: A method for urban soundscape surveys—questionnaires with acoustical and contextual information. Appl Sci 10(7):2397. https://doi.org/10.3390/app10072397

Musacchia G, Schroeder CE (2009) Neuronal mechanisms, response dynamics and perceptual functions of multisensory interactions in auditory cortex. Hear Res 258:72. https://doi.org/10.1016/j.heares.2009.06.018

Oberman T, Obad Šcitaroci BB, Jambrošic K (2018) Towards a virtual soundwalk. In: Handbook of research on perception-driven approaches to urban assessment and design. IGI Global, Hershey, pp 317–343

Oberman T, Jambrošić KJ, Horvat M et al (2020) Using virtual soundwalk approach for assessing sound art soundscape interventions in public spaces. Appl Sci 10(6):2102. https://doi.org/10.3390/APP10062102

Pheasant RJ, Watts GR, Horoshenkov KV (2009) Validation of a Tranquillity rating prediction tool. Acta Acust united with Acust 95:1024–1031. https://doi.org/10.3813/AAA.918234

Puyana Romero V, Maffei L, Brambilla G, Ciaburro G (2016) Modelling the soundscape quality of urban waterfronts by artificial neural networks. Appl Acoust 111:121. https://doi.org/10.1016/j.apacoust.2016.04.019

Rathbun SL, Stein ML (2000) Interpolation of spatial data: some theory for kriging. J Am Stat Assoc 95:1010. https://doi.org/10.2307/2669494

Rogalla MM, Rauser I, Schulze K et al (2020) Mice tune out not in: violation of prediction drives auditory saliency. Proc R Soc B Biol Sci 287:20192001. https://doi.org/10.1098/rspb.2019.2001

Russell JA (1980) A circumplex model of affect. J Pers Soc Psychol 39:1161. https://doi.org/10.1037/h0077714

Sadaghiani S, Kleinschmidt A (2016) Brain networks and α-oscillations: structural and functional foundations of cognitive control. Trends Cogn Sci 20:805–817. https://doi.org/10.1016/J.TICS.2016.09.004

Salamon J, Bello JP (2017) Deep convolutional neural networks and data augmentation for environmental sound classification. IEEE Signal Process Lett 24:279–283. https://doi.org/10.1109/LSP.2017.2657381

Schönwiesner M, Zatorre RJ (2009) Spectro-temporal modulation transfer function of single voxels in the human auditory cortex measured with high-resolution fMRI. Proc Natl Acad Sci U S A 106:14611–14616. https://doi.org/10.1073/PNAS.0907682106

Schulte-Fortkamp B (2010) The daily rhythm of the soundscape "Nauener Platz" in Berlin. J Acoust Soc Am 127(3):1774. https://doi.org/10.1121/1.3383871

Schulte-Fortkamp B, Fiebig A (2006) Soundscape analysis in a residential area: an evaluation of noise and people's mind. Acta Acust united with Acust 92(6):875–880

Sedley W, Gander PE, Kumar S et al (2016) Neural signatures of perceptual inference. elife 5:e11476. https://doi.org/10.7554/ELIFE.11476

Segura J, Felici S, Cobos M et al (2015) Psychoacoustic annoyance monitoring with WASN for assessment in urban areas. In: 138th audio engineering society convention 2015. Audio Engineering Society, Warsaw

Sikora M, Russo M, Derek J, Jurčević A (2018) Soundscape of an archaeological site recreated with audio augmented reality. ACM Trans Multimed Comput Commun Appl 14(3):1–22. https://doi.org/10.1145/3230652

Steele D, Bild E, Tarlao C, Guastavino C (2019) Soundtracking the public space: outcomes of the musikiosk soundscape intervention. Int J Environ Res Public Health 16(10):1–38. https://doi.org/10.3390/ijerph16101865

Sun K, Sanchez GME, De Coensel B et al (2018) Personal audiovisual aptitude influences the interaction between landscape and soundscape appraisal. Front Psychol 9. https://doi.org/10.3389/fpsyg.2018.00780

Sun K, De Coensel B, Filipan K et al (2019) Classification of soundscapes of urban public open spaces. Landsc Urban Plan 189:139. https://doi.org/10.1016/j.landurbplan.2019.04.016

Tenbrink T (2015) Cognitive discourse analysis: accessing cognitive representations and processes through language data. Lang Cogn 7(1):98–137. https://doi.org/10.1017/langcog.2014.19

Thoret E, Varnet L, Boubenec Y et al (2020) Characterizing amplitude and frequency modulation cues in natural soundscapes: a pilot study on four habitats of a biosphere reserve. J Acoust Soc Am 147(5):3260. https://doi.org/10.1121/10.0001174

Turner SF, Cardinal LB, Burton RM (2015) Research design for mixed methods: a triangulation-based framework and roadmap. Organ Res Methods 20:243–267. https://doi.org/10.1177/1094428115610808

Tversky B (1993) Cognitive maps, cognitive collages, and spatial mental models. In: Lecture notes in computer science (including subseries lecture notes in artificial intelligence and lecture notes in bioinformatics). Springer, Berlin, Heidelberg, pp 14–24

van den Bosch KAM, Welch D, Andringa TC (2018) The evolution of soundscape appraisal through enactive cognition. Front Psychol 9:1129. https://doi.org/10.3389/FPSYG.2018.01129/BIBTEX

Van Renterghem T, Dekoninck L, Botteldooren D (2020a) Multi-stage sound planning methodology for urban redevelopment. Sustain Cities Soc 62:102362. https://doi.org/10.1016/J.SCS.2020.102362

Van Renterghem T, Vanhecke K, Filipan K et al (2020b) Interactive soundscape augmentation by natural sounds in a noise polluted urban park. Landsc Urban Plan 194:103705. https://doi.org/10.1016/j.landurbplan.2019.103705

Viollon S, Lavandier C, Drake C (2002) Influence of visual setting on sound ratings in an urban environment. Appl Acoust 63(5):493. https://doi.org/10.1016/S0003-682X(01)00053-6

Watts GR, Pheasant RJ, Horoshenkov KV (2011) Predicting perceived tranquillity in urban parks and open spaces. Environ Plan B Plan Des 38(4):585–594. https://doi.org/10.1068/B36131

Webb EJ, Campbell DT, Schwartz RD, Sechrest L (1966) Unobtrusive measures: nonreactive research in the social sciences. Rand McNally & Co, Chicago

Wei W, Van Renterghem T, De Coensel B, Botteldooren D (2016) Dynamic noise mapping: a map-based interpolation between noise measurements with high temporal resolution. Appl Acoust 101:127–140. https://doi.org/10.1016/j.apacoust.2015.08.005

Yu L, Kang J (2009) Modeling subjective evaluation of soundscape quality in urban open spaces: an artificial neural network approach. J Acoust Soc Am 126(3):1163. https://doi.org/10.1121/1.3183377

Zannin PHT, Engel MS, Fiedler PEK, Bunn F (2013) Characterization of environmental noise based on noise measurements, noise mapping and interviews: a case study at a university campus in Brazil. Cities 31:317–327. https://doi.org/10.1016/J.CITIES.2012.09.008

Zhang XL, Wu J (2013) Deep belief networks based voice activity detection. IEEE Trans Audio Speech Lang Process 21(4). https://doi.org/10.1109/TASL.2012.2229986

# Chapter 9
# Soundscape and Health

**Peter Lercher and Angel M. Dzhambov**

**Abstract** This chapter assesses the current state of research that relates the acoustic environment to health-related quality of life and severe health effects. An integrated approach is used to consider and characterize the acoustic environment and its associated physical, structural, social, and cultural contexts. The conceptual perspective of health-related soundscape research is introduced followed by a brief review of the established health effects of noise exposure. Limitations of acoustic environment assessments and alternative indicators are discussed. Current research is evaluated along with the potential for soundscape research to complement classic health assessments. The central section provides a brief description of auditory system pathways, relevant reviews, and a selection of research projects with broad, health-related perspectives. In addition, the assessment of quietness/tranquility and their restorative potentials are considered. A discussion of studies that cover multisensory experiences and combined exposures (noise, vibration, air pollution) is followed by the integration of research that links the acoustic environment with greenspace and health. The chapter concludes with a summary of the current status of soundscape research and further needs in integrated soundscape research.

**Keywords** Acoustic environment · Sound perception · Greenspace · Well-being · Quality of life

---

P. Lercher (✉)
Institute for Highway Engineering and Transport Planning, Graz University of Technology, Graz, Austria
e-mail: peter.lercher@tugraz.at

A. M. Dzhambov
Department of Hygiene, Faculty of Public Health, Medical University of Plovdiv, Plovdiv, Bulgaria
e-mail: angel.dzhambov@mu-plovdiv.bg



B. Schulte-Fortkamp et al. (eds.), *Soundscapes: Humans and Their Acoustic Environment*, Springer Handbook of Auditory Research 76,
https://doi.org/10.1007/978-3-031-22779-0_9

## 9.1 Introduction: A Conceptual Perspective on Soundscape and Health

Human health and well-being are shaped by a myriad of environmental influences working together at different spatial scales (home, neighborhood, population), across contexts (leisure, work, school), and throughout life (Wild 2012). To understand the role of the acoustic environment in human health requires conceptualization that is not based on a stand-alone exposure but rather is considered as an integral component interwoven in the fabric of physical, social, and biological factors that dynamically interact with humans and give rise to perceptual phenomena. Human responses are triggered by sensory information and social cues provided by the environment (enviroscape), which are then subject to cognitive and psychological processing (psychscape) (Job and Hatfield 2001). Figure 9.1 provides an overview of the conceptual space that the acoustic environment and its perceptual counterpart occupy in the larger ecological dimension.

The following sections of this chapter address several focal points of interest outlined in this model, namely the processing of acoustic information by the auditory system (Sect. 9.3.2), cognitive appraisal of sounds in a multisensory and contextual perspective (Sect. 9.4.3), and operationalization of perceived acoustic environments (Sect. 9.4.6). Both empirical and theoretical work on soundscape and relevant constructs are summarized to provide an overview of this transdisciplinary field and its implications for supporting human health.

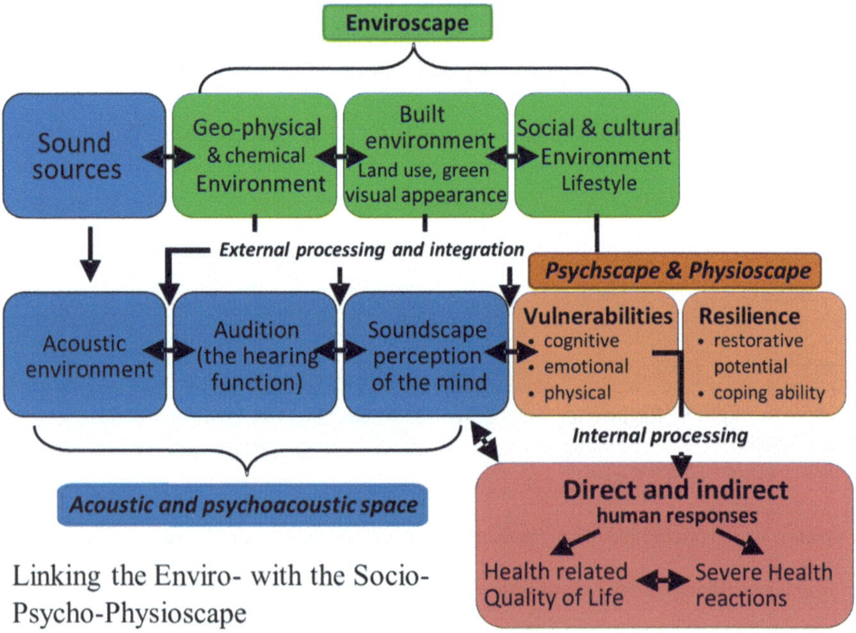


**Fig. 9.1** Conceptual model of soundscape and health. (Modified after Lercher et al. 2013)

## 9.2 Status of Research on Environmental Noise and Health

### *9.2.1 Noise as a Harmful Environmental Factor*

Environmental noise has been regarded as a major public health problem since the early twentieth century (Bijsterveld 2008). Currently, 133 million (75%) Europeans living in major agglomerations are exposed to traffic noise ≥50 dBA day-evening-night noise level ($L_{den}$) and 145 million (82%) are exposed to ≥40 nighttime noise level ($L_{night}$), which exceeds the health safety limits set by the World Health Organization (WHO) (Houthuijs et al. 2015). Conservatively, 48,000 new cases of ischemic heart disease and 12,000 premature deaths annually may be attributed to long-term exposure to environmental noise (European Environment Agency 2020). The WHO commissioned a series of systematic reviews on common health outcomes of environmental noise to inform the revision of community noise guidelines for Europe with state-of-the-art evidence on the health effects of noise and updated recommendations on acceptable exposure levels. They focused on noise originating from transportation (road traffic, railway, and aircraft), wind turbines, and occurring in leisure settings in which people spend the majority of their time (Jarosińska et al. 2018). Exposure-response functions could be constructed only for some of the outcomes based on the available evidence at the time. For the first time, source-specific evidence-based guideline values were proposed (Table 9.1) (WHO 2018).

The new guideline values for sound exposure of the respective sources were lowered compared with the previous guidelines (Berglund et al. 1999), the guidelines of the Environmental noise directive (European Union 2002), and the former WHO night-time noise guideline (Kim and van Berg 2010). Specifically, the exposure values for aircraft and railway sounds were set to substantially lower levels, and there was no evidence that railway noise led to a lower percentage of annoyed

**Table 9.1** Guideline values of environmental noise levels (L) to prevent negative health effects (WHO 2018)

| Noise source | WHO Guideline values |
|---|---|
| *Road traffic noise* | |
| Day/evening/night ($L_{den}$) | 53 dBA |
| Night ($L_{night}$) | 45 dBA |
| *Railway noise* | |
| Day/evening/night ($L_{den}$) | 54 dBA |
| Night ($L_{night}$) | 44 dBA |
| *Aircraft noise* | |
| Day/evening/night ($L_{den}$) | 45 dBA |
| Night ($L_{night}$) | 40 dBA |
| *Wind turbine noise* | |
| Day/evening/night ($L_{den}$) | 45 dBA |

people than road traffic noise at equal exposure levels (i.e., a railway bonus), as previously believed by some experts.

Following the completion of the WHO evidence reviews (2016–2020), new studies and updated meta-analyses surfaced. This newer evidence not only supported the guideline values but added further credence to the possible adverse effects of environmental noise on mental health in adults (Dzhambov and Lercher 2019a; Lan et al. 2020), in children (Schubert et al. 2019), and in birth outcomes (Dzhambov and Lercher 2019b). Nevertheless, some limitations remain due to the applied methodology and the focus on transportation noise. The major caveats included concerns about the exposure assessment techniques, paucity of data on the effect of multiple/combined sources and exposures, and the effect of contextual factors.

### *9.2.2 Noise as an Annoyance*

As mentioned previously, one of the primary outcomes considered by the WHO is noise annoyance. It is defined as "a multifaceted psychological concept, covering immediate behavioral noise effects aspects, like disturbance and interfering with intended activities, and evaluative aspects like "nuisance," "disturbance," "unpleasantness," and "getting on one's nerves" (Guski et al. 1999; Heinonen-Guzejev 2008). However, research into noise annoyance often relies on classical acoustic indicators based on equivalent sound level ($L_{den}$, $L_{eq}$), which may lead to either over- or under-estimates of the health effects at the public health level in the broad sense of the WHO definition (see Sect. 9.4.6 for alternative acoustic indicators). This is a clear challenge specifically in areas with idiosyncratic landscape fabric, social environments, and lifestyles. An illustration of this point is the exposure-response curve for road traffic noise and high annoyance (Fig. 9.2). Particularly noticeable is the distinct difference in the annoyance responses between alpine (within the blue-dashed circle) and Asian studies (all red symbols) included in the WHO review of evidence on noise as an annoyance (Fig. 9.2). The annoyance responses (highly annoyed) from all five surveys in alpine valleys lie far above the new WHO summary curve. The percent difference in those highly annoyed is often 20% or greater across the full sound exposure range. In contrast, the responses in the Asian studies, even the largest study from Hong Kong (big red triangles), not only lie below the new summary curve but are also below the old curve (red line) used in European regulations (Fig. 9.2). Even when cities with similar built-up structures are compared, Asian and European cities differ in the annoyance responses by approximately 10% at lower sound levels and differ even more at higher sound levels. Sensitivity analyses suggested that the variability was not due to methodological differences (e.g., sampling, questionnaire items, and noise metrics) in the selected studies (Guski et al. 2017). This state-of-the-art information helps to raise the level of awareness of health effects from environmental noise at the supra-national level. However, such high variability in human responses to similar sound-level intensities can hamper measures tailored specifically for remediation at the community level in

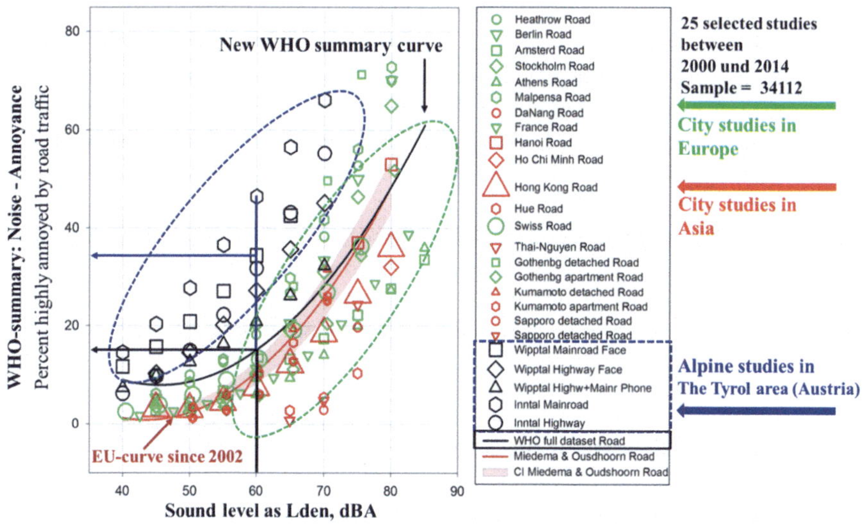


**Fig. 9.2** Exposure-response curves for severe annoyance by road traffic as a function of day-evening-night sound level. Ldn day-night equivalent sound level, dBA decibels with A-filter. (Guski et al. 2017)

environmental health impact assessments. Furthermore, considerable uncertainty may enter noise action plans that will complicate mitigation measures and future planning. The potentially high costs that may be involved with under- and over-estimations of the overall health burden across geographies, landscapes, cultures, city structures, and city designs must be kept in mind.

Similar variations are apparent when one compares measured noise levels in European cities or satisfaction with noise in European opinion surveys (Lercher 2019). The results on both scales (overall sound level and satisfaction with noise climate) differ widely, and the correlations between the subjective and objective indicators of the noise climate are poor. The major factors responsible for the large variability and the poor correlation between classic acoustic indicators and subjectively measured annoyance (or satisfaction with noise levels) remain unexplored.

The "enviroscape" boxes in Fig. 9.1 point to the main drivers of such observed heterogeneity. One obvious factor is the diversity of the geo-physical and built environments. The amphitheater effect of sound propagation to the slopes of the alpine valleys under specific meteorological and seasonal conditions is one example. The large heterogeneity of the macro-, meso-, and micro-scale structures of cities worldwide is another known reason for major differences in sound levels. Variations in urban morphology (Han et al. 2018), population density (Yuan et al. 2019), traffic density and composition, and land use patterns (King et al. 2012) introduce further factors that may hinder accurate planning in urban areas. Due to the potential sources of error in noise modeling in cities (Rey Gozalo et al. 2019), the same modeling software cannot be used reliably to calculate noise levels across spatially heterogeneous urban environments (Hornikx 2016).

Another aspect to consider is the amount of urban green and blue space available for recreation, which differs widely across the globe (Yuan et al. 2019) and within seemingly homogenous urban agglomerations in Europe (Kabisch et al. 2016). Green space or general vegetation level in the area (see Sect. 9.4.4.3) is associated with not only lower sound levels (Margaritis and Kang 2016) but also with improvements in physical health (Astell-Burt et al. 2014), mental health (Geneshka et al. 2021; Yang et al. 2021), and sleep (Shin et al. 2020). The surveys in the alpine area were taken as an example to illustrate that disregarding important co-factors may result in large variability of responses as observed in Fig. 9.2. Moreover, these co-factors can also have cumulative effects on both sound levels and health, which calls for a more holistic perspective of acoustic environments and their health impacts.

## 9.3 Status of Research on Soundscape and Health

### *9.3.1 Introduction to the Concept of Soundscape*

Beyond the outcomes of discomfort and potential negative effects, a genuine and innovative research effort has been directed toward a positive human experience of the acoustic environment, such as perceived pleasantness, vibrancy, quietness, and restoration. The concept of soundscape represents the way that people perceive the full range of sounds in a place, with a focus on desirable sounds as experienced at a given time (Brown 2012; ISO 12913-1 2014). As the International Organization for Standardization (ISO) standard 12913-1 (2014) distinguishes the perceptual construct (the soundscape) from the physical phenomenon (the acoustic environment), a logical consequence is that human perception of the acoustic environment (Fig. 9.1: "the soundscape perception of the mind") is at the center of the approach. This implies an investigation of a much broader set of human responses, including source-related attitudes, beliefs, preferences, judgments, behaviors, and experiences (e.g., interference, emotional states, well-being, and quality of life). However, unlike the health effects of traffic noise, relatively little is known about the contribution of soundscape to human health. The current field of health research within the larger soundscape community is rather diverse in terms of the use of acoustic indicators to operationalize soundscape, the choice of health outcomes, and analysis strategies.

The relevance of the soundscape approach as complementary to noise annoyance assessments was first outlined in a special session at Internoise 1997 and later summarized at the International Commission on Biological Effects of Noise conference in Rotterdam (Lercher and Schulte-Fortkamp 2003). The Swedish "Soundscape Support to Health Project" was the first large effort that provided new insight into the potential of the soundscape approach to support public health and planning (Berglund 2006; Gidlöf-Gunnarsson and Öhrström 2010). Another important research contribution followed with "The Positive Soundscape Project" (Jennings and Cain 2013; Watts et al. 2013). European Cooperation in Science and Technology (COST) Action TD0804 "European Landscapes and Soundscapes" was initiated in 2009 and resulted in a book focused on soundscapes and the built environment with relevant health-related chapters (e.g., van Kamp et al. 2016; Lercher et al. 2016; also see summary by Kang et al. 2016a). A new European Research Council (ERC)-funded SSID (soundscape indices) project is designed to characterize urban soundscapes based on large data collections. This will be achieved with audio-visual recordings and soundscape judgments by questionnaires. Both assessments will be paired with extensive acoustic data collection (Mitchell et al. 2020). In addition, the United Nations Environment Programme (UNEP) published the report "Listening to cities: from noisy environments to positive soundscapes" that drew attention to the long-term physical and mental health impacts of noise pollution and included measures that can be implemented to create positive and restorative soundscapes in urban areas (Aletta 2022). This was a considerable change in how the acoustic environment was viewed: sound as a potentially valuable resource for enhancing appreciation of places by people. Such a perspective also helps to widen the view of planning and design of environments and remedial actions. Furthermore, this approach coincides with aspects of acoustic environments that are important to improve and restore health and to cope with the daily challenges of environmental stress. Examples of relevant population studies that used both soundscape and acoustic approaches consistent with the approach in Fig. 9.1 are presented in the following subsections.

### 9.3.2 Auditory System Mechanisms and Multi-sensory Integration

Auditory experience induces structural and functional changes in the central auditory system and, therefore, may affect features of central auditory system plasticity. These potential changes include synaptic growth, changes in receptor and transmitter activity, and reorganization of frequency tuning (Kraus and Canlon 2012). This central plasticity can be seen as an adaptive process that increases signal detection but can also play a maladaptive role in the development of tinnitus (Shore et al. 2016). Perhaps of more importance in the context of soundscape recognition and responses is the fact that important non-auditory brain regions (the limbic system) are involved in the processing of auditory input. The amygdala and the hippocampus receive either direct or indirect neuronal input from the central auditory system and integrate them in processing emotions, threats, memories, and in fear conditioning (Kraus and Canlon 2012). In turn, limbic regions provide direct or indirect feedback to the central auditory system that can affect neuronal activity and modulate plasticity. Interestingly, the amygdala operates independently from the sensory modality (Ghazanfar and Schroeder 2006) and processes visual, auditory, olfactory, and gustatory information (Gerdes et al. 2014; Andersson et al. 2018), which supports the idea that sound perception is intertwined with other senses (Wu et al. 2015). For example, audio-visual pairings with pleasant sounds and pictures were rated as more pleasant than pleasant pictures presented alone (Gerdes et al. 2014). However, the extent to which audio-visual incongruence affects perceived soundscape dimensions requires further effort in this area. This lack of established knowledge hampers the study and interpretation of multi-sensory and combined exposure in epidemiologic studies.

Another issue that is not fully understood is the role of individual noise sensitivity (Dzhambov 2015), which is seen as a stable personality trait (Stansfeld 1992) that is possibly hereditary (Heinonen-Guzejev et al. 2005), that is conceptually independent of noise intensity or type of noise but is predictive of noise annoyance (Van Kamp et al. 2004). Noise sensitivity was an independent predictor of poor mental health/health-related quality of life (Lercher 1996; Shepherd et al. 2010) and future risk of illness (Stansfeld and Shipley 2015). Several studies have also investigated neurophysiological correlates of noise sensitivity (Lee et al. 2012; Kliuchko et al. 2016). Noise-sensitive people were impaired in the process of sensory gating such that they actively inhibited the response to repetitive stimuli (gating out) and showed increased responses to stimuli with novel features (gating in) (Kliuchko et al. 2016; Shepherd et al. 2016). Hence, individual noise sensitivity is an important aspect influencing auditory perceptions and, by extension, soundscape quality.

### *9.3.3 "Subjective" and "Objective" Measurements*

Both objective and perceived acoustic quality are relevant for human health and well-being. However, classic acoustic operationalization is not ideally suited to answering questions about the health effects of soundscapes. What we need to have for both planning and remedial action is a more balanced total assessment of local soundscapes that covers both potentially adverse sound level conditions and also the promotive and supportive aspects of the enviroscape as related to the health of people over time (Lercher et al. 2016; van Kamp et al. 2016). Therefore, alternative sound exposure indicators are needed to better reflect the temporal and frequency components of noise sources and to reflect the acoustical appraisal in various contexts (Botteldooren et al. 2016) in which the interplay of multiple sensory modalities shape human responses to the environment. Acoustic indicators must better reflect psycho-physiological appraisal (Hume and Ahtamad 2013), namely the grade and type of neurophysiologic stimulation (stressful, neutral, or restorative). In addition, indicators should account for the person-environment fit, that is, whether soundscape components interfere or support the intentions or human expectations for a particular place. Finally, indicators are needed to measure the holistic potential of a place to support health, which is contingent on the availability of control/coping strategies in a setting as well as meanings and emotional content attached to the sonic environment.

The classical labels "subjective" and "objective" suggest a dichotomy that is not justified and can lead to misconceptions about the quality of both types of measurements. The addition of quotation marks is suggested to reflect the lack of well-defined usage for those terms in noise and health and soundscape research (cf., Kompier 2005). The information we can gain from relating two subjective measures in a soundscape or noise study (e.g., pleasantness/annoyance and general symptoms or the perceived stress or impact on sleep as assigned to a scale on a questionnaire) is limited, as discussed in an early review by Kasl (1984). Both types of indicators should be employed in research as they can complement each other (Pendrill 2014) for explaining the multi-faceted relationship and potential pathways between the wider acoustic environment and health. Kasl (1984) concisely summarized the issue: "If the objective measure is confirmed as a risk factor for disease outcome, and the subjective measure is associated with both the objective measure and the disease outcome (and the latter two associations are stronger than the first), then we have a substantial suggestion of a disease effect of environmental exposure, operating primarily through the intervening process of subjective perception or reaction."

Whether an observed reaction is due to the environment or to some extent (e.g., moderation or mediation) due to pre-existing vulnerabilities or an underlying disease of the subject is difficult to distinguish. Moreover, reactions may not be caused directly by the sound, rather a reaction may be mediated by intermediate variables, such as health status or sleep disturbance. Likewise, neglecting other general factors (e.g., age, gender, socioeconomic status, and noise sensitivity) can distort results if linearity of effects is assumed in the analysis. In their review, Van Gerven et al. (2009) found an inverted U-shaped annoyance reaction to noise, peaking around age 45 years with the lowest numbers representing the youngest and the oldest groups independent of noise sensitivity and sound levels. When specific diseases are being considered as the outcomes in the noise-health relationship, the pathways can be even more complicated. For example, a study on noise and blood pressure found that people with hypertension reported the least annoyance across the full noise exposure range (45–70 dBA), although they belonged to a highly vulnerable group regarding the effects of noise (Lercher et al. 2011). This seemingly paradoxical result can be explained by two biopsychological factors. First, persons with hypertension do not complain as much about home or work stress (Nyklíček et al. 1996). This behavior is attributed to "emotional dampening," which is a lowered capability to recognize emotions correctly in social and environmental contexts (McCubbin et al. 2018). Second, elevations in blood pressure may lead to reduced arousal and pain sensitivity through baroreceptor activation (Bruehl et al. 2018). Both factors can lead to a distorted subjective assessment of an experienced noise burden.

Furthermore, successful or unsuccessful coping with existing exposure can strongly impact perceptions (Lercher 1996; Botteldooren and Lercher 2004). A challenging result was found in noise-sensitive people: although they rated their sensitivity as high and exhibited greater health worries and more sleep disturbance, they were less engaged in active coping measures to reduce the burden, such as closing windows, complaining to authorities, or moving to another sleeping room (Lercher and Kofler 1996). Other common life stressors (e.g., work, family, social, major life events) are operating concurrently and must also be evaluated in any assessment (Gomez-Bernal et al. 2019).

## 9.4 Soundscape and Health

### *9.4.1 Outline of the Review*

Using the keywords "soundscape and health" in Google Scholar reveals around 21,000 results. Most of these studies were conducted in narrowly defined settings, such as city parks, nature parks, green/historic areas, city squares, pedestrian areas, courtyards, shopping malls/markets, crossings, bridges, etc. (Engel et al. 2018). Other research investigated more specific sounds, such as birds, traffic sources, human speech, loud music, and sounds from rivers and fountains (Engel et al. 2018).

Only a small number of studies took place in residential areas. Those studies applied a broad variety of more or less standardized open or closed questionnaire items and semantic differentials (scales with two polar adjectives at the ends) or other scales (Engel et al. 2018) to measure the acoustic comfort, pleasantness, quietness, vibrancy, or other qualities of the perceived soundscape, mostly as momentary assessments. The measurements often took place during soundwalks or in direct interviews with people (most of which were non-representative convenience samples of small size) in the area of interest (Berglund and Nilsson (2006)). Classical acoustic indicators were used in many studies, but few employed psychoacoustic methods. Most studies did not make further attempts to link both measures (classic and soundscape) to clearly defined health or positive soundscape endpoints.

A review, intended to summarize current knowledge of the effect of positive soundscapes on positive health, found only seven studies (Aletta et al. 2018). Three were field studies and two used annoyance ratings. Negative soundscape assessments were excluded. Lower annoyance in these studies was interpreted as a positive soundscape outcome. This is a reasonable approach as it would allow a direct comparison with results from standard noise surveys given that another genuine soundscape measure was available. Although all selected studies had both subjective and objective assessments of the acoustic environment, no cross-evaluation of both measures with the health outcomes was reported. Such information is a critical requirement for the implementation of soundscape research into general practice. Moreover, this type of cross-validation is needed for an appropriate interpretation of the results.

The following overview prioritizes selected field studies on public health effects reasonably representative of the general population. Those interested in information covering lab and small-area studies can find further information in a comprehensive book by Kang and Schulte-Fortkamp (2016), summary articles (Aletta et al. 2018; Engel et al. 2018), and systematic reviews (Erfanian et al. 2019; Li and Lau 2020).

### 9.4.2 Selected Studies on the Health Effect of Soundscapes

One of the first studies that included features of the building and neighborhood environment took place in children. Third and fourth graders from 26 schools (total $n = 1280$) rated their perceived environment and levels of annoyance with standard, Likert-type verbal scales. Noise exposure, obtained by modeling and on-site measurements, was assigned to residential addresses using a graphical information system. The annoyance responses varied (Figs. 9.3 and 9.4) to a highly significant degree according to attitudes (place attachment) and behavior, and they were slightly less correlated with built environmental variables (density, building shape) and the traffic source (road versus rail). Moreover, the assessment of their environment as quiet was associated with lower blood pressure (Fig. 9.5).

The results for rail traffic noise similarly showed strong modification of the effect by attitude and behavior but not by the built environment variables. The data indicated that reduction in the annoyance response was large when children could play freely and make noise in their neighborhood. Thus, it was not surprising that those children liked to live there.

Another early soundscape study was conducted in the framework of the "Soundscape Support to Health Project" in four residential areas of Stockholm (Nilsson and Berglund 2006). All four areas were exposed to road traffic noise (on both sides of the building, only one side, or shielded). Residents listened to a soundscape for periods of 30 s at six different pre-selected listening places. Four of those were indoors (noisy and shielded façade with open/closed window) and two were

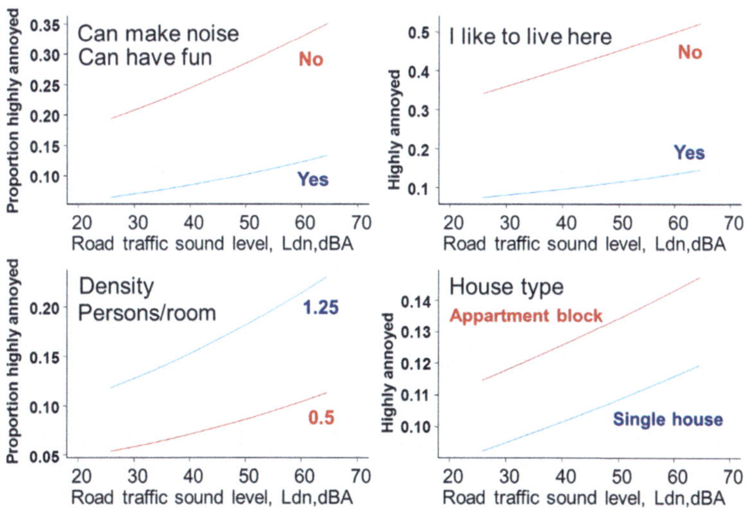


**Fig. 9.3** The relationship between day-night road traffic sound level, the built environment, and child attitude and behavior. Ldn day-night equivalent sound level, dBA decibels with A-filter. (Lercher et al. 2000)

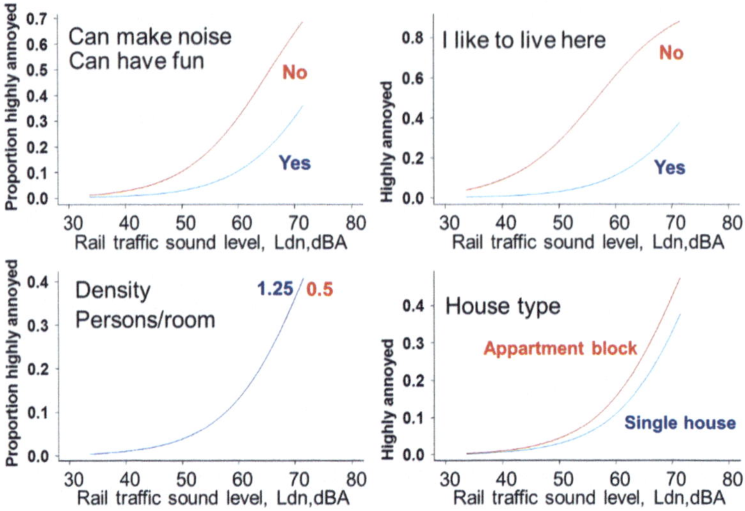


**Fig. 9.4** The relationship between day-night rail traffic sound level, the built environment, and child attitude and behavior. $L_{dn}$ day-night equivalent sound level, dBA decibels with A-filter. (Lercher et al. 2000)

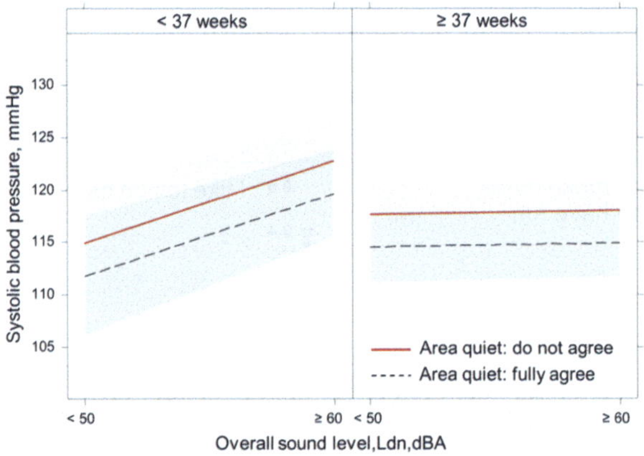


**Fig. 9.5** The effect of perceived quietness on blood pressure in children with a significant interaction between sound level and short gestation (<37 weeks). $L_{dn}$ day-night equivalent sound level, dBA decibels with A-filter. (Lercher et al. 2013)

outdoor building sites (noisy, 55–65 dBA; and shielded, 5–15 dBA less). Simultaneously, binaural and monaural recordings of the corresponding 30-s acoustic soundscapes were obtained. The quality of the soundscape at each listening place was assessed by means of a visual analog scale. The soundwalks were conducted individually. Each participant rated 12 perceptual-emotional attributes (soothing, pleasant, light, dull, eventful, exciting, stressful, hard, intrusive, annoying, noisy, and loud). The participants were instructed to rate the match (Swedish "passade" or

German "Passung") of the attribute with the 30-s soundscape heard at each of the six listening places. They found the strongest collinear relationships for sound level in dBA and perceived loudness ($r^2 = 0.81$) and for annoyance with sound level ($r^2 = 0.58$). As expected, the positive attributes of pleasant and soothing were inversely associated with the sound level. Soothing ($r^2 = 0.45$) was more strongly associated than pleasantness ($r^2 = 0.08$). The soundscape quality was rated as less adverse (i.e., less annoying, loud, intrusive) and more soothing or pleasant at shielded sites compared with traffic noise-exposed sides of the same building. The quality of indoor soundscapes (closed window) at exposed sides of buildings was rated as less good (less pleasant/soothing) than outdoor soundscapes at shielded sides of buildings. For a good soundscape quality outdoors (pleasantness/soothing), sound levels should be below 50 dBA $L_{Aeq,30s}$. However, soundscapes below this sound level are not always of good quality. A principal component analysis of the 12-attribute profiles of the 30-s soundscapes was able to identify four comprehensible classes of characteristic perceptual qualities ("soundscape signatures").

The results of another in-depth Swedish study conducted with adults pointed in a similar direction (Gidlöf-Gunnarsson and Öhrström 2010). Four subsamples from a larger sample ($N = 956$) were selected which formed two sound level categories: 58–62 dBA and 63–68 dBA for a courtyard space. All participants had access to a "quiet" side. The quality of this quiet side was assessed with respect to "courtyard utilization" and "naturalness" as low or high. The authors observed a significant effect with an index that included all noise-disturbed outdoor activities. Furthermore, significant differences were found in high and low courtyard-usage groups for more natural and human sound sources. This information was used as a positive indicator for the quality of the courtyard. Courtyard quality also significantly influenced the ratings of residential satisfaction (52% versus 42% in high and low courtyard categories).

There are only a few studies on physical health outcomes with both subjective and objective measures of the soundscape and sufficient consideration of important confounders in the regression models used. For instance, an Austrian study sampled 1280 children with a school survey that considered the mother's education and the noise levels at the child's home. The same perceived environment and annoyance ratings shown in Figs. 9.3 and 9.4 were used. Independent of noise level, when children rated their area as quiet, their blood pressure was significantly lower than when their area was rated as not quiet (Fig. 9.5). Notably, there was a significant variation with length of pregnancy and noise level: children born prematurely show an additional blood pressure increase when the home sound level is higher.

### *9.4.3 Selected Examples of Multisensory Experiences and Combined Exposures*

The sensory experience of our environment includes all sensory modalities and perceptible contexts like spatiality, building structure, urban morphology, or rural topography. Hong and Jeon (2017) performed multiple linear regression analyses to predict pleasantness and eventfulness scores using acoustic, morphological, and spatiotemporal indicators. Among the tested models, the pleasantness model showed a significantly higher explanatory power (50%) than the eventfulness model (13%). Open spaces and water features explained about 50% of the variance in the pleasantness model. In univariate analyses, bird sounds were more correlated with the pleasantness score than were water features; however, open space also was correlated with bird song. The eventfulness scores were mainly related to human voices and music features in business and high-density commercial areas and seemed less suited to characterizations of residential areas.

Odor is another example of a less examined sensory component of our environment (Xiao et al. 2018). Natural scents and traffic-related air pollution provide incongruent messages to the brain, which can up- or down-rate perceived health-related quality or even affect cognition. A multisensory stress study used a novel approach to evaluate three senses (audio, visual, and smell) simultaneously (Hedblom et al. 2019). While the urban scene contributed most to the stress experience measured by skin conductance levels, the park and forest exposure contributed the least. In the stress period, odor pleasantness significantly predicted stress response, auditory pleasantness demonstrated only marginal significance, and visual pleasantness had no significance. In the recovery period, the same results were observed. Parks and forests with singing birds and natural smells could lower stress levels within a minute of stressor offset.

Another factor rarely considered in soundscape studies is the kinesthetic sense of vibration. Heavy trucks, buses, tramways, and underground rail systems emit structure-borne sound, low-frequency noise, and perceptible vibrations into nearby buildings (Crocker 2007). Sufficient evidence is available on the combined effects of noise and vibration on levels of annoyance (Öhrström 1997; Trollé et al. 2015) and sleep disturbance (Persson Waye et al. 2019) in transportation studies.

A Swedish study (Pedersen 2015) investigated the effect of multiple environmental stressors (various noise sources, vibration, odors, light, place characteristics) in urban dwellings of a town with no major noise sources (no major highways or railways, airport, or noisy industrial facilities). Outcomes measured were quality of life (health status, life satisfaction), taking into consideration sensitivity and reported daily stress. A structural equation model was specified, using five latent variables: quality of life (general health, life satisfaction), sensitivity (to noise, odor, and vibration), stress (stress in daily life, need for stress recovery), residential satisfaction (dwelling and neighborhood), and place relation (place attachment and restorative possibilities). Also included were area (as a proxy for exposure) and the annoyance score from all exposures (Fig. 9.6). The relationship between annoyance and quality of life in the structural equation model was mediated by residential satisfaction, which in turn was largely influenced by place attachment. Sensitivity was correlated to some degree with stress. Stress had a direct negative impact on quality of life and was also correlated to place relation. The study demonstrated that although most respondents were not severely annoyed, many single annoyances can sum up to a substantial impairment of health-related quality of life.

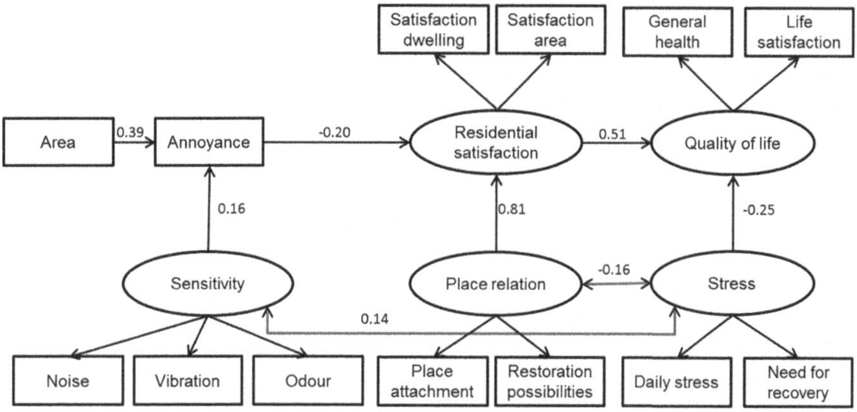


**Fig. 9.6** Structural equation model of the relationship between area, annoyance, and quality of life with mediating and moderating variables. Regression estimates are shown next to each path. (Pedersen 2015)

A study conducted in alpine valleys investigated the annoyance response related to multiple sound sources (rail, highway, main road) and considered perceptional, emotional, and behavioral factors and coping efforts (Lercher et al. 2017). In a multivariate regression, the perception of dust/soot and vibration substantially increased the annoyance response. The perception effects varied depending on the traffic source. The amount of coping needed (e.g., closing windows and avoid sitting outdoors) to adapt to the stress load from several traffic sources turned out to be another factor of equal importance in the model. The higher burden of a multi-source exposure in the environment was further indicated by higher annoyance responses at lower sound levels (45–55 dBA, $L_{den}$), a higher dissatisfaction about the area soundscape, and higher anger ratings toward the traffic. In a sensitivity analysis with emergence indicators instead of $L_{den}$, main road emergences contributed significantly to the overall annoyance beyond the overall emergences from the other sources.

A Canadian study (Oiamo et al. 2015), that used a structural equation modeling framework, found that odor and noise annoyances had a significant effect on physical and mental factors of health and well-being, as measured by the 12-Item Short Form Survey questionnaire. However, the observed effect was small compared with other variables in the model (e.g., age, noise sensitivity, or illness). Noise annoyance explained 6% and 8% of the overall variance in functional mental and physical health, respectively. Odor annoyance contributed even less. After adjustments were made for background variables (noise sensitivity, risk perception, residential exposure to traffic noise, and air pollution), a study area indicator and the deprivation index did not show a significant effect. The researchers hypothesized that higher coping activities (closing windows, spending less time outdoors, decreasing social activities) are responsible for the smaller effect.

The last two field studies support what Lepore and Evans (1996) described earlier: the observed additive effects may indicate that multiple environmental burdens demand different coping resources or strategies, which both drains our resources and compromises our restoration options.

## *9.4.4 Greenspace, Soundscape, Restoration, and Health*

### 9.4.4.1 Theoretical Premise

Through the lens of core theories attempting to explain the benefits conferred by contact with a natural environment, soundscape appears to be an intrinsic component of the salutary experiences in nature. The biophilia hypothesis (Kellert and Wilson 1993) and the stress reduction theory (Ulrich 1984) posit that humans have an evolutionary predisposition to benefit from even short encounters with nature, which may reduce psychophysiological stress. According to the complementary attention restoration theory (Kaplan and Kaplan 1989), coping with everyday stressors and adaptive demands taxes neurocognitive and attentional resources, which require periodic restoration. Restoration more generally refers to renewal, recovery, or reestablishment of adaptive capacities diminished in ongoing efforts to meet adaptive demands at the individual, dyad/group (family), or community/population levels (von Lindern et al. 2017; Hartig 2021). A growing body of theoretical and empirical work has established nature sounds as an integral component of the overall quality of natural or natural element-dominated settings, with the potential to augment the positive emotional and physiological effects of these settings (van Kamp et al. 2016; Ratcliffe 2021). However, while wanted sounds can boost the restorative quality of a setting, unwanted sounds, like traffic noise, have the capacity to harm health both by being a stressor and by acting as a constraint on neighborhood restorative experiences. A seminal study by von Lindern et al. (2016) found that higher noise and air pollution levels were associated with higher annoyance and, through that, also were associated with lower residential satisfaction and perceived health. Dzhambov et al. (2017, 2018a, b) replicated these findings and found neighborhood greenspace has less anthropogenic noise sources and vegetation might psychologically buffer noise and reduce annoyance.

Large population studies and earlier reviews found various protective associations between urban green space exposure and health outcomes, such as mental health, cognition, self-reported health, cardiometabolic outcomes, and birthweight (Geneshka et al. 2021; Yang et al. 2021). The pathways behind these effects were grouped by Hartig et al. (2014) and Markevych et al. (2017) into (1) reducing harm (reducing/modifying exposure/perception of noise, exposure to particulate matter, or excessive heat); (2) restoring capacities (attention restoration, stress reduction); and (3) building new capacities (health promotion through social support/cohesion, physical activities). Still, only a handful of these studies have investigated to what extent restoration and acoustic perceptions (e.g., lower noise annoyance) accounted for the beneficial associations between natural environments and health (Dzhambov et al. 2020). Some studies, however, specifically investigated the association between green and blue spaces and noise perception.

#### 9.4.4.2 Greenspace, Acoustic Environment, and Health

Although green and blue spaces exist and act within the context of other main characteristics of the ambient environment (landscape, built environment, traffic exposures), they have rarely been studied in a quantitative way with sufficient adjustment for potential confounders and sensitivity analyses (moderation or mediation tests). Due to the increased availability of noise mapping data, more epidemiological studies have evaluated the relative importance of the acoustic space (with classical sound indicators) compared with other influential factors in multivariable models. A review by Dzhambov et al. (2020) found 11 studies that tested noise as the potential mediator of the overall relationship between greenspace and various health outcomes. Most studies analyzed physical activity and air pollution (over 40 studies). Mediation effects were found in 44% of studies on physical activity, 45% on noise, 52% on social interaction/cohesion, and 60% on air pollution and temperature. Higher percentages were found for actual perceived greenness/use (63%) and when perceived restorativeness was used as an indicator variable (89%). Overall, among the health outcomes tested, positive findings were observed mainly with respect to cognition (80%), physical health (50%), and mental health/stress (35%).

#### 9.4.4.3 Greenspace, Acoustic Environment, and Annoyance

Studies have repeatedly shown that vegetation can physically reduce sound directly by redistribution (reflection, diffraction, scattering) and absorption or indirectly through acoustically soft soil underneath vegetation or due to aspects of the microclimate (Van Renterghem 2014). Worth noting is that these phenomena depend on the configuration of green elements (types of vegetation, width, and height of the vegetation belt). In addition, whether reductions in sound level are apparent depends on how sound frequency attenuation or annoyance is investigated (Fang and Ling 2003). For example, when belts of higher trees are used as shields against a noise source, these can be perceived also as a visual barrier, which can lead to higher annoyance levels (Tamura 1997). On the other hand, studies found higher annoyance when the sound source was visible (Pedersen and Larsman 2008).

Besides physical reduction of noise levels, green and blue spaces have the potential to improve perceived soundscape quality, for example, by reducing noise annoyance. Multiple studies have observed lower noise annoyance for people living closer to greenspace or having a green view out of their home's windows (Dzhambov 2017; Van Renterghem 2019). For example, Gidlöf-Gunnarsson and Öhrström (2007) found that accessibility to nearby green spaces determined the amount of annoyance reduction, and Li et al. (2010) observed varying effects that depended on the visual setting: wetland parks and garden parks reduced noise annoyance to a greater degree than grassy hills. In another study, Van Renterghem, Botteldooren (2016) found that noise annoyance was lower only when vegetation was visible through the living room window. Bluescapes (Leung et al. 2014) and water features embedded in urban design may generate pleasant water sounds (e.g., stream, waterfall, sea) that mask noise and reduce the discomfort and annoyance (Kang 2012). Overall, it appears that the combined audio-visual experience provided by natural elements of the setting improved soundscape quality. On the one hand, the visual and aesthetic characteristics seem critical for the psychological "buffering" effect (Lercher 1996; Dzhambov and Dimitrova 2014). On the other hand, nature sounds can mask unwanted sounds when they are presented at the same time, and partial visual screening of the noise source seems to reduce annoyance. In addition, the masking effect of nature sounds is context-specific as the masking effect is greater when the masker and the noise signal are the same frequency; however, traffic noise is in the lower register while bird songs and small water bodies, for example, generate mid-to-high-frequency sounds (Galbrun and Ali 2013).

While most studies on this phenomenon were (quasi-) experimental or small case studies, a handful of epidemiological studies also investigated the association between geographic measures of greenspace and noise annoyance. A Polish study found a small but significant adjusted effect of greenspace within a 300 m buffer zone (Koprowska et al. 2018). A Bulgarian study used different greenspace indicators: normalized difference vegetation index, NDVI; tree cover density; percentage of green space in circular buffers of 100–500 m; and the Euclidean distance to the nearest structured green space (Dzhambov et al. 2018a). In multivariate models that controlled for potential individual and area-level confounding factors, the authors found that NDVI and percentage of green space were associated with lower noise annoyance in all buffers, while higher tree cover was only effective in the 100 m buffer zone. The effects of NDVI and percentage of green space were mediated by higher perceived greenspace and lower $L_{day}$ (daytime equivalent sound level). Both pathways seemed equally important. Another Swiss population-based survey provided evidence that increasing residential greenspace (NDVI and percent green space in the area) reduced the levels of annoyance for road traffic and railway noise (equivalent to about 3–6 dBA); however, it observed less clear beneficial effects of having a green view from home and having better accessibility to green spaces, which effects were contingent on the degree of urbanicity of the area (Schäffer et al. 2020). Interestingly, greenspace was correlated with increased annoyance with aircraft noise in that study. Research conducted in urban areas has shown that urban form (radial versus linear), building shape, and building density strongly influence calculated noise levels (Margaritis and Kang 2016; Yuan et al. 2019). Therefore, greener cities are not necessarily perceived as quieter.

### 9.4.5 *Lockdown Soundscapes*

Changes in the use of public spaces and transportation networks resulted from the public health measures enacted by many governments during the initial stages of the SARS-CoV-2 virus pandemic. That gave birth to what is now referred to as "lockdown soundscapes" (Aletta 2022; Asensio et al. 2020). The resulting reduction in mobility led to temporary decreases in traffic volume and noise levels within cities. On average, this reduction was in the region of only 6–10 dBA (Aletta 2022). While these changes to the objective acoustic situation were not trivial, an even more profound change was seen in human-occupied spaces in which the relative prevalence of unwanted and desired sounds shifted in the home environment during confinement periods (Dzhambov et al. 2021). Decrease in anthropogenic noise in cities unmasked nature sounds from birds, water, and wind (Derryberry et al. 2020; Sakagami 2020). In one study, Bulgarian university students who spent over 20 hours/day at home reported poorer self-rated health when they experienced mechanical sounds more frequently; nature sounds were related to higher restorative quality of the home, and through that, with better self-rated health (Dzhambov et al. 2021). That finding fits the evidence of the beneficial effects of nature on mental health in times of the COVID-19 pandemic (Pouso et al. 2021) and suggests that restorative indoor soundscapes may be conducive to promoting health.

### 9.4.6 *Operationalization of Soundscape and Health-Relevance of Acoustic Indicators*

#### 9.4.6.1 Characterization of the Soundscape by Appropriate Indicators

The WHO's evaluation of the effects of environmental noise on health (described in Sect. 9.2.1) relied on standard physical descriptors of noise ($L_{den}$, $L_{night}$, $L_{eq}$, $L_{16h}$, or $L_{24h;}$ see Table 9.2 for definitions). Since the year 2000, various approaches have been proposed for the characterization of soundscapes. For brevity, only improved indicators suitable for long-term assessment of chronic effects on health and well-being are considered in this chapter. After an extensive review of acoustic indicators, the following theoretical quality requirements were postulated to characterize sounds appropriately (slightly modified after Can 2015).

- Covering large noise amplitudes
- Against background sounds in rural or shielded areas (25–35 dBA), peak sound levels often intrude up to 90 dBA.
- Covering large spectrum variations
- A-weighting (based on hearing experiments with single frequencies!) underestimates seriously the potential impact of certain sounds (e.g., trucks/buses/trams, wind turbines).

**Table 9.2** Abbreviations used in the chapter

| Abbreviation | Definition |
|---|---|
| A-filter | Sound level meter correction procedure that weights and integrates registered sounds waves depending on their frequency in a way similar to how the human auditory apparatus registers sounds |
| CoG | Spectrum center of gravity (1/3-octave spectrum) |
| COST | European Cooperation in Science and Technology |
| dBA | Decibels with an A-filter applied |
| $L_{16h}$ | 16-hour equivalent sound level |
| $L_{24h}$ | 24-hour equivalent sound level |
| $L_{A10}$ | A-weighted equivalent sound level exceeded for 10% of the measurement period |
| $L_{A10}$-$L_{A90}$ | Difference between A-weighted sound level exceeded 10% and 90% of the time |
| $L_{A50}$ | A-weighted equivalent sound level exceeded for 50% of the measurement period |
| $L_{A95}$ | A-weighted equivalent sound level exceeded for 95% of the measurement period |
| $L_{Aeq}$ | A-weighted equivalent sound level |
| $L_{Ceq}$ | C-weighted equivalent sound level |
| $L_{Ceq}$-$L_{Aeq}$ | Difference between C- and A-weighted sound level |
| $L_{day}$ | Daytime equivalent sound level |
| $L_{den}$ | Day-evening-night equivalent sound level |
| $L_{dn}$ | Day-night equivalent sound level |
| $L_{max}$ | Maximum sound pressure level |
| $L_{night}$ | Nighttime equivalent sound level |
| ML | Music-likeness |
| N | Zwicker loudness (sone), ISO 532 B |
| NCN | Number of noise events ($L_{Aeq,1s}$ 3 dB above $L_{A50}$ for at least 3 seconds) |
| NDVI | Normalized difference vegetation index |
| WHO | World Health Organization |

- Covering large temporal variations
- Typical 1 s and fast weighting ($\tau$ = 125 ms) does not always capture the large temporal fluctuations of sound sources, especially when impulse components occur.
- Covering large spatial variations
- Some noise sources can affect inhabitants over more than 1 km, depending on source spectrum, built-up environment, topography, and meteorology. Standard noise mappings underestimate such exposures and, therefore, the number of exposed, which leads to misclassification of health associations in epidemiological studies.

The overarching goal is to have valid sound exposure assessments for critical time segments (e.g., evening, night), critical spatial situations (e.g., street canyons, backyards, slopes, etc.), and various temporal and frequency characteristics. Several research groups worked toward these requirements by adapting existing indices or creating new ones.

#### 9.4.6.2 Improving Classical Acoustic Measures

Fiebig and Schomer (2015) pointed out that a major precondition for the development of supplemental metrics is that the metric does not correlate well ($r < 0.7$) with the standard indicator $L_{den}$. A number of researcher groups took on the challenge. Nilsson et al. (2007) applied a triplet approach with indices from overall sound level ($L_{Aeq}$, $L_{A50}$, $L_{A95}$, N; see Table 9.2 for definitions), the average spectrum ($L_{Ceq}$-$L_{Aeq}$, CoG), and the time pattern of the soundscape (ML music likeness; NCN, number of noise events; $L_{A10}$-$L_{A90}$). The notice-event model (De Coensel et al. 2009) focused on the temporal structure of the sound (fluctuation and emergence) and accounted for the subject's activity and time spent at home. The intermittency ratio (IR) attempted to account for the temporal characteristic of several sound sources (Wunderli et al. 2016) and aimed to enrich the $L_{den}$ (Idea: $L_{eq+}$).

Other metrics, such as the Community Tolerance Level, implicitly include the effects of low-frequency noise annoyance and noise-induced rattles (Schomer et al. 2013). Can et al. (2015) evaluated a three-step procedure for city soundscapes to improve the description of the variability of urban acoustic environments at multiple spatial scales. The same group tried to accomplish this with sound-level time series along a typical walking path (corresponding to an average pedestrian trip) and parallel testing in the laboratory (Aumond et al. 2017). Alternatively, dense low-cost sensor networks and participative data collection procedures via smartphone applications (Picaut et al. 2019) were developed to cover longer time periods and provide additional spatial resolution. Detailed in-depth statistical analysis of these sound-level time series can provide input for better sound modeling and for capturing local differences of soundscapes. Unfortunately, these approaches mostly lack comparisons and validations against other sound indicators and health outcomes.

Eventually, several research groups tried to find appropriate sound indicators to characterize quiet areas with or without reference to pleasantness, calmness, or tranquility ratings. De Coensel and Botteldooren (2006) developed a multi-criteria acoustical approach and found statistical noise levels in the range $L_{A95}$ to $L_{A50}$ were the best predictors for quietness. In a larger study with participant ratings that used a linear model, Cassina et al. (2017) found that perceived tranquility was negatively correlated to $L_{A10}$ and to sound sources or visual elements that were evaluated negatively. On the other hand, sound sources or visual elements that were rated positively were correlated positively with perceived tranquility. The inclusion of the visual perspective is a clear improvement.

#### 9.4.6.3 Psychoacoustic Measures

Psychoacoustic measurements were considered as alternatives to classical equivalent noise indicators and can consist of single or aggregated measures (see Genuit, Schulte-Fortkamp, and Fiebig, Chap. 6). Zwicker's (1991) metric, "unbiased annoyance," was based on loudness, sharpness, and fluctuation strength. Another construct, called the "intrusiveness" of the sound, combined the difference between the loudness and the background sound with sharpness and the amount of distortion of informational content (Preis 1987). These and later similar approaches were derived from laboratory work with short presentations of single sounds and without context. Moreover, an evaluation against other acoustic measures rarely took place in this laboratory work, and only a few examples of field studies have been published (e.g., Genuit and Fiebig 2006; Yang and Kang 2013; Rey Gozalo et al. 2015). For specific sources such as motorbikes and scooters, it has been shown that Lmax and roughness distinguish better against car sounds with respect to annoyance ratings (Paviotti and Vogiatzis 2012).

Beginning in 2010, a research group in Lyon developed more specific psychoacoustic indices and additional models with laboratory studies based on recorded sounds (Klein et al. 2015) and then tested these against annoyance ratings in field studies. This group also studied mixed sound sources. In a combined laboratory (Morel et al. 2016) and field study (Gille and Marquis-Favre 2019), partial and total annoyance models were developed based on psychoacoustic indices and noise sensitivity. A systematic review demonstrated that human perceptions of sound were governed by a general judgment based on sensitivity to overall magnitude and the temporal and spectral components of the perceived sounds (Ma et al. 2018). Three perceptual components were derived by principal component analysis from nine adjective pairs: noisy versus quiet, relaxed versus tense, pleasant versus unpleasant, deep versus metallic, high versus low, sharp versus dull, quiet versus loud, light versus heavy, and weak versus strong. These adjective pairs can be used as items in an index to analyze correlations among the physical stimuli from multiple sound sources and human perceptions about their acoustic environments. However, most studies (even smaller ones) published in peer-reviewed journals lack validation against other measures and are not yet in compliance with the current technical specification (ISO/TS 12913-2) that requires the use of binaural measurement methods to maintain necessary spatial information.

#### 9.4.6.4 Genuine Soundscape Measures

Most general soundscape measures use a two-dimensional soundscape model with the affective poles of pleasantness–eventfulness (Axelsson et al. 2010), and/or calmness–vibrancy (Cain et al. 2013). Beyond these affective qualities a supplementary dimension “appropriateness of a soundscape to a place” was suggested (Aletta et al. 2016). However, the development of predictive models for residential areas is still in its infancy. There is room for better agreement on procedures and standardization (Fiebig 2018). Moreover, few specialized measures that target quietness/tranquility of the experience (Pheasant et al. 2010b; Watts and Marafa 2017) or the restorativeness of the soundscape (Payne 2013) have been developed. A systematic review of prediction models for the experience of urban soundscapes has been published by Lionello et al. (2020).

#### 9.4.6.5 Improved Sound and Soundscape Mapping Procedures

A Norwegian study (Klæboe et al. 2006) included the larger neighborhood soundscape (75 m area) as a potential modifier that affects noise annoyance ratings. The hypothesis was that noisy neighborhoods, backyards, and walkways have an additional effect on noise perception beyond the noise exposure at the façade (typical exposure measure). Adjustments for these differences were labeled "neighborhood soundscape effect."

A French study (Tenailleau et al. 2015) documented an even larger effect when they compared 50-m radius buffers with 400-m buffers. The calculated $L_{Aeq, 24\,h}$ level values varied across buildings from −9.4 to +22.3 dBA. Other research confirmed the large impact of the search radius on noise exposure assessments (Meyer et al. 2017). A 1000–2000 m radius was recommended to avoid errors. The SONORUS project observed underestimations up to 13 dB(A) for sound exposure in a neighborhood's quiet area with standard noise mapping procedures (Kropp et al. 2016).

New promising approaches for "dynamic noise mapping" are being published that consider saliency (Filipan et al. 2019) and the temporal and spatial resolution of sound exposure for study participants (Wei et al. 2016). Others consider not only the prime sources (road traffic) but also sounds from nature and humans (Aumond et al. 2017). Any gain in the prediction of health effects from these approaches must still be compared to classical sound mapping procedures as evaluated in broader population studies.

Most soundscape mapping has been carried out across short distances or within smaller subsections of cities, street canyons, and specific traffic situations (Kang et al. 2016a, b; Aumond et al. 2017). For field studies, proper propagation methods to the participant's home over larger distances are required to better match both real sound exposure and human perception at these distances, which is often influenced by other sources, meteorology, and topography.

#### 9.4.6.6 Soundscape Indicators and Audiovisual Assessments Related to Tranquility and Restoration

Humans are multisensory by nature (Ghazanfar and Schroeder 2006) and the environment is never experienced in isolation. However, the sensory assessment of our environment has been conducted mostly through unimodal studies, especially vision and audition. In the framework of "The Positive Soundscape Project," several studies were conducted in both the laboratory and real landscapes with sufficient participants and a representative range of landscape types. These studies demonstrated to what extend the auditory soundscape or visual landscape influenced the perception of tranquility in a real, multisensory environment (Pheasant et al. 2010a). With the mean tranquility rating as the dependent variable, a regression analysis was performed using weighted mean loudness, percentage of natural features present at each location (excluding sky), equivalent A-weighted continuous sound pressure level ($L_{Aeq}$), and maximum sound pressure level ($L_{max}$) as independent variables in the bi-modal analysis. The results indicate that bi-modal tranquility was rated higher than the average of the uni-modal estimates in which the visual features were omitted.

In a parallel study (Hunter et al. 2010), functional magnetic resonance imaging was used to examine neural responses to visually distinct scenes (beach images versus freeway images) that were experienced as tranquil (beach) or non-tranquil (freeway). Both scenes had the same auditory components (similar auditory spectral and temporal characteristics). Results showed that the visual context can modulate activity in regions of the auditory cortex implicated in the generation of subjective states and that the same sound may be associated with different perceptions depending on the pattern of activity stimulated between the auditory cortex and other brain regions. Both results strongly indicate that bi-modal stimuli are essential for a full characterization of tranquil space.

The same research group developed the Tranquility Rating Prediction Tool (TRAPT) in laboratory studies to identify the auditory and visual factors that make restorative or tranquil environments valued in that sense (Pheasant et al. 2010b). This assessment tool was further improved in field studies by including two factors (presence of litter and water sounds) that can significantly moderate the results of the tranquility assessments. In a subsequent field study, a high correlation was found between the predicted tranquility ratings and the level of relaxation as reported from site visitors (Watts et al. 2013). This tool can help to not only protect quiet/tranquil areas but also can support planning in urban and rural areas with respect to their restorative effect and can account for potential visual impact. It was further validated cross-culturally (Hong Kong, China, other countries) with smaller samples and showed a good measure of agreement in field and laboratory studies (Watts and Marafa 2017).

Another metric (Perceived Restorativeness Soundscape Scale, PRSS) used selected items from widely used perceived restorativeness scales (Hartig et al. 1997) to become sound specific rather than as an assessment of environments in general (Payne 2013). This seems reasonable as nature sounds (e.g., wind, rustling trees, water, and animals) are known to be preferred over anthropogenic sounds and natural sounds contribute to restoration beyond the visual aspect of nature (Ratcliffe et al. 2020). The final nine items resulted in a very reliable one-factor structure scale. The PRSS was able to significantly differentiate between soundscapes from different environmental types (urban, urban park, and rural) and of the same environment type (urban parks). Both instruments (TRAPT, PRSS) would have the potential to complement classic exposure measures in epidemiological studies with long-term health outcomes, but validation in residential areas is needed.

Another tool, the Green Soundscape Index is defined as the ratio of the perceived number of natural sounds compared with traffic sounds (Kogan et al. 2018). The lack of consideration of visual components of nature means this index is less useful than the assessment tools discussed previously.

### 9.4.7 *Caveats in Extant Soundscape Research and Recommendations to Move Forward*

Often overlooked in soundscape research is the fact that people spend only a short time of their overall life experience in public spaces that have been studied the most (parks, places, shopping malls, etc.). Some of the momentary perceptual outcomes investigated (e.g., vibrancy) are not appropriate for residential areas (Aletta and Kang 2018) where people spend most of their time, including nighttime when they are more sensitive to sounds. In addition, momentary pleasantness ratings are only intermediate products on the path toward health and need to be related to more distinct long-term health outcomes (standard single-item health ratings, established multi-item questionnaires, or specific mental health inventories adhering to international disease classifications). Assessments should be combined with appropriate consideration of the general environment (Kogan et al. 2017) and general satisfaction (Olsen et al. 2019), which will require adjustments for personal preferences and vulnerabilities (Lercher 2003; Oiamo et al. 2015). Moreover, to implement research findings in regulation, noise management, and noise prevention, further work is needed to link the perceptual constructs with improved acoustic indicators in a standardized manner (Fiebig 2018) beyond ISO/TS 12913-2: 2018. Based on the premises described previously (Kang and Schulte-Fortkamp 2016), the soundscape approach should be able to complement health assessments by integrating the above-listed requirements for a holistic approach to plan and implement measures that ensure efficient and sustainable solutions and foster good health.

## 9.5 Summary and Conclusions

Up to 2023, most soundscape studies have addressed mostly subjective sound qualities (e.g., pleasantness, vibrancy), acoustic comfort, or preferences at rather low spatial scale levels, such as parks, recreational areas, shopping centers, and squares. In addition, most studies consisted of small, non-representative population groups or samples chosen for convenience, and the judgments made were based on rather short time spans of exposure. In studies that used objective measurements of the acoustic space, the techniques used were rather diverse, but they did not yet follow the new ISO soundscape guidelines and the objective measurements were not used in analyses to track potential pathways. Instead, often only intercorrelations are reported and a serious discussion and mutual reference to other studies is often lacking.

Previously, associations with long-term effects in residential areas and more serious health outcomes were not considered and integrated assessments of multiple environmental stressors measured both subjectively and objectively were lacking. Studies that focused only on subjective perception and subjective health assessments did not capture the more hidden effects (and were not subject to individual recognition) of the ambient acoustic and non-acoustic space on overall health and well-being (see Sects. 9.2.1, 9.2.2, and 9.4.2). Specifically, the possible moderating and mediating pathways between the factors of the surrounding environment could not be revealed. Knowing the major pathways to health is necessary to guide both prevention and remedial action in sustainable planning and is required for strategic and environmental health impact assessments. If those pathways toward health are hidden or not well known, invalid conclusions can result that can negatively impact long-term planning in complex environments.

Soundscape approaches have provided useful input for small-scale environmental assessment and planning. In future, these approaches should be more closely integrated with ongoing or future large epidemiologic studies to combine the strength of the two approaches. To achieve this, the studies reviewed in Sects. 9.4.2, 9.4.3, 9.4.4, and 9.4.5 indicate that soundscape research must integrate not only the most relevant evidence-based factors but, in addition, must investigate multiple pathways via moderation and mediation analyses while considering important confounders. Using indicators of cumulative environmental exposures may provide more realistic assessments of the potential health burden.

**Compliance with Ethics Requirements** Peter Lercher declares that he has no conflict of interest.

Angel M. Dzhambov declares that he has no conflict of interest.

## References

Aletta F (2022) Listening to cities: from noisy environments to positive soundscapes. In: Frontiers 2022: noise, blazes and mismatches. United Nations Environment Programme. ISBN: 978-92-807-3917-6

Aletta F, Kang J (2018) Towards an urban vibrancy model: a soundscape approach. Int J Environ Res Public Health 15:1712. https://doi.org/10.3390/ijerph15081712

Aletta F, Kang J, Axelsson O (2016) Soundscape descriptors and a conceptual framework for developing predictive soundscape models. Landsc Urban Plan 149:65–74

Aletta F, Oberman T, Kang J (2018) Associations between positive health-related effects and soundscapes perceptual constructs: a systematic review. Int J Environ Res Public Health 15:2392. https://doi.org/10.3390/ijerph15112392

Andersson L, Sandberg P, Olofsson JK, Nordin S (2018) Effects of task demands on olfactory, auditory, and visual event-related potentials suggest similar top-down modulation across senses. Chem Senses 43:129–134. https://doi.org/10.1093/chemse/bjx082

Asensio C, Aumond P, Can A, Gascó L et al (2020) A taxonomy proposal for the assessment of the changes in soundscape resulting from the COVID-19 lockdown. Int J Environ Res Public Health 17(12):4205. https://doi.org/10.3390/ijerph17124205

Astell-Burt T, Feng X, Kolt GS (2014) Is neighborhood green space associated with a lower risk of type 2 diabetes? Evidence from 267,072 Australians. Diabetes Care 37:197–201. https://doi.org/10.2337/dc13-1325

Aumond P, Can A, De Coensel B, Botteldooren D et al (2017) Modeling soundscape pleasantness using perceptual assessments and acoustic measurements along paths in urban context. Acta Acust united with Acust 103:430–443. https://doi.org/10.3813/AAA.919073

Axelsson Ö, Nilsson ME, Berglund B (2010) A principal components model of soundscape perception. J Acoust Soc Am 128:2836–2846. https://doi.org/10.1121/1.3493436

Berglund B (2006) From the WHO Guidelines for community noise to healthy soundscapes. In: Proceedings of the Institute of Acoustics. pp 1–9

Berglund B, Nilsson ME (2006) On a tool for measuring soundscape quality in urban residential areas. Acta Acust united with Acust 92:938–944(7)

Berglund B, Lindvall T, Schwela DH, World Health Organization (1999) Guidelines for community noise. World Health Organization, Geneva

Bijsterveld K (2008) Listening to technology. In: Mechanical sound: technology, culture, and public problems of noise in the twentieth century. The MIT Press, Cambridge

Botteldooren D, Lercher P (2004) Soft-computing base analyses of the relationship between annoyance and coping with noise and odor. J Acoust Soc Am 115:2974–2985. https://doi.org/10.1121/1.1719024

Botteldooren D, Andringa T, Aspuru I, Brown L et al (2016) From sonic environment to soundscape. In: Kang J, Schulte-Fortkamp B (eds) Soundscape and the built environment. CRC Press, Boca Raton, pp 17–41

Brown AL (2012) A review of progress in soundscapes and an approach to soundscape planning. Int J Acoust Vib 17(2):73–81. https://doi.org/10.20855/ijav.2012.17.2302

Bruehl S, Olsen RB, Tronstad C et al (2018) Chronic pain-related changes in cardiovascular regulation and impact on comorbid hypertension in a general population: the Tromsø study. Pain 159:119–127. https://doi.org/10.1097/j.pain.0000000000001070

Cain R, Jennings P, Poxon J (2013) The development and application of the emotional dimensions of a soundscape. Appl Acoust 74:232–239. https://doi.org/10.1016/j.apacoust.2011.11.006

Can A (2015) Noise pollution indicators. In: Environmental indicators. Springer Netherlands, Dordrecht, pp 501–513

Can A, Guillaume G, Gauvreau B (2015) Noise indicators to diagnose urban sound environments at multiple spatial scales. Acta Acust united with Acust 101:964–974. https://doi.org/10.3813/AAA.918891

Cassina L, Fredianelli L, Menichini I et al (2017) Audio-visual preferences and tranquility ratings in urban areas. Environments 5:1. https://doi.org/10.3390/environments5010001

Crocker MJ (2007). Handbook of noise and vibration control. John Wiley & Sons Ltd., Hoboken

De Coensel B, Botteldooren D (2006) The quiet rural soundscape and how to characterize it. Acta Acust united with Acust 92:887–897

De Coensel B, Botteldooren D, De Muer T et al (2009) A model for the perception of environmental sound based on notice-events. J Acoust Soc Am 126:656–665. https://doi.org/10.1121/1.3158601

Derryberry EP, Phillips JN, Derryberry GE, Blum MJ, Luther D (2020) Singing in a silent spring: birds respond to a half-century soundscape reversion during the COVID-19 shutdown. Science 370:575–579

Dzhambov AM (2015) Noise sensitivity: a neurophenomenological perspective. Med Hypotheses 85:650–655. https://doi.org/10.1016/j.mehy.2015.08.006

Dzhambov AM (2017) More residential greenspace is associated with lower noise annoyance: results from a quantitative synthesis of the literature. In: Łucjan C, Gérard D (eds) Traffic noise: exposure, health effects and mitigation. Nova Science Publishers, New York, pp 77–104

Dzhambov AM, Dimitrova D (2014) Urban green spaces′ effectiveness as a psychological buffer for the negative health impact of noise pollution: a systematic review. Noise Health 16:157–165. https://doi.org/10.4103/1463-1741.134916

Dzhambov AM, Lercher P (2019a) Road traffic noise exposure and birth outcomes: an updated systematic review and meta-analysis. Int J Environ Res Public Health 16:2522. https://doi.org/10.3390/ijerph16142522

Dzhambov AM, Lercher P (2019b) Road traffic noise exposure and depression/anxiety: an updated systematic review and meta-analysis. Int J Environ Res Public Health 16:4134. https://doi.org/10.3390/ijerph16214134

Dzhambov A, Markevych I, Tilov B, Dimitrova D (2017) Residential greenspace modifies the effect of road traffic noise exposure on mental health in youth. Eur J Pub Health 27(ckx187):279. https://doi.org/10.1093/eurpub/ckx187.279

Dzhambov A, Hartig T, Markevych I et al (2018a) Urban residential greenspace and mental health in youth: different approaches to testing multiple pathways yield different conclusions. Environ Res 160:47–59. https://doi.org/10.1016/j.envres.2017.09.015

Dzhambov AM, Markevych I, Tilov B et al (2018b) Lower noise annoyance associated with GIS-derived greenspace: pathways through perceived greenspace and residential noise. Int J Environ Res Public Health 15:1533. https://doi.org/10.3390/ijerph15071533

Dzhambov AM, Browning MHEM, Markevych I et al (2020) Analytical approaches to testing pathways linking greenspace to health: a scoping review of the empirical literature. Environ Res 186:109613. https://doi.org/10.1016/j.envres.2020.109613

Dzhambov AM, Lercher P, Stoyanov D, Petrova N, Novakov S, Dimitrova DD (2021) University students' self-rated health in relation to perceived acoustic environment during the COVID-19 home quarantine. Int J Environ Res Public Health 18:2538. https://doi.org/10.3390/ijerph18052538

Engel MS, Fiebig A, Pfaffenbach C, Fels J (2018) A review of socio-acoustic surveys for soundscape studies. Curr Pollut Rep 4:220–239. https://doi.org/10.1007/s40726-018-0094-8

Erfanian M, Mitchell AJ, Kang J, Aletta F (2019) The psychophysiological implications of soundscape: a systematic review of empirical literature and a research agenda. Int J Environ Res Public Health 16:3533. https://doi.org/10.3390/ijerph16193533

European Environment Agency (2020) Environmental noise in Europe — 2020. Publications Office of the European Union, Luxembourg. https://doi.org/10.2800/686249

European Union (2002) Directive 2002/49/EC of the European Parliament and of the council of 25 June 2002 relating to the assessment and management of environmental noise. Off J Eur Communities 189:0012–0026

Fang C-F, Ling D-L (2003) Investigation of the noise reduction provided by tree belts. Landsc Urban Plan 63:187–195. https://doi.org/10.1016/S0169-2046(02)00190-1

Fiebig A (2018) Recent developments in the standardization of soundscape. In: INTER-NOISE and NOISE-CON congress and conference proceedings. pp 4469–4479

Fiebig A, Schomer P (2015) Supplemental metrics for day/night average sound level and day/evening/night average sound level. I-INCE publication no. 2015-1. 111p

Filipan K, De Coensel B, Aumond P et al (2019) Auditory sensory saliency as a better predictor of change than sound amplitude in pleasantness assessment of reproduced urban soundscapes. Build Environ 148:730–741. https://doi.org/10.1016/j.buildenv.2018.10.054

Galbrun L, Ali TT (2013) Acoustical and perceptual assessment of water sounds and their use over road traffic noise. J Acoust Soc Am 133:227–237

Geneshka M, Coventry P, Cruz J, Gilbody S (2021) Relationship between green and blue spaces with mental and physical health: a systematic review of longitudinal observational studies. Int J Environ Res Public Health 18(17):9010

Genuit K, Fiebig A (2006) Psychoacoustics and its benefit for the soundscape approach. Acta Acust united with Acust 92:952–958

Gerdes ABM, Wieser MJ, Alpers GW (2014) Emotional pictures and sounds: a review of multimodal interactions of emotion cues in multiple domains. Front Psychol 5. https://doi.org/10.3389/fpsyg.2014.01351

Ghazanfar A, Schroeder C (2006) Is neocortex essentially multisensory? Trends Cogn Sci 10:278–285. https://doi.org/10.1016/j.tics.2006.04.008

Gidlöf-Gunnarsson A, Öhrström E (2007) Noise and well-being in urban residential environments: the potential role of perceived availability to nearby green areas. Landsc Urban Plan 83:115–126. https://doi.org/10.1016/j.landurbplan.2007.03.003

Gidlöf-Gunnarsson A, Öhrström E (2010) Attractive "quiet" courtyards: a potential modifier of urban residents' responses to road traffic noise? Int J Environ Res Public Health 7:3359–3375. https://doi.org/10.3390/ijerph7093359

Gille L-A, Marquis-Favre C (2019) Estimation of field psychoacoustic indices and predictive annoyance models for road traffic noise combined with aircraft noise. J Acoust Soc Am 145:2294–2304. https://doi.org/10.1121/1.5097573

Gomez-Bernal F, Madva EN, Puckett J et al (2019) Relationships between life stressors, health behaviors, and chronic medical conditions in mid-life adults: a narrative review. Psychosomatics 60:153–163. https://doi.org/10.1016/j.psym.2018.12.007

Guski R, Felscher-Suhr U, Schuemer R (1999) The concept of noise annoyance: how international experts see it. J Sound Vib 223:513–527

Guski R, Schreckenberg D, Schuemer R (2017) WHO environmental noise guidelines for the European region: a systematic review on environmental noise and annoyance. Int J Environ Res Public Health 14:1539. https://doi.org/10.3390/ijerph14121539

Han X, Huang X, Liang H, Ma S et al (2018) Analysis of the relationships between environmental noise and urban morphology. Environ Pollut 233:755–763. https://doi.org/10.1016/j.envpol.2017.10.126

Hartig T (2021) Restoration in nature: beyond the conventional narrative. In: Schutte AR, Torquati JC, Stevens JR (eds) Nature and psychology. Nebraska symposium on motivation, vol 67. Springer, Cham, pp 89–151. https://doi.org/10.1007/978-3-030-69020-5_5

Hartig T, Korpela K, Evans GW, Gärling T (1997) A measure of restorative quality in environments. Scand Hous Plan Res 14:175–194. https://doi.org/10.1080/02815739708730435

Hartig T, Mitchell R, de Vries S, Frumkin H (2014) Nature and health. Annu Rev Public Health 35:207–228. https://doi.org/10.1146/annurev-publhealth-032013-182443

Hedblom M, Gunnarsson B, Iravani B et al (2019) Reduction of physiological stress by urban green space in a multisensory virtual experiment. Sci Rep 9:10113. https://doi.org/10.1038/s41598-019-46099-7

Heinonen-Guzejev M (2008) Noise sensitivity – medical, psychological and genetic aspects [dissertation]. University of Helsinki: Helsinki University Printing House Helsinki

Heinonen-Guzejev M, Vuorinen HS, Mussalo-Rauhamaa H et al (2005) Genetic component of noise sensitivity. Twin Res Hum Genet 8:245–249. https://doi.org/10.1375/twin.8.3.245

Hong JY, Jeon JY (2017) Exploring spatial relationships among soundscape variables in urban areas: a spatial statistical modelling approach. Landsc Urban Plan 157:352–364. https://doi.org/10.1016/j.landurbplan.2016.08.006

Hornikx M (2016) Ten questions concerning computational urban acoustics. Build Environ 106:409–421. https://doi.org/10.1016/j.buildenv.2016.06.028

Houthuijs D, Swart W, van Beek A, van Kempen E, de Leeuw F (2015) Health impact assessment for noise in Europe: expected consequences of the limitations of the available noise exposure data. ETC/ACM Technical Paper 2014/9 February. Available at: http://acm.eionet.europa.eu/reports/docs/ETCACM_TP_2014_9_HIA-noise_europe.pdf

Hume K, Ahtamad M (2013) Physiological responses to and subjective estimates of soundscape elements. Appl Acoust 74:275–281. https://doi.org/10.1016/J.APACOUST.2011.10.009

Hunter MD, Eickhoff SB, Pheasant RJ et al (2010) The state of tranquility: subjective perception is shaped by contextual modulation of auditory connectivity. NeuroImage 53:611–618. https://doi.org/10.1016/j.neuroimage.2010.06.053

International Organization for Standardization (ISO). (2014). ISO/FDIS 12913-1: 2014 Acoustics—Soundscape—Part 1: Definition and conceptual framework. Geneva: ISO.

Jarosińska D, Héroux M-È, Wilkhu P et al (2018) Development of the WHO environmental noise guidelines for the European region: an introduction. Int J Environ Res Public Health 15:813. https://doi.org/10.3390/ijerph15040813

Jennings P, Cain R (2013) A framework for improving urban soundscapes. Appl Acoust 74:293–299. https://doi.org/10.1016/J.APACOUST.2011.12.003

Job RFS, Hatfield J (2001) The impact of soundscape, enviroscape, and psychscape on reaction to noise: implications for evaluation and regulation of noise effects. Noise Control Eng J 49:120. https://doi.org/10.3397/1.2839647

Kabisch N, Strohbach M, Haase D, Kronenberg J (2016) Urban green space availability in European cities. Ecol Indic 70:586–596. https://doi.org/10.1016/j.ecolind.2016.02.029
Kang J (2012) On the diversity of urban waterscape. In: Proceedings of the Acoustics 2012 Conference, Nantes
Kang J, Schulte-Fortkamp B (eds) (2016) Soundscape and the built environment. CRC Press, Boca Raton
Kang J, Aletta F, Gjestland TT et al (2016a) Ten questions on the soundscapes of the built environment. Build Environ 108:284–294. https://doi.org/10.1016/j.buildenv.2016.08.011
Kang J, Schulte-Fortkamp B, Fiebig A, Botteldooren D (2016b) Mapping of soundscape. In: Kang J, Schulte-Fortkamp B (eds) Soundscape and the built environment. CRC Press, Boca Raton, pp 161–196
Kaplan S, Kaplan R (1989) The experience of nature. Cambridge University Press, Cambridge
Kasl SV (1984) Stress and health. Annu Rev Public Health 5:319–341. https://doi.org/10.1146/annurev.pu.05.050184.001535
Kellert SR, Wilson EO (1993) The biophilia hypothesis. Island Press, Washington, DC
Kim R, van den Berg M (2010) Summary of night noise guidelines for Europe. Noise & Health 12(47): 61–63
King G, Roland-Mieszkowski M, Jason T, Rainham DG (2012) Noise levels associated with urban land use. J Urban Health 89:1017–1030. https://doi.org/10.1007/s11524-012-9721-7
Klæboe R, Engelien E, Steinnes M (2006) Context sensitive noise impact mapping. Appl Acoust 67:620–642. https://doi.org/10.1016/j.apacoust.2005.12.002
Klein A, Marquis-Favre C, Weber R, Trollé A (2015) Spectral and modulation indices for annoyance-relevant features of urban road single-vehicle pass-by noises. J Acoust Soc Am 137:1238–1250. https://doi.org/10.1121/1.4913769
Kliuchko M, Heinonen-Guzejev M, Vuust P et al (2016) A window into the brain mechanisms associated with noise sensitivity. Sci Rep 6:39236. https://doi.org/10.1038/srep39236
Kogan P, Turra B, Arenas JP, Hinalaf M (2017) A comprehensive methodology for the multidimensional and synchronic data collecting in soundscape. Sci Total Environ 580:1068–1077. https://doi.org/10.1016/j.scitotenv.2016.12.061
Kogan P, Arenas JP, Bermejo F et al (2018) A green soundscape index (GSI): the potential of assessing the perceived balance between natural sound and traffic noise. Sci Total Environ 642:463–472. https://doi.org/10.1016/j.scitotenv.2018.06.023
Kompier M (2005) Assessing the psychosocial work environment—"subjective" versus "objective" measurement. Scand J Work Environ Health 31:405–408. https://doi.org/10.5271/sjweh.946
Koprowska K, Łaszkiewicz E, Kronenberg J, Marcińczak S (2018) Subjective perception of noise exposure in relation to urban green space availability. Urban For Urban Green 31:93–102. https://doi.org/10.1016/j.ufug.2018.01.018
Kraus KS, Canlon B (2012) Neuronal connectivity and interactions between the auditory and limbic systems. Effects of noise and tinnitus. Hear Res 288:34–46. https://doi.org/10.1016/j.heares.2012.02.009
Kropp W, Forssén J, Estévez-Mauriz L (2016) Urban sound planning: the sonorus project. Gothenburg
Lan Y, Roberts H, Kwan M-P, Helbich M (2020) Transportation noise exposure and anxiety: a systematic review and meta-analysis. Environ Res 110118. https://doi.org/10.1016/j.envres.2020.110118
Lee JSY, Jenny SY, Hautus MJ, Shepherd D (2012) Neural correlates of noise annoyance and sensitivity. J Acoust Soc New Zeal 25:4–11
Lepore SJ, Evans GW (1996) Coping with multiple stressors in the environment. In: Zeidner M, Endler NS (eds) Handbook of coping: theory, research, applications. Wiley, pp 350–377
Lercher P (1996) Environmental noise and health: an integrated research perspective. Environ Int 22:117–129. https://doi.org/10.1016/0160-4120(95)00109-3
Lercher P (2003) Which health outcomes should be measured in health related environmental quality of life studies? Landsc Urban Plan 65:63–72. https://doi.org/10.1016/S0169-2046(02)00238-4

Lercher P (2019) Noise in cities: urban and transport planning determinants and health in cities. In: Integrating human health into urban and transport planning. Springer International Publishing, Cham, pp 443–481
Lercher P, Kofler WW (1996) Behavioral and health responses associated with road traffic noise exposure along alpine through-traffic routes. Sci Total Environ 189–190:85–89. https://doi.org/10.1016/0048-9697(96)05194-7
Lercher P, Schulte-Fortkamp B (2003) The relevance of soundscape research to the assessment of noise annoyance at the community level. In: Proceedings of the eighth international congress on noise as a public health problem, Rotterdam, The Netherlands, pp 225–231
Lercher P, Schulte-Fortkamp B (2013) Soundscape of European cities and landscapes–harmonising. In: Kang J, Chourmouziadou K, Sakantamis K et al (eds) Soundcape COST TUD action TD0804. Soundscape-COST, Oxford, UK, pp 120–127
Lercher P, Brauchle G, Kofler W et al (2000) The assessment of noise annoyance in schoolchildren and their mothers. In: Proceedings of the 29th international congress and exhibition on noise control engineering. Société Française d'Acoustique, Nice, pp 2318–2322
Lercher P, Botteldooren D, Widmann U et al (2011) Cardiovascular effects of environmental noise: research in Austria. Noise Health 13:234–250
Lercher P, Evans GW, Widmann U (2013) The ecological context of soundscapes for children's blood pressure. J Acoust Soc Am 134:773–781. https://doi.org/10.1121/1.4807808
Lercher P, van Kamp I, von Lindern E, Botteldooren D (2016) Perceived soundscapes and health-related quality of life, context, restoration, and personal characteristics: case studies. In: Kang J, Schulte-Fortkamp B (eds) Soundscape and the built environment. CRC Press, Boca Raton, pp 133–160
Lercher P, De Coensel B, Dekonink L, Botteldooren D (2017) Community response to multiple sound sources: integrating acoustic and contextual approaches in the analysis. Int J Environ Res Public Health 14:663. https://doi.org/10.3390/ijerph14060663
Leung TM, Chau CK, Tang SK, Pun LSC (2014) On the study of effects of views to water spaces on noise annoyance perceptions at home. Melbourne, Internoise
Li H, Lau S-K (2020) A review of audio-visual interaction on soundscape assessment in urban built environments. Appl Acoust 166:107372. https://doi.org/10.1016/j.apacoust.2020.107372
Li HN, Chau CK, Tang SK (2010) Can surrounding greenery reduce noise annoyance at home? Sci Total Environ 408:4376–4384
Lionello M, Aletta F, Kang J (2020) A systematic review of prediction models for the experience of urban soundscapes. Appl Acoust 170:107479. https://doi.org/10.1016/j.apacoust.2020.107479
Ma KW, Wong HM, Mak CM (2018) A systematic review of human perceptual dimensions of sound: meta-analysis of semantic differential method applications to indoor and outdoor sounds. Build Environ 133:123–150. https://doi.org/10.1016/j.buildenv.2018.02.021
Margaritis E, Kang J (2016) Relationship between urban green spaces and other features of urban morphology with traffic noise distribution. Urban For Urban Green 15:174–185. https://doi.org/10.1016/j.ufug.2015.12.009
Markevych I, Schoierer J, Hartig T et al (2017) Exploring pathways linking greenspace to health: theoretical and methodological guidance. Environ Res 158:301–317. https://doi.org/10.1016/j.envres.2017.06.028
McCubbin JA, Nathan A, Hibdon MA et al (2018) Blood pressure, emotional dampening, and risk behavior. Psychosom Med 80:544–550. https://doi.org/10.1097/PSY.0000000000000598
Meyer R, Lavandier C, Gauvreau B, Benetto E (2017) Influence of the search radius in a noise prediction software on population exposure and human health impact assessments. Appl Acoust 127:63–73. https://doi.org/10.1016/j.apacoust.2017.05.028
Mitchell A, Oberman T, Aletta F et al (2020) The Soundscape Indices (SSID) protocol: a method for urban soundscape surveys—questionnaires with acoustical and contextual information. Appl Sci 10:2397. https://doi.org/10.3390/app10072397

Morel J, Marquis-Favre C, Gille L-A (2016) Noise annoyance assessment of various urban road vehicle pass-by noises in isolation and combined with industrial noise: a laboratory study. Appl Acoust 101:47–57. https://doi.org/10.1016/j.apacoust.2015.07.017

Nilsson ME, Berglund B (2006) Soundscape quality in suburban green areas and city parks. Acta Acust united with Acust 92:903–911(9)

Nilsson M, Botteldooren D, De Coensel B (2007) Acoustic indicators of soundscape quality and noise annoyance in outdoor urban areas. In: 19th International Congress on Acoustics Madrid, 2–7 September 2007. pp 2–7

Nyklíček I, Vingerhoets AJJM, Van Heck GL (1996) Hypertension and objective and self-reported stressor exposure: a review. J Psychosom Res 40:585–601. https://doi.org/10.1016/0022-3999(95)00647-8

Öhrström E (1997) Effects of exposure to railway noise—a comparison between areas with and without vibration. J Sound Vib 205:555–560. https://doi.org/10.1006/jsvi.1997.1025

Oiamo TH, Baxter J, Grgicak-Mannion A, Xu X et al (2015) Place effects on noise annoyance: cumulative exposures, odour annoyance and noise sensitivity as mediators of environmental context. Atmos Environ 116:183–193. https://doi.org/10.1016/j.atmosenv.2015.06.024

Olsen JR, Nicholls N, Mitchell R (2019) Are urban landscapes associated with reported life satisfaction and inequalities in life satisfaction at the city level? A cross-sectional study of 66 European cities. Soc Sci Med 226:263–274. https://doi.org/10.1016/j.socscimed.2019.03.009

Paviotti M, Vogiatzis K (2012) On the outdoor annoyance from scooter and motorbike noise in the urban environment. Sci Total Environ 430:223–230

Payne SR (2013) The production of a perceived restorativeness soundscape scale. Appl Acoust 74:255–263. https://doi.org/10.1016/J.APACOUST.2011.11.005

Pedersen E (2015) City dweller responses to multiple stressors intruding into their homes: noise, light, odour, and vibration. Int J Environ Res Public Health 12:3246–3263. https://doi.org/10.3390/ijerph120303246

Pedersen E, Larsman P (2008) The impact of visual factors on noise annoyance among people living in the vicinity of wind turbines. J Environ Psychol 28:379–389

Pendrill L (2014) Man as a measurement instrument. NCSLI Meas 9:24–35. https://doi.org/10.1080/19315775.2014.11721702

Persson Waye K, Smith MG, Hussain-Alkhateeb L et al (2019) Assessing the exposure-response relationship of sleep disturbance and vibration in field and laboratory settings. Environ Pollut 245:558–567. https://doi.org/10.1016/j.envpol.2018.09.082

Pheasant RJ, Fisher MN, Watts GR et al (2010a) The importance of auditory-visual interaction in the construction of 'tranquil space.'. J Environ Psychol 30:501–509. https://doi.org/10.1016/J.JENVP.2010.03.006

Pheasant RJ, Horoshenkov KV, Watts GR (2010b) Tranquility rating prediction tool (TRAPT). Acoust Bull 35:18–24

Picaut J, Fortin N, Bocher E et al (2019) An open-science crowdsourcing approach for producing community noise maps using smartphones. Build Environ 148:20–33. https://doi.org/10.1016/j.buildenv.2018.10.049

Pouso S, Borja Á, Fleming LE, Gómez-Baggethun E, White MP, Uyarra MC (2021) Contact with blue-green spaces during the COVID-19 pandemic lockdown beneficial for mental health. Sci Total Environ 756:143984. https://doi.org/10.1016/j.scitotenv.2020.143984

Preis A (1987) Intrusive sounds. Appl Acoust 20:101–127. https://doi.org/10.1016/0003-682X(87)90057-0

Ratcliffe E (2021) Sound and soundscape in restorative natural environments: a narrative literature review. Front Psychol 12:570563

Ratcliffe E, Gatersleben B, Sowden PT (2020) Predicting the perceived restorative potential of bird sounds through acoustics and aesthetics. Environ Behav 52:371–400. https://doi.org/10.1177/0013916518806952

Rey Gozalo G, Trujillo Carmona J, Barrigón Morillas JM et al (2015) Relationship between objective acoustic indices and subjective assessments for the quality of soundscapes. Appl Acoust 97:1–10. https://doi.org/10.1016/j.apacoust.2015.03.020
Rey Gozalo G, Gómez Escobar V, Barrigón Morillas JM et al (2019) Statistical attribution of errors in urban noise modeling. Appl Acoust 153:20–29. https://doi.org/10.1016/j.apacoust.2019.04.001
Sakagami K (2020) A note on variation of the acoustic environment in a quiet residential area in Kobe (Japan): seasonal changes in noise levels including COVID-related variation. Urban Sci 4:63
Schäffer B, Brink M, Schlatter F, Vienneau D, Wunderli JM (2020) Residential green is associated with reduced annoyance to road traffic and railway noise but increased annoyance to aircraft noise exposure. Environ Int 143:105885. https://doi.org/10.1016/j.envint.2020.105885
Schomer P, Mestre V, Schulte-Fortkamp B, Boyle J (2013) Respondents' answers to community attitudinal surveys represent impressions of soundscapes and not merely reactions to the physical noise. J Acoust Soc Am 134:767–772. https://doi.org/10.1121/1.4808075
Schubert M, Hegewald J, Freiberg A et al (2019) Behavioral and emotional disorders and transportation noise among children and adolescents: a systematic review and meta-analysis. Int J Environ Res Public Health 16:3336. https://doi.org/10.3390/ijerph16183336
Shepherd D, Welch D, Dirks KN, Mathews R (2010) Exploring the relationship between noise sensitivity, annoyance and health-related quality of life in a sample of adults exposed to environmental noise. Int J Environ Res Public Health 7:3579–3594. https://doi.org/10.3390/ijerph7103580
Shepherd D, Hautus MJ, Lee SY, Mulgrew J (2016) Electrophysiological approaches to noise sensitivity. J Clin Exp Neuropsychol 38:900–912. https://doi.org/10.1080/13803395.2016.1176995
Shin JC, Parab KV, An R, Grigsby-Toussaint DS (2020) Greenspace exposure and sleep: a systematic review. Environ Res 182:109081. https://doi.org/10.1016/j.envres.2019.109081
Shore SE, Roberts LE, Langguth B (2016) Maladaptive plasticity in tinnitus — triggers, mechanisms and treatment. Nat Rev Neurol 12:150–160. https://doi.org/10.1038/nrneurol.2016.12
Stansfeld SA (1992) Noise, noise sensitivity and psychiatric disorder: epidemiological and psychophysiological studies. Psychol Med Monogr Suppl 22:1–44. https://doi.org/10.1017/S0264180100001119
Stansfeld SA, Shipley M (2015) Noise sensitivity and future risk of illness and mortality. Sci Total Environ 520:114–119. https://doi.org/10.1016/j.scitotenv.2015.03.053
Tamura A (1997) Effects of landscaping on the feeling of annoyance of a space. In: Schick A, Klatte M (eds) Contributions to psychological acoustics. BIS, Oldenburg, pp 135–161
Tenailleau QM, Bernard N, Pujol S et al (2015) Assessing residential exposure to urban noise using environmental models: does the size of the local living neighborhood matter? J Expo Sci Environ Epidemiol 25:89–96
Trollé A, Marquis-Favre C, Parizet É (2015) Perception and annoyance due to vibrations in dwellings generated from ground transportation: a review. Low Freq Noise, Vib Act Control 34:413–458. https://doi.org/10.1260/0263-0923.34.4.413
Ulrich R (1984) View through a window may influence recovery from surgery. Science (80-) 224:420–421. https://doi.org/10.1126/science.6143402
Van Gerven PWM, Vos H, Van Boxtel MPJ et al (2009) Annoyance from environmental noise across the lifespan. J Acoust Soc Am 126:187–194. https://doi.org/10.1121/1.3147510
van Kamp I, Job RFS, Hatfield J et al (2004) The role of noise sensitivity in the noise–response relation: a comparison of three international airport studies. J Acoust Soc Am 116:3471–3479. https://doi.org/10.1121/1.1810291
van Kamp I, Klæboe R, Brown A, Lercher P (2016) Soundscapes, human restoration, and quality of life. In: Soundscape and the built environment. CRC Press, Boca Raton, pp 43–68
Van Renterghem T (2014) Guidelines for optimizing road traffic noise shielding by non-deep tree belts. Ecol Eng 69:276–286. https://doi.org/10.1016/j.ecoleng.2014.04.029

Van Renterghem T (2019) Towards explaining the positive effect of vegetation on the perception of environmental noise. Urban For Urban Green 40:133–144
Van Renterghem T, Botteldooren D (2016) View on outdoor vegetation reduces noise annoyance for dwellers near busy roads. Landsc Urban Plan 148:203–215. https://doi.org/10.1016/j.landurbplan.2015.12.018
von Lindern E, Hartig T, Lercher P (2016) Traffic-related exposures, constrained restoration, and health in the residential context. Health Place 39:92–100. https://doi.org/10.1016/j.healthplace.2015.12.003
von Lindern E, Lymeus F, Hartig T (2017) The restorative environment: a complementary concept for salutogenesis studies. In: The handbook of salutogenesis. Springer International Publishing, Cham, pp 181–195
Watts G, Marafa L (2017) Validation of the Tranquility Rating Prediction Tool (TRAPT): comparative studies in UK and Hong Kong. Noise Mapp 4:67–74. https://doi.org/10.1515/noise-2017-0005
Watts G, Miah A, Pheasant R (2013) Tranquility and soundscapes in urban green spaces—predicted and actual assessments from a questionnaire survey. Environ Plan B Plan Des 40:170–181. https://doi.org/10.1068/b38061
Wei W, Van Renterghem T, De Coensel B, Botteldooren D (2016) Dynamic noise mapping: a map-based interpolation between noise measurements with high temporal resolution. Appl Acoust 101:127–140. https://doi.org/10.1016/j.apacoust.2015.08.005
WHO (2018) Environmental noise guidelines for the European region. Copenhagen
Wild CP (2012) The exposome: from concept to utility. Int J Epidemiol 41:24–32. https://doi.org/10.1093/ije/dyr236
Wu C, Stefanescu RA, Martel DT, Shore SE (2015) Listening to another sense: somatosensory integration in the auditory system. Cell Tissue Res 361:233–250. https://doi.org/10.1007/s00441-014-2074-7
Wunderli JM, Pieren R, Habermacher M et al (2016) Intermittency ratio: a metric reflecting short-term temporal variations of transportation noise exposure. J Expo Sci Environ Epidemiol 26:575–585. https://doi.org/10.1038/jes.2015.56
Xiao J, Tait M, Kang J (2018) A perceptual model of smellscape pleasantness. Cities 76:105–115. https://doi.org/10.1016/j.cities.2018.01.013
Yang M, Kang J (2013) Psychoacoustical evaluation of natural and urban sounds in soundscapes. J Acoust Soc Am 134:840–851
Yang B-Y, Zhao T, Hu L-X, Browning MHEM, Heinrich J, Dharmage SC, Jalaludin B, Knibbs LD, Liu X-X, Luo Y-N, James P, Li S, Huang W-Z, Chen G, Zeng X-W, Hu L-W, Yu Y, Dong G-H (2021) Greenspace and human health: an umbrella review. The Innovation. https://doi.org/10.1016/j.xinn.2021.100164
Yuan M, Yin C, Sun Y, Chen W (2019) Examining the associations between urban built environment and noise pollution in high-density high-rise urban areas: a case study in Wuhan, China. Sustain Cities Soc 50:101678. https://doi.org/10.1016/j.scs.2019.101678
Zwicker E (1991) A proposal for defining and calculating the unbiased annoyance. In: Schick A, Höge H, Lazarus-Mainka G (eds) Contributions to psychological acoustics. BIS, Oldenburg

# Chapter 10
# Hospital Soundscapes

**Ilene Busch-Vishniac and Erica Ryherd**

**Abstract** Hospital soundscapes are challenging because there are many noise sources that contribute to the soundscape at all hours, which can potentially affect a vulnerable population. Traditional measures of sound in buildings tend not to capture the essential quality of the hospital soundscape, and interventions that have been perceived to produce improvements often show nearly no impact on such acoustic measures. Statistical approaches to hospital sound characterization offer a better means of correlating objective measures to subjective responses.

Hospital soundscapes affect staff and patients, potentially increasing stress in staff and anxiety in patients. Studies of hospital soundscapes using sophisticated statistical approaches suggest a key determinant of patient and staff satisfaction with the hospital soundscape is how calm or relaxing it seems to be. Interventions that might improve hospital soundscapes include the implementation of quiet times, architectural designs that reduce reverberation, addition of sound absorption, the use of earbuds or headphones, and the use of nature sounds to mask some less appreciated hospital sounds.

**Keywords** Noise control · Occurrence rate · Noise and sleep · Noise and psychological response · Noise and physiological response · Noise interventions

I. Busch-Vishniac (✉)
BeoGrin Consulting, Baltimore, MD, USA

E. Ryherd
Durham School of Architectural Engineering and Construction, University of Nebraska, Omaha, NE, USA
e-mail: eryherd@unl.edu



B. Schulte-Fortkamp et al. (eds.), *Soundscapes: Humans and Their Acoustic Environment*, Springer Handbook of Auditory Research 76,
https://doi.org/10.1007/978-3-031-22779-0_10

## 10.1 Introduction

In many situations, the approach to noise control is to eliminate the offending sounds. As we have come to better understand human reaction to sound, we have also come to appreciate that people and sounds interact in complicated ways and that the eradication of noise is not always the best approach to produce environments that are viewed as positive experiences. Imagine, for instance, a silent playground—would we think this is desirable? It is the complex interaction between people and sound that we define as the soundscape, including the types of sounds that exist in an environment and the responses, physiological and emotional, they produce in people exposed to them (International Organization for Standardization 2014).

Hospitals are a very interesting and challenging environment in which to consider the soundscape. Hospitals are densely packed with people and noise sources, are places where auditory communication can be critical and urgent, and are places where staff are cognizant that their noisy operations might negatively affect their mission by interrupting sleep, interfering with speech communication, and producing irritation and anxiety in patients.

In this chapter, we discuss the compelling reasons to be concerned about hospital soundscapes and the challenges the environment presents in Sect. 10.2, then delve into what is known about the acoustic characterization of hospital environments and their psychoacoustic impacts in Sect. 10.3. In Sect. 10.4, we discuss the impact of the acoustic environment on patients, families, and staff. Finally, we present some work that has been done on interventions to improve hospital soundscapes in Sect. 10.5. Throughout, we will carefully describe both what is well researched and what areas are ripe for further study. This chapter is not intended to be an all-encompassing literature review that discusses and cites all papers published on hospital acoustics. More detailed literature reviews can be found in various articles (Cvach 2012; Hsu 2012; Ryherd et al. 2012).

## 10.2 Why Do We Care About Hospital Soundscapes?

There are a number of reasons to care about the soundscape in hospitals, ranging from economic to health mission-driven reasons, and an equally daunting set of challenges to improving hospital soundscapes given the constraints imposed by infection control and the seemingly endless production of new medical devices for the bedside.

### 10.2.1 Exposure to Hospital Noise—The Demographics

One reason to consider the hospital soundscape is simply the sheer volume of people who are in hospitals at any given moment. It is possible to get an estimate of how many people are being treated as in-patients in hospitals on any day by looking at the data collected by the Organisation for Economic Co-operation and Development (OECD) (2019). There are 31 countries for which OECD has information for the year 2016 on the total number of discharges of in-patients and on the average patient stay. Combining this with populations by country permits us to estimate the fraction of people in hospitals as a patient on a typical day. The numbers range from a low of 0.04% of the population (Mexico) to 1% (Japan) with a median of 0.3% among these countries. While patients are the most vulnerable population in hospitals at any moment, they certainly aren't the only people in hospitals. Hospital staff and visitors bump the density by a minimum of a factor of two to a median estimate of about 0.6%. The point is that a large number of people in the world are in a hospital on any given day, many on a recurring basis because they work in the hospital, and thus it behooves us to consider how the hospital environment affects them, including the sonic environment (i.e., the soundscape).

### 10.2.2 Economic Importance

In the United States, over 21 million people are currently employed in the labor sector identified as Healthcare and Social Assistance and it is the fastest-growing sector of the economy (Bureau of Labor Statistics 2018a). Hospitals are a large part of the healthcare cost and labor force, so it is clear that there are economic reasons to pursue making the experience of people in hospitals as pleasant, effective, and efficient as possible.

The appropriateness of a soundscape depends on the population occupying the space and its intended uses. While the hospital staff population is roughly the same age distribution as that seen in office buildings all over the world, the patient population in hospitals tends to be skewed toward the very young (under 5 years of age) and the old (over 65 and especially over 90 years of age) (Eurostat Data 2009). The very nature of the hospital mission mandates that we understand the soundscape in hospitals so as to create acoustic environments that promote healing rather than soundscapes that tend to delay healing and prolong hospital stays. While studies relating hospital stays and hospital noise are rare, a study by Fife and Rappaport (1976) showed that patients in hospital during a major construction phase (and its associated noise) had longer stays than were the norm when the hospital was modestly less noisy. There is an economic cost to prolonging hospital stays unnecessarily and although we are not yet at the point of being able to estimate this cost, we know enough to presume that it is non-zero and possibly quite significant.

### 10.2.3 Hospital Design and Noise Interact

Hospitals have some of the most stringent requirements found in buildings in order to deal with the large flow of people, the density of equipment, and the need for the control of pathogens. Hospitals have very high air flow rates, have hard surfaces that are frequently cleaned, have people and equipment constantly in motion, and operate through the transmission of a large amount of information orally. These characteristics pose challenges for crafting a soundscape that is pleasant. Hard, cleanable surfaces tend to be low in sound absorption while high airflow rates can result in increased systems noise, and the constant conversations of people in halls and rooms at all hours of the day disrupt the normal circadian rhythm of patients. A further complication is that hospitals have traditionally been designed to be efficient for staff with less regard for the impact of those designs on patients and on the soundscape, although this is changing in modern hospital designs.

### 10.2.4 Patient Views on the Hospital Soundscape

There is certainly evidence that patients are unhappy with the soundscape in hospitals. While noise in hospitals has been among the top few complaints of patients worldwide for decades, even surpassing complaints about the food by more than a factor of two (Fick and Vance 2006), the recent creation of patient satisfaction surveys has shown how pervasive and irritating this problem has become. From the moment the Hospital Consumer Assessment of Health Providers and Systems survey was created in the United States (known as the HCAHPS), the single question about noise universally received the lowest average score (Jha et al. 2008). This has recently changed due to modifications in the HCAHPS survey, but noise remains as the second lowest score in the survey. Given that government financial support is tied to performance in the HCAHPS survey, hospital administrators now care greatly about the soundscape in their hospitals. This is evident from the surveys by the Beryl Institute of hospital administrators about their top patient concerns, an example of which is demonstrated in the Fig. 10.1 word cloud (Wolf 2013).

What has been largely missing have been efforts to rigorously relate the sound in hospitals to impacts on patients, staff, and visitors as measured by more than complaints. We know, for instance, that the turnover rate of staff in hospitals is higher than seen in most other business sectors (Bureau of Labor Statistics 2018b), but we don't know the extent to which that relates to the soundscape. Nor do we know enough about the physiological and psychological impacts on patients of the soundscape in hospitals. We will delve further into what is known about the relationship between sound and human interactions in hospitals in later sections of this chapter.

Finally, work that has been done on hospital noise shows that new hospitals are not necessarily quieter than old hospitals (Madaras 2017). Madaras (2017) looked at HCAHPS scores for quiet at night in new hospitals (in operation under 3 years)

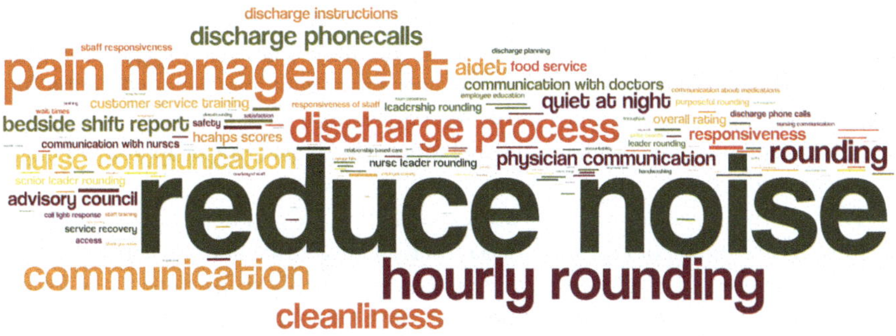


**Fig. 10.1** Top patient management concerns from Beryl Institute, 2013

compared to the national average of all hospitals and found that results were either the same or different by the smallest amount possible. This suggests that the hospital soundscape has not been adequately considered in hospital design stages and that we need to have a better foundation of knowledge from which to create soundscapes conducive to healing, leading to healing architectures as described in Nickl and Nickl-Weller (2013).

## 10.3 The Characterization of Hospital Noise

Noise in hospitals has been recognized as a problem for patients for nearly as long as hospitals have existed. Indeed, in 1859 Florence Nightingale noted "unnecessary noise, then, is the most cruel absence of care which can be inflicted either on sick or well" (Nightingale 1969). Papers discussing noise in hospital date back as far as the 1860s and continue to appear in significant numbers annually.

### *10.3.1 Sound Pressure Levels*

Hospital noise has tended to be characterized in much the way most noisy environments have been objectively described using the *A*-weighted equivalent sound pressure level, the $L_{eq}(A)$. The $L_{eq}$ is defined as the steady-state sound pressure level containing the same sound energy as the original time-varying signal over a given time interval. The $L_{eq}(A)$ is viewed as an appropriate measure of noise in a space as it reflects the average sound energy present as filtered through a weighting network that roughly reflects the acuity of human hearing. This metric is thus a time and frequency average.

Busch-Vishniac et al. (2005) traced the trends in measured $L_{eq}(A)$ for hospitals from 1960 to 2005. The approach was to include all reports of hospital noise that used the $L_{eq}(A)$ and that could be trusted to have averaged correctly, regardless of the type of unit monitored or the country in which the hospital was located. Most of the papers on hospital noise follow the medical literature convention of presenting a mean and standard deviation and unfortunately most seem to calculate these by literally averaging the level values rather than converting back to energy, averaging, and then converting once again to decibel units. This is an error in approach that is prevalent still today.

The reported $L_{eq}(A)$ up to 2005 was expanded by Ryherd et al. (2011) to 2010. The results are shown in Fig. 10.1 for daytime and nighttime hours. There are a few items to note in these graphs. First, there is a monotonically rising trend to the levels as a function of year, indicating that hospitals have gotten progressively noisier. One can speculate on causes for this increase: hospitals first started to be air conditioned in the 1960s; required airflow rates have continued to increase to drive pathogens from the air; the use of hard surfaces has gained popularity as a means to reduce infection; the amount of equipment in use at bedsides has constantly risen with time and each new machine or instrument introduces some noise; the density of patients has increased greatly as time has marched on.

Second, virtually none of the recorded $L_{eq}(A)$ fall below the recommended maxima defined by the World Health Organization (WHO). The WHO guidelines recommend a maximum $L_{eq}(A)$ of 35 dBA in patient treatment and observation rooms and 30 dBA in ward rooms (Berglund and Lindvall 1999). However, the data in Fig. 10.2 are typically occupied background noise levels and the WHO guidelines are applicable for unoccupied (e.g., building system) noise levels. The WHO work in this area is intended to define the maximum level for which there is no evidence of an adverse impact on people. Clearly, the literature supports the assertion that current levels of sound in hospital are sufficient to have a negative impact on patients.

Third, the distribution of $L_{eq}(A)$ reported in the literature is tighter than one might expect given the wide variation of countries, ages of hospitals, and types of units measured (intensive care, medical/surgical, pediatric, and even psychiatric). Given that the buildings are very different and that they have widely varying utilities and equipment, one can surmise that the major source of noise in hospitals, at least as

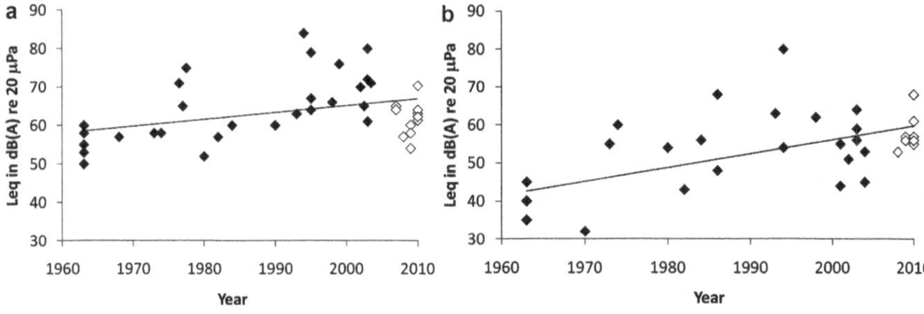


**Fig. 10.2** A-weighted equivalent sound pressure levels as a function of year of publication during: (**a**) daytime hours, and (**b**) nighttime hours. (Reproduced with permission (Busch-Vishniac et al. 2005; Ryherd et al. 2011))

measured by the $L_{eq}(A)$, seems to be hospital occupants or standard and common medical equipment. There are two additional pieces of information that support this speculation—studies of occupied versus unoccupied hospital rooms, and studies of sound levels versus frequency in hospitals.

Ryherd et al. (2011) looked at sound levels with occupied and unoccupied rooms in seven different units of a hospital and found that occupied rooms showed higher $L_{eq}(A)$ by 6–15 dB. This identifies the dominant sound sources as those that exist only when patients are present—conversation and other human-origin sounds, and equipment noise for items only used when patients are present. Results were fairly uniform although the equipment used in the units was not, suggesting that it is human sound sources that dominate in a typical hospital.

## 10.3.2 *Sound Spectra*

Frequency spectra of hospital noise also support the hypothesis that humans are the major source of sound. Figure 10.3 shows a typical graph of sound versus frequency in octave bands in a variety of hospital units in a single hospital. The preponderance of low-frequency sound is typical of HVAC system noise. In the mid range, the level is flatter and it is this range where speech dominates and other human occupancy sounds also contribute. This is distinct from a typical unoccupied space, in which the level decreases monotonically with frequency.

Besides human-associated noises, there are many additional sources of noise that contribute to an overall hospital soundscape often described as pandemonium. Siebein and Skelton (2009) classed hospital sound sources into five distinct categories: occupational; medical equipment; conversational; building equipment; and

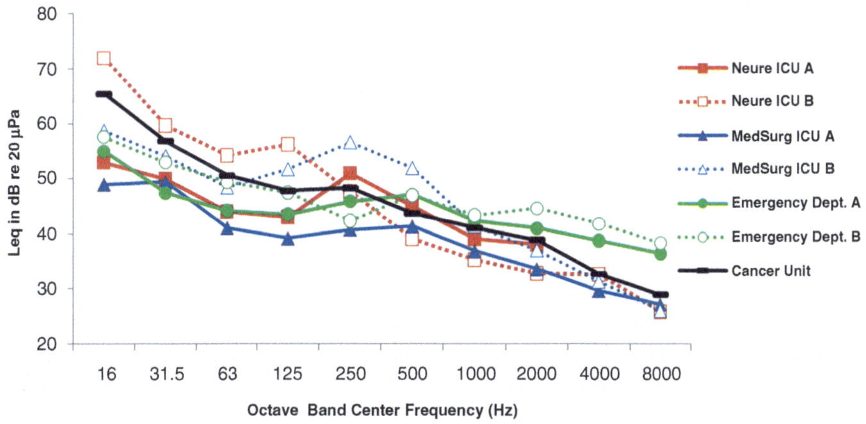


**Fig. 10.3** Average unoccupied background noise levels in octave bands for various types of units (i.e., Neurological and Medical-Surgical Intensive Care Units, Emergency Departments, and Cancer Units). (Reproduced with permission (Ryherd et al. 2011))

exterior noise. Occupational sources include telephones, overhead paging systems, and food carts. Medical equipment includes EKG monitors, ventilators, and various pumps. Many of these devices have alarms which sound frequently but for such short durations that they would not be well captured by a measure which averages over time such as a $L_{eq}$. Building equipment sound sources include HVAC systems and floor cleaning equipment. Exterior noise sources are related to automobile traffic, helicopter landings and takeoffs, and sirens from ambulance arrivals. Compared to a typical household, the hospital sound sources differ in a few ways: there are more of them, they don't seem to quiet down at night, and some of them produce alarms.

## *10.3.3 Sound Levels Over Time*

Hospitals are unusual environments in terms of sound versus time—typically displaying a great deal of variation of amplitude on short time scales but very little over longer time scales. Consider Fig. 10.4, which shows a typical sound recording versus time on a patient unit over the course of 6 h at night. Viewing the entire period, the average seems to be fairly constant and indeed little difference between nighttime and daytime hours was found (Ryherd et al. 2008). By contrast, looking at the sound versus time on short time scales shows it is peaky.

### 10.3.3.1 Loudness Measures

Other standard measures shown in Fig. 10.4 include the minimum, maximum, and peak sound pressure levels ($L_{min}$, $L_{max}$, and $L_{peak}$). The $L_{min}$ and the $L_{max}$ are the minimum and maximum root-mean-squared sound pressure levels observed over a specified time-averaging period: 125 ms (fast), 1 s (slow), and 35 ms (impulse) time constants. The peak sound pressure level ($L_{peak}$) is the highest amplitude sound pressure level instantaneously sensed. It is traditional to use the $L_{peak}$ to describe sounds of very short duration and to use the C-weighting scale, which is nearly linear and derived from human perception of loud sounds.

The $L_{eq}(A)$ is not terribly well suited to environments in which the sound is peaky or contains pure tone alarms. Further, the $L_{min}$, $L_{max}$, or $L_{peak}$ values are highly influenced by single events. For this reason, many efforts to improve hospital soundscapes have found these measures alone do not predict response to interventions that produce noticeable changes in the soundscape. In the period from 2000 to 2020, researchers have made attempts to find measures of the sound in hospital settings that correlate better with subjective reactions as measured in surveys such as the HCAHPS.

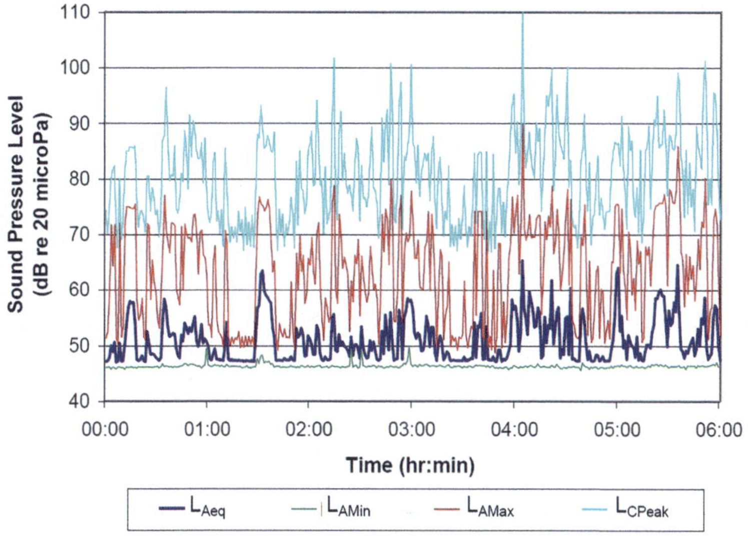


**Fig. 10.4** A-weighted equivalent, minimum, and maximum ($L_{Aeq}$, $L_{AMin}$, $L_{AMax}$) and C-weighted peak ($L_{CPeak}$) sound pressure levels measured over 1 min intervals overnight. (Reproduced with permission (Ryherd et al. 2008)

#### 10.3.3.2 OR(*N*) as a Measure

The occurrence rate is a newer measure that shows promise. It is derived from traditional statistical distribution analysis techniques such as standard noise percentile level analysis. Percentile or exceedance levels ($L_n$) are historically used to describe how often certain levels are exceeded, based on the running sound pressure level. For example, $L_{90}$ reflects the sound level exceeded 90% of the time. The occurrence rate expands upon the traditional $L_n$ by applying a statistical distribution analysis specifically to $L_{min}$, $L_{max}$, and $L_{peak}$ metrics. The occurrence rate therefore gives a better sense of the distribution of $L_{min}$, $L_{max}$, and $L_{peak}$ sound levels. The occurrence rate, which we will write as OR(*N*), thus shows the fraction the time that a measurement of sound pressure level has a value that exceeds *N* dB. For example, OR $(90)_{peak}$ indicates the fraction of the time that peak sound levels exceed 90 dBC. A typical example of an OR(*N*) graph is shown in Fig. 10.5. Note that, by definition, the graph starts at 100% at the lowest sound pressure level and decreases monotonically to 0% at the highest values of *N*.

The OR(*N*) was hinted at as an acoustic measure by Ryherd et al. (2008), Williams et al. (2007), and Kracht et al. (2007), who used similar techniques to describe the environments of adult and neonatal intensive care units, and to quantify noise from operating room surgeries, respectively. It was more formally defined in two papers by Okcu (2011) and Okcu et al. (2012) that examined the reaction of nursing staff in two different units of a hospital with similar staff activities and acuity levels. The units showed nearly the same noise levels using the standard $L_{eq}$, $L_{max}$,

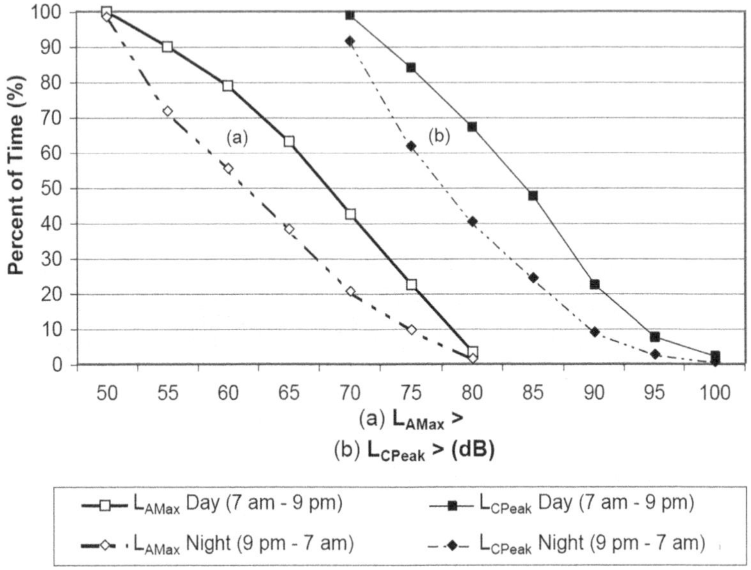


**Fig. 10.5** Statistical distribution of peak and maximum levels. *Y* axis represents the percent of time, or occurrence rate (OR) that (a) $L_{AMax}$ and (b) $L_{CPeak}$ exceed *N* values shown on the *x* axis. (Reproduced with permission (Ryherd et al. 2008))

$L_{min}$, and $L_{peak}$ measures, but staff views of how loud/annoying the environment was and of the impact of the noise on staff performance, health and anxiety were quite different. The unit in which staff had a harsher view of their soundscape had a much higher OR(90), with sound peaks in excess of 90 dBC more than 50% of the time as opposed to just over 20% of the time in the other unit (Okcu et al. 2012).

The OR(*N*) measure of noise in hospitals is growing in popularity and several papers have used this measure and linked it to staff outcomes. For example, Sbihi et al. (2011) studied three long-term care facilities and found peak occurrence rates were correlated with staff perception of noise-related health effects including distraction, stress, fatigue, and tension headache. Okcu et al. (2011, 2012) linked nurse loudness and annoyance perception to mid-level occurrence rates.

Theoretically, occurrence rate analysis can be applied to any acoustic metric. For example, Ryherd et al. (2013) used occurrence rate to determine how often speech intelligibility fell within certain thresholds such as "poor," "marginal," and "good." They found that a unit retrofitted with sound absorption had higher speech intelligibility ratings for a larger percentage of time compared to an identical untreated unit.

While the OR(*N*) is a significant improvement, modified versions might be even more useful in predicting startle responses of patients, visitors, and staff. For example, the OR(*N*) looks at the peak or max sound pressure level only and the result is independent of the average or minimum level present at the time of measurement. If one is concerned about sharp changes in sound energy of the sort that might awaken or startle a patient, then the interest might more appropriately be on the range of noise levels encountered. Work by Bliefnick (2018) developed secondary occurrence rate metrics to account for the range of sound levels experienced and found correlations between the occurrence rate range and patient satisfaction. Follow-up laboratory listening studies found correlations between the occurrence rate range and annoyance. However, some conflicting results indicate this area needs additional research.

### *10.3.4 Speech in Hospitals*

Hospitals are unusual in that an extraordinary amount of information is communicated orally—when medical orders are issued, when staff and patients converse, and when patients and visitors interact. In spite of this reliance on oral communication, little work has focused on the quality of these speech interactions. Kwon et al. (2007), Godfrey and Feth (2011), and Ryherd et al. (2013) measured the Speech Intelligibility Index (SII) in seven different hospitals from 2007 to 2013. In not one of these hospitals did the SII predict good speech communication. Indeed, the environments typically were found to be marginal or poor for speech communication. Further, Ryherd et al. (2013) found SII to be correlated with nurse perception of communication problems. This has serious repercussions. Patients might not follow medical directions because they didn't hear or comprehend them, a problem exacerbated by patients being less able to focus due to their illness or medications they are taking. Speech communication problems also contributed to medication errors prompting rules to be changed to require written drug orders in hospitals. Characterizing hospital speech communication issues, their impact on patients, staff, and visitors, and a means of improving speech communication are areas ripe for further research work.

### *10.3.5 Alarms*

Hospital soundscapes typically include a large number of alarms. Most of the studies reporting on alarm frequency in hospitals have focused on intensive care units, where alarms sounding 150–500 times per day per patient are the norm (Cvach et al. 2013; Whalen et al. 2014). Even on units with lower acuity, such as medical/surgical units in community hospitals, alarm rates of about 100 per patient per day are common.

Alarm noise poses quite a challenge for hospital soundscapes. Sounding alarms have traditionally been used to indicate an urgent situation that staff must address, and the sounds are specifically intended to grab attention. Further, for a variety of legal and medical reasons, there is a desire to err on the side of false positive alarms rather than false negative alarms. This has led to a situation in which over 90% of clinical alarms in hospitals result in no action being taken (Cvach 2012).

In spite of this enormous rate of excess alarms, there are consistent problems in hospitals with alarm failures resulting in deaths and loss of function of patients. The Emergency Care Research Institute (ECRI) listed clinical alarms as the top medical technology hazard in 2013 and 2014, and number 2 in 2015 (ECRI Institute 2013). During those years, alarm failures accounted for about 200 deaths in the United States annually. An analysis of the literature and databases on alarm errors by Busch-Vishniac (2015) estimated that about 3% of the time, alarms that sound are not responded to in a timely fashion, and another 9% of the time, alarms that should sound according to current standards do not.

The impact of alarms in the hospital soundscape is very different for staff and for patients and their visitors. Staff are expected to respond to alarms but they sound so frequently that they tend to produce a response referred to as "alarm fatigue" in which caregivers become desensitized to alarms, leading to sometimes missing critical clinical alarms. In one study more than half (56%) of nurses admitted they sometimes tune out alarms (Okcu et al. 2012) and in another study almost half (49%) revealed they sometimes adjust the alarm levels so that they would not hear them (Ryherd et al. 2008). For staff, then, alarms contribute to a very stressful environment.

For patients and their visitors, alarms sounding at or near the bedside can produce anxiety and disrupt conversation and sleep. Alarms are routinely listed by patients as one of the most disturbing noise sources.

The hospital soundscape could be improved by reducing the number of clinical alarms and changing the sound they produce. There are a number of issues with the use of alarms that merit further study. For instance, currently alarms sound not because of a medical diagnosis (as in cardiac arrest) but rather because a physiological measure, such as oxygen saturation level, exceeds or goes below a threshold. Further, we have not made clinical alarms particularly smart. For instance, a study of patients by Gorges et al. (2009) indicated that incorporating a 14 s delay before alarming would eliminate most of the alarms. We have virtually no information available at this point on whether the presence of alarms sounding at a patient bedside has an impact on the medical outcomes for that patient—yet hospital staff would agree that there is no medical reason for patients to hear the alarms going off.

There are a host of questions that should be studied related to the alarms themselves. Current alarms contain little information: they can be hard to localize because they tend to be pure tones and they aren't standardized so the alarm source can't be easily recognized. An early study of alarms by Lawless (1994) that was equivalent to a "name that tune" test found that hospital staff were largely unable to identify the equipment alarming based on the sound. Since that time, the number of alarms has exploded making the problem worse. Even more basic is the question of when alarms should sound rather than providing an electronic notice or visual alarm, and what sounds might be used to provide information beyond merely a location. Work on the latter question is considered by Edworthy and her collaborators (e.g., Edworthy and Hellier 2006).

## 10.4 The Impact of Hospital Soundscapes on Staff and Patients

Most of the work on noise in hospitals has sought to characterize the sound environment. Because of the difficulty in conducting studies with human subjects, especially a vulnerable group in hospital, there are a limited number of rigorous studies of the impacts of the soundscape on staff and patients. The studies that exist tend to examine how a specific physiological or psychological characteristic or behavior correlates with a measure of the overall loudness of the environment. It is only starting in the 2000s that the methods of analysis developed to study the complex sound/reaction interactions of humans have been applied to hospital soundscapes.

In what follows, we describe the early work to identify the impacts of hospital soundscapes on staff and patients, and then present the work using soundscape analytical approaches. We begin with staff studies and then move to patient studies, with special note on the impact of the soundscape on neonates.

### *10.4.1 The Impact of the Hospital Soundscapes on Staff*

The literature examining the impacts of hospital noise on staff outcomes is relatively sparse compared to the number of studies done with hospital patients and there are some conflicting findings. Overall research generally points to the importance of the sound environment on staff stress, job performance, and occupational health (Ryherd et al. 2012). Examples of the potential impacts of hospital soundscapes on staff outcomes are depicted in Fig. 10.6.

### *10.4.2 Staff Stress and Auditory Monitoring*

The most studied group of staff in hospitals is nurses, who are not only a large fraction of the staff but also the people who work in closest contact with patients. Nursing is a difficult occupation and stress and burnout have been identified as significant job issues, leading to higher rates of substance abuse, depression, suicide, and reduced satisfaction with care among patients/clients (Dyrbye et al. 2017). In addition to having a negative impact on health and wellbeing, stress leads to a high turnover rate for staff in hospitals, especially registered nurses, which negatively affects the quality of patient care and the operating costs of hospitals.

Because stress has been identified as a top concern for hospital staff, most studies of noise impacts in hospitals have focused on the noise/stress nexus. Two of the earliest studies relating noise to stress in nurses were reported by Topf (1988) and Topf and Dillon (1988) who studied 100 critical care nurses and showed that

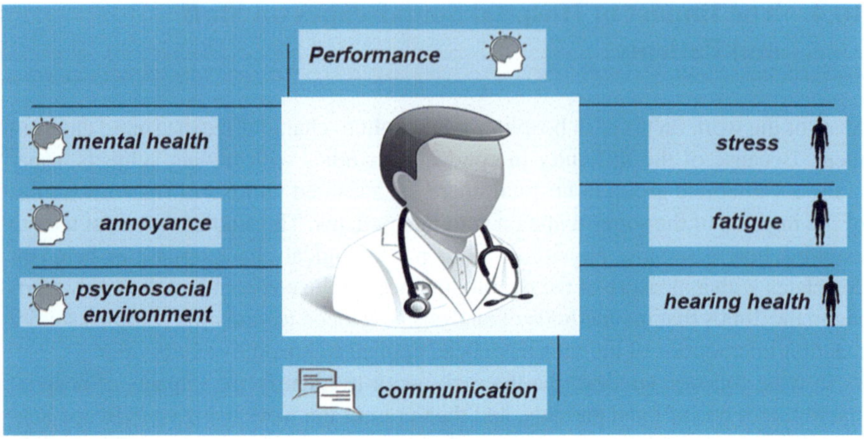


**Fig. 10.6** Potential impacts of hospital soundscapes on staff outcomes. (Ryherd et al. 2012)

self-reported sensitivity to noise on the unit correlated with self-reports of stress-related health issues such as headaches on the job. Morrison et al. (2003) later conducted a more nuanced study in which nurses in a pediatric intensive care unit were individually followed for a period of 3 h, during which they were surveyed about their level of stress and annoyance while their heart rate, cortisol levels in saliva (a known stress indicator) and sound level exposure were monitored. They found a correlation between self-assessed annoyance and sound level, between self-reported stress and sound level, and between cortisol levels in saliva and sound level.

Ryherd et al. (2008) surveyed 47 nurses in a neurological intensive care unit and found that 91% reported that noise negatively affects them in their daily work environment. Many of those surveyed reported symptoms of noise-induced stress including irritation (66%), fatigue (66%), concentration problems (43%), and tension headaches (40%). Mahapatra (2011) surveyed 65 staff members in two Emergency Departments and found that 96% of physicians, 89% of nurses, and 91% of other staff (e.g., nurse practitioners, emergency medical technicians, and patient relations staff) felt that their workplaces were "somewhat" to "extremely" noisy. Another study by Applebaum and Fowler (2010) examined the impact of odor, noise, light, and color on nursing stress by surveying nurses in medical and surgical suites in a 500-bed level I trauma center. They found that among the characteristics considered, only noise was significantly related to perceived stress. Further, this study reported that perceived stress was significantly related to job satisfaction and turnover intention, thus indirectly linking noise to job satisfaction and nursing turnover.

From these studies, we conclude there is a correlation between noise and stress in nursing staff in hospitals. Less is known at this point about the contribution to stress from specific sources of sound and the extent to which a reduction in stress might be obtained through soundscape interventions. Without this information, it is difficult to determine soundscape interventions that might improve the working environment for hospital staff.

The Mahapatra (2011) Emergency Department study provides some insight, as subjects were asked to evaluate whether various noise sources disturbed their concentration. Items that most often were reported as "moderately" to "extremely" disturbing are shown in Fig. 10.7. The majority of disturbing sounds were mechanical or human generated. Subjects reported visitor conversation, patient sounds, emergency procedures, operational sounds of medical equipment, building and service sounds, and exterior sounds to be "not at all" to "somewhat" disturbing.

Simply eliminating noise is not an option for hospital staff. For nurses, auditory monitoring of patients is a key part of their job. A conceptual overview of the components of auditory monitoring is shown in Fig. 10.8. We know that staff rely on auditory cues and they must be able to hear calls for help, listen to body sounds and discriminate normal from abnormal, hear sounds indicating threats to patient safety (as in slips and falls), and notice and respond to clinical alarms (Okcu et al. 2008). Further, nurses report that effective auditory monitoring requires recognition, localization, and immediate reaction to these auditory cues. Thus, hospital soundscapes are a good example of a situation where the required solutions are much more complex than simply elimination of noise.

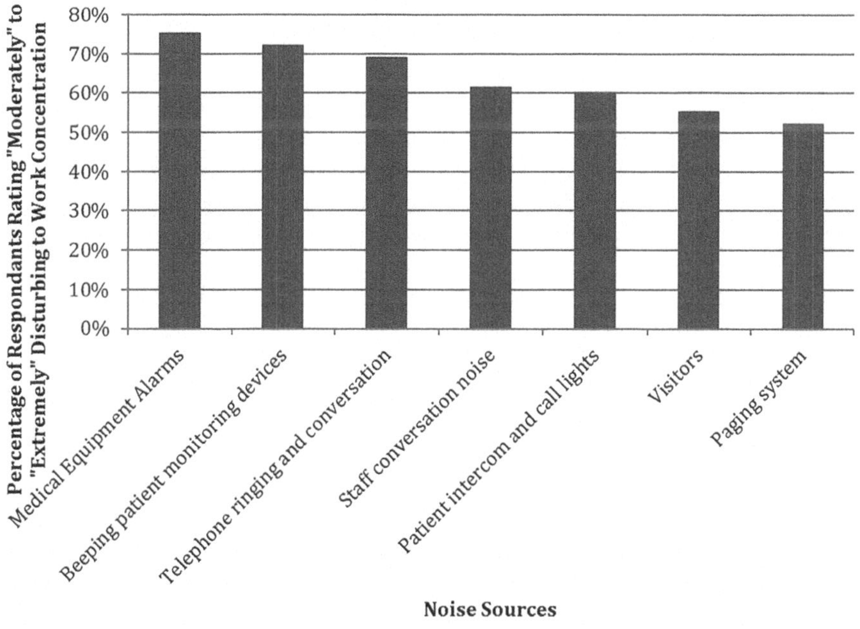


**Fig. 10.7** Perceived work concentration disturbance due to various noise sources in the Emergency Department. (Mahapatra 2011)

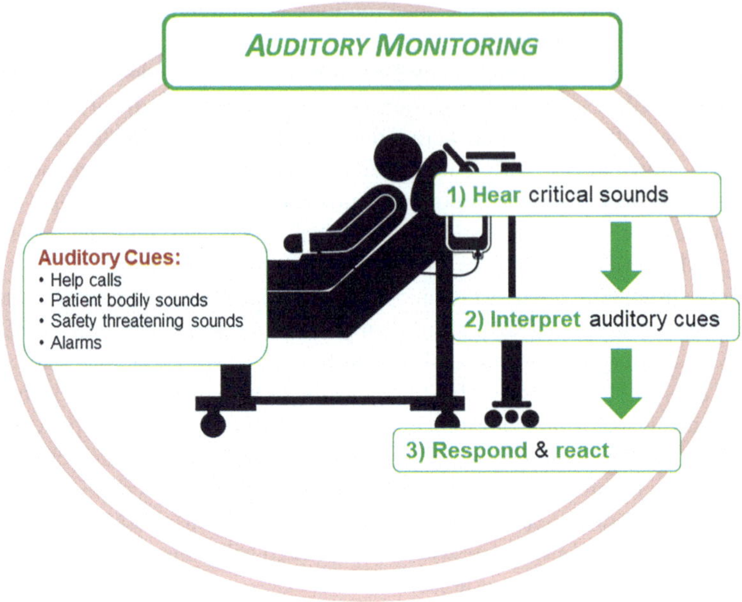


**Fig. 10.8** Components of caregiver auditory monitoring (Okcu et al. 2008)

### *10.4.3 Hearing Loss and Staff Performance*

There are two additional impacts of the soundscape worth mentioning: noise-induced hearing loss and impact on performance. Hospitals generally aren't sufficiently loud that there is concern about noise-induced hearing loss. However, operating rooms can be hearing hazards. Operating rooms have very high air flow rates because surgical site infections decrease with the number of room air changes per hour. They are also equipment dense, with each device capable of producing alarms or making other noises. Kracht et al. (2007) looked at the sound levels of typical surgeries at Johns Hopkins Hospital (Baltimore, MD) and categorized them by the type of surgery involved (e.g., cardiology, neurology, etc.). While none of the surgeries produced $L_{eq}(A)$ values which would cause hearing loss concern, many of them showed the presence of high peak sounds, as characterized by an occurrence rate-type measure shown in Fig. 10.9. Neurosurgery and orthopedic operations were found to have peak levels over 100 dBC more than 40% of the time with peaks occasionally exceeding 120 dBC. A handful of studies specifically examined hearing health among orthopedic surgeons and staff (Willett 1991; Holmes Jr. et al. 1996). Though results were mixed, findings point to potential risks for occupational noise-induced hearing loss among this population due to the high levels produced by orthopedic instruments.

Operating rooms often include music at the request of the surgical team. According to Spotify, 90% of doctors listen to music in the operating room

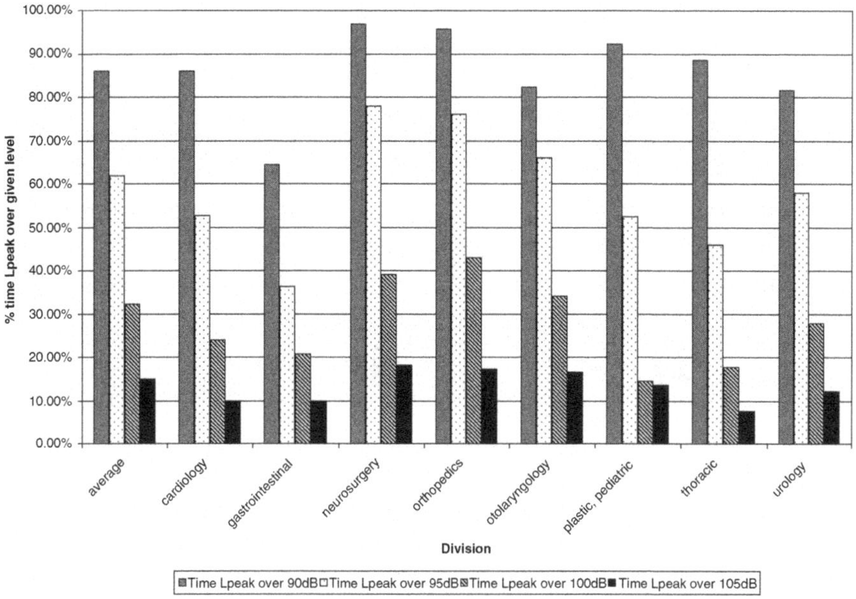


**Fig. 10.9** Fraction of time $L_{\text{peak}}$ exceeds 90, 95, 100, and 105 dB (unweighted) by category of surgery. (Reproduced with permission (Kracht et al. 2007))

(Ahmed 2019), although a review by Vahed and Kabiri (2016) cited a rate of 62–72%. There are articles that suggest that music can relax surgeons and improve their performance, though the research findings in this area are sparse and conflicting (Moorthy et al. 2004; Zun and Downey 2005).

There have been concerns that the hospital soundscape with its intensity, its many sound sources, its alarms, and its dynamic nature could have a negative impact on task performance. However, the research literature is not clear, with some studies finding no significant difference in performance between quiet and noisy conditions (Hawksworth et al. 1998; Moorthy et al. 2004), while others have found the impact of noise on performance depends on individual preference for quiet or noise and that the impact mostly seems to affect short term memory and mental efficiency (Park and Song 1994; Murthy et al. 1995). There is certainly room for further investigation of task performance in hospitals and how it relates to the soundscape.

### *10.4.4 The Impact of the Hospital Soundscapes on Patients*

The impact of hospital soundscapes on patients is quite different from staff because patients in hospitals are present round-the-clock and rely on the hospital to provide all of their required services. Additionally, patients are a vulnerable population, often anxious about their condition and trying to recover. Further, while staff

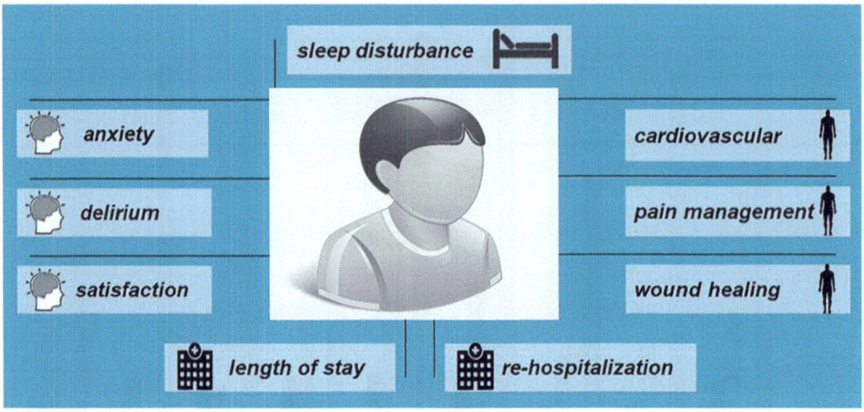


**Fig. 10.10** Potential impacts of hospital soundscapes on patient outcomes. (Hsu et al. 2012)

members have modest control over the noise produced in a unit, patients have almost none, with the exception of conversations with visitors and choosing whether to watch TV. This lack of control tends to negatively affect the patient's experience with the hospital environment.

A host of potential reactions have been investigated over the years, including sleep disturbance, physiological responses (e.g., cardiovascular response, hospital stay, pain management, wound healing, other physiological reactions), and psychological reactions (e.g., general perception, delirium, satisfaction) (Hsu et al. 2012). Example potential impacts of hospital soundscapes on patient outcomes are depicted in Fig. 10.10. Results generally show that hospital soundscapes impact patients.

#### 10.4.4.1 Sleep

There is a significant body of literature on the impact of noise in hospitals on patients, much of which focuses on sleep. Disrupted sleep is known to relate to changes in blood pressure, weight gain, heart disease, pain, stress, and inflammation. Therefore, a key issue is whether hospital soundscapes promote or inhibit patient sleep.

A study by Gabor et al. (2003) set the stage for what we understand today about the relationship between noise and sleep for patients. Prior to this study, most of the research focused on indirect correlations of sleep and noise. In this study, healthy subjects and subjects on ventilators were monitored for noise arousals and awakenings. Results showed some commonality between healthy and ventilated patients and some differences. For healthy subjects, the majority of arousals were caused by sound peaks. For ventilated patients, the *minority* of arousals (about 20%) were caused by sound. For both healthy subjects and patients, alarm sounds were less disruptive to sleep than conversation or staff activities. Overall, the majority of awakenings were of unknown cause. When asked the next day about the noise sources causing arousals, subject perceptions did not generally match reality well.

Another study of the noise-sleep link for hospital sounds that is widely cited is the Sound Sleep Study (Buxton et al. 2012). The aim of this study was to determine the influence of typical hospital noises on various sleep stages as measured under controlled conditions in a sleep lab. Buxton et al. (2012) found that as sounds got louder they were more likely to cause arousal. They also found that heart rate increases correlated with arousals, particularly during REM sleep. Among the sounds, alarms and ringing phones were most likely to cause arousal, then conversations and overhead paging. Once again, self-reports of noise sources were not well correlated with actual noise sources to which they were exposed.

These two studies provide information about what hospital sounds disrupt sleep. A number of other studies confirm the potential impacts of the hospital soundscape on patient sleep, through direct and indirect measures of polysomnography, electroencephalography (EEG), structured questionnaires, and interviews (Hsu et al. 2012). For example, Persson Waye et al. (2013) found that sleep was more fragmented with less slow-wave sleep, more arousals, and more time awake among subjects exposed to typical ICU noise as compared to a quieter, reference night. Berg (2001) linked the addition of sound-absorbing ceiling tiles to a significant reduction in EEG arousals for laboratory subjects exposed to a variety of specific noise sources. What these and other sleep studies do not do is provide information about whether the type of sleep that patients achieve in the hospital is prolonging hospital stays, delaying medical improvements, or preventing recovery. While a study of medical outcomes linked to sleep quality in hospitals would be extremely difficult to conduct, this is certainly an area that would benefit from the knowledge that would be gained in such studies.

The HCAHPS question about sound, asking patients to rate whether the area around their room is always, usually, sometimes, or never quiet at night, is aimed at determining whether patients believe their sleep is disturbed by noise. The results of HCAHPS surveys are available online and Locke and Pope (2017) compared the responses to the noise question in 2010 and 2014. They found in 2010 that 70% of patients said their room was always quiet at night, 25% said it was usually quiet at night, and 5% said it was sometimes or never quiet at night. The 2014 data shows a drop in the fraction of patients saying their room is always quiet at night (down to 62%), with an increase to 29% finding their room is usually quiet and an increased 9% saying their room is sometimes or never quiet at night. This result shows patient perceptions moving in an undesirable direction.

#### 10.4.4.2 Physiology

In addition to sleep, there have been a few studies of patient physiological measures as a function of the level of noise in hospitals to which they are exposed. For instance, Hagerman et al. (2005) reported on a study of patients and staff in a unit in which they could change the soundscape by changing the material used for the ceiling in the central area and patient rooms. They compared a reflective ceiling to an absorptive ceiling, a change that dropped the reverberation time from 0.8 s to 0.4 s and dropped the background $L_{eq}$ by 5–6 dB. The study found that speech intelligibility improved and staff felt fewer demands and less irritation (Blomkvist et al. 2005). However, there were initially no significant differences observed in patient heart rate, heart rate variability (a stress measure), or blood pressures. In subsequent analysis, acute myocardial infarction and unstable angina pectoris patient groups were found to have significantly lower values of pulse amplitude at night with the absorptive ceiling. There was also a higher rate of rehospitalizations at 3 months for the group of patients exposed to a reflective ceiling and the patients exposed to a sound-absorbing ceiling considered staff attitudes to be better.

Hsu et al. (2011, 2012) found that sound levels at various thresholds are correlated with increases in patient heart rate, respiration rate, and systolic and diastolic blood pressure, and a decrease in blood oxygen saturation level. While the patient cohort was relatively small, it was possible to evaluate the risk likelihood of these physiologic changes. For instance, this study found that exposures to levels over 50 dBA meant patients had a 22% risk of having a higher heart rate.

Again, what is generally missing from these and other studies of physiological measures of patients is a determination of whether the physiological changes result in a significant change to medical outcomes. Also, since many studies have relied on the $L_{eq}$ to describe the sound environment, it isn't clear whether there are particular sorts of sounds that are more likely than others to affect patient physiological measures. Some insight was provided by Hsu et al. (2012), which linked a variety of psychoacoustic metrics (i.e., loudness, sharpness, fluctuation strength, and roughness) to patient physiology in addition to the more traditional metrics. However, more research on the relationships between detailed characteristics of soundscapes and patient physiology across a broad variety of patient populations is needed.

There is also a significant body of literature on the impact of sound on neonates in hospitals. Wachman and Lahav (2011) presented a review of the literature in this area, addressing how neonatal intensive care unit soundscapes impact the cardiovascular, respiratory, auditory, and nervous systems of preterm neonates. These authors note that while the survival rate of very low birth-weight neonates has dramatically improved, as these children have reached school age, they seem to be displaying a high incidence of neurodevelopmental problems. There is a concern that the NICU environment might be responsible, in part, for these problems—an issue still unresolved. Wachman and Lahav (2011) show that noise in the NICU can increase neonatal blood pressure and heart rate, depress respiration rate, reduce sleep time, and make babies fussier.

#### 10.4.4.3 Psychology

Finally, we note that there have been some studies of patient psychological responses to the hospital soundscape. These studies have considered the impact of music on patients, and the perception of wellbeing as it relates to noise.

There are a number of studies on the impact of music on patients, their visitors, and staff. For instance, Perez-Cruz and Nguyen (2012) exposed patients, caregivers, and healthcare providers to background music and surveyed their reaction. Overwhelmingly, all groups preferred having the music present with no significant difference between the groups. McClurkin and Smith (2016) studied the impact of music on preoperative patients to understand whether music can reduce the need for anti-anxiety medications prior to surgery. They determined that listening to as little as 15 min of music prior to surgery was sufficient to reduce anxiety. Iyendo (2016) presented a very complete review of the work on music and its healing properties particularly related to hospital environments. In addition to research in this area, there are companies that have produced soothing sound products for hospitals and clinics to use.

A study by Johansson et al. (2012) examined the link between patient perception of sounds, ICU delirium, and noise. While patients generally prefer a quieter environment, if rooms were too quiet it could create feelings of being abandoned. Positive sounds such as quietly working staff created feelings of safety, security, and familiarity. Conversely, negative sounds such as sick patients or medical equipment created feelings of fear, helplessness, and anxiety. This work points to the importance of considering holistic soundscapes that reduce negative noises while promoting positive sounds.

Finally, Cunha and Silva (2015) studied the relationship between the hospital soundscape and a patient's perception of wellbeing. They had subjects from three units in a hospital (post-anesthesia care, coronary intensive care, and intermediate surgical care) take two surveys: the Environmental Comfort Questionnaire (EMQ) to assess noise perception and the Positive and Negative Affect Schedule (PANAS) to measure emotion. They compared the results of these surveys with sound levels measured and found statistically significant correlations between wellbeing and noise levels, with higher noise levels leading to lowered sense of wellbeing. This is important because a patient's sense of wellbeing tends to be a good indicator of health-related benefits they are enjoying. Another study by Bliefnick (2018) utilized PANAS in hospital occurrence rate listening tests. Positive mood was found to significantly decrease after 30 subjects listened to simulated hospital soundscapes for 30 min. Though follow-up studies are needed, these results might indicate that simply being immersed in hospital soundscapes may negatively impact mood.

Overall, we know a fair amount about the correlation between noise and patient physiological and psychological reaction, but we have merely scratched the surface of what we could know. In the next section, we will discuss studies that specifically use soundscape analytical approaches. These studies are producing results that have already started to guide hospital interventions to produce soundscapes more conducive to healing.

### 10.4.5 Studies of the Hospital Soundscape Using Soundscape/ Analytical Approaches

One of the earlier efforts to use analytical techniques developed for psychology and now applied to soundscape analysis was reported by Mourshed and Zhao (2012). They developed a list of 16 design factors in hospitals that had been previously shown to be important for workers in healthcare facilities, both from a perspective of satisfaction with the facility and delivery of safe, high-quality care for patients. They visited hospitals in China to determine what options were open to them for architectural changes and administered a questionnaire to doctors, nurses, technicians, and administrative staff that focused on topics they could change. They analyzed their results using principal component analysis and followed up with selective interviews to confirm results.

The results obtained in this study showed three significant dimensions in hospital designs, which they labeled as spatial, environmental, and maintenance. Overall, they found that cleanliness was the top concern of hospital staff, followed by air quality, then noise, then thermal comfort. While the goal of this work was not aimed at understanding the hospital soundscape in detail, but rather to understand architectural design options in hospitals, this work makes it clear that noise (and thus the soundscape) in hospitals is one of the top concerns of staff as well as patients. As we will discuss below, changing hospital layouts and designs is one significant means of altering the hospital soundscapes.

An impressive body of work on hospital soundscapes using sophisticated analytical approaches was conducted by Mackrill et al. (2013a, b, 2014). This study of a cardiothoracic ward had multiple parts and was aimed to identify the positive and negative aspects of the hospital soundscape as described by nurses and patients. The idea was that interventions would then be based upon the results of the study, preserving positive aspects of the soundscape while mitigating or eliminating negative aspects. In the first part of the work, Mackrill et al. (2013a) used semi-structured interviews on topics covering the hospital general environment, sound as part of that environment, and future designs. Results of this study identified sound sources most often mentioned and whether the sounds were viewed positively or negatively.

The next part of this study used recordings of the sounds most likely to be mentioned by the staff and patients in a laboratory listening study (Mackrill et al. 2013b). Subjects were asked to listen to each sound sample and to describe how it made them feel. Then a principle component analysis was used to determine significant dimensions for assessment of sound sources in their ward. They found two perceptual dimensions, the first of which they labeled as relaxation and the second as interest and understanding. The relaxation dimension described 56.8% of the variance seen in their results and the interest and understanding dimension 13.2%. These results are not surprising. Overall, staff and patients seek a soundscape that is relaxing. Further, they are more willing to forgive sound intrusions if they understand why they exist and view them as necessary.

The last part of the work reported to date by Mackrill et al. (2014) involves the potential to improve the soundscape by introducing masking sounds or sounds of nature. That work will be discussed in the following section of this chapter.

Subsequent to the publication of the work by Mackrill, there have been other studies using similar approaches. The study by Azzahra et al. (2017), for instance, asked nurses in an intensive care unit to rate the soundscape on a variety of pre-established scales such as pleasant to unpleasant, and anxious to calm. Results were analyzed using principal component analysis and three significant dimensions identified. The first dimension was labeled information, accounted for 31% of the variance, and related to the scales uninformative/informative, unclear/clear, and complex/simple. The second dimension was labeled calmness, accounted for 31% of the variance, and related to the scales pleasant/unpleasant, anxious/calm, and uncomfortable/comfortable. The final significant dimension found was labeled dynamics, accounted for 23% of the variance, and related to the scales of loud/quiet, soft/hard, and flat/sharp.

The results obtained by Azzahra et al. (2017) agree well with those found by Mackrill et al. (2013b) in that they both identify information and calmness (or relaxation) as important dimensions. The information dimension isn't typically found in urban soundscape analysis. Azzahra et al. (2017) hypothesize that this demonstrates that information content is critical in hospital environments for patients and staff.

Work by Hasegawa and Ryherd (2019) utilized sophisticated statistical approaches applied to both occupant response surveys and acoustic measurements in hospital settings. Principle component analysis utilized in staff perception of specific noise sources revealed three inherent categories for noise source annoyance (facility noise, human/speech activity noise, and alarm noise) that were also grouped by frequency content (broadband, speech band, and narrow band, respectively). Statistical clustering analyses allowed for measured background noise to be post-processed and classed into active (louder) and quieter periods. This approach may provide better insight into the distributions of typical "occupied" and "unoccupied" noise levels experienced in units.

Sophisticated statistical approaches to the hospital soundscape are providing a nuanced insight into the soundscape in hospitals. These approaches are informing us about the links between sounds in hospitals and perceptions of staff and patients, identifying which sounds are most concerning, and suggesting means of mitigating negative aspects of the soundscape. There remains much research to be done using these techniques, for instance introducing interventions and testing the impact to see whether the results support the original soundscape analysis.

## 10.5 Interventions

One measure of how successfully we have come to understand hospital soundscapes is how well we have produced interventions that improve them. By this measure, we have had only modest success. In this section, we discuss hospital soundscape interventions from a research perspective. Our aim is to connect interventions to soundscape research (i.e., we focus on evidence-based interventions). Further, we aim to establish a framework for consideration of interventions and identification of potentially fruitful avenues for further work.

Hospital soundscape interventions are much like noise control work in nearly any venue in that they can be categorized in terms of the classical source-path-receiver model. In this model, noise control can be accomplished by changing sound sources in some way to mitigate their impact, by impeding the path the sound follows from sound sources to an observer, or by protecting the person observing the sound (the receiver). Among these approaches, noise control at the source is normally viewed as most effective and efficient, although it is often impossible to take this approach to intervention. Noise control along-the-path from the source to the receiver is a very common approach, often involving the use of sound barriers or acoustical absorption. Noise control at the receiver is normally reserved for situations that don't yield to other approaches as it requires equipment for each individual.

Hospital soundscape interventions at the source include decreasing alarm numbers, lowering voices, and the implementation of quiet times. Along-the-path approaches include addition of sound absorption, closing doors, and adjusting architectural layouts. Noise control approaches at the receiver include adding masking or natural sounds locally and using earplugs, earphones, or headphones. Each of these is discussed below.

### *10.5.1 Source Interventions*

It is no surprise interventions to change hospital soundscapes at sound sources have focused on alarms and conversations as these are routinely cited as some of the most disrupting sounds in hospitals. One approach that has succeeded to a modest extent is to reduce the number of alarms sounding or, at a minimum, to reduce their impact at the patient bedside. For instance, Cvach et al. (2013) have discussed how to reduce the number of nuisance alarms without compromising safety standards and have successfully done so on a number of units of Johns Hopkins Hospital. However, even with reductions, alarms in intensive care units sound often enough to remain terribly bothersome. Additional reductions in alarm numbers in the near future are difficult to imagine because of the potential medical and legal repercussions of an alarm not sounding in an urgent situation.

There are actions that could be taken that preserve patient safety but change the soundscape by mitigating alarms as a sound source. For example, many hospitals now collect alarms at monitors at the nursing station in a unit. Alarms show on the monitor (visual alarms) as well as sounding there. Alarms at bedside, then, are largely redundant and continue to exist to ensure that a nurse not at the nursing station is aware of an alarm. A solution to that problem might be to refer alarms to a device carried by the nurse assigned to a patient—a tablet computer, a phone, or equivalent. Even if these devices are to still produce sounding alarms, they can be set to insonify a much smaller number of people than currently exposed to alarms, likely improving the hospital soundscape for everyone. Another option is to use vibrating alarms. One pilot study found better identification rates using vibro-tactile alarms compared to auditory or combination auditory/vibro-tactile system (Ng et al. 2005). A final option is to change the way alarms sound. One study by Stanford et al. (1985) engineered alarms to mimic human vowel sounds. They found that the new alarms could be detected with at least 93% accuracy, even in the presence of masking noise. Although some improved alarm technologies exist and have been incorporated in newer facilities, progress is slow to implement these on a wide scale even though nurses seem open to the idea. For example, Ryherd et al. (2008) found that 62% of ICU nurses surveyed felt audible alarms could be replaced with visual alarms, though only 26% thought vibrating alarms were a feasible option. Interestingly, although many of the nurses were willing to change alarm systems, more than half (55%) did not think their managers were open to changing the alarm environment.

In addition to alarms, conversations are clearly seen as a major negative aspect of the hospital soundscape. There have been two approaches to mitigating conversational noise: campaigns to produce lowered voices, and the creation of designated quiet times. Campaigns to produce lowered voices are common and largely not terribly useful. Much of the conversational noise during the day is from visitors and patients themselves and requests for quiet don't tend to work on this cohort. Further, the turnover in hospital staff on wards is sufficiently high that quieting by changing behavior requires constant reinforcement. Some interventions have gone so far as to install devices that provide a visual indication of sound getting loud, but unfortunately staff tend to habituate to these visual alarms just as they do to the audio alarms. The bottom line is that asking people to change their behavior by talking more softly rarely works long term.

Contrary to lowered voice campaigns, the implementation of quiet times in hospitals has been shown to be effective. Quiet times are designated blocks of time (often two consecutive hours each day) during which operations are intentionally set up to produce a quieter environment. Typically, lights are dimmed, doors are closed, and fewer procedures are scheduled. Detailed protocols can be developed that incorporate behavioral, environmental, and scheduling components as shown in Fig. 10.11.

Both staff and patients appreciate these times of rest. Weber showed, for instance, that over 90% of nurses felt quiet time was useful to them, their patients, and the families of their patients, with some additional positive benefits to infant physiology (Weber et al. 2016; Weber 2018). Similarly, Adatia et al. (2014) showed that quiet times were useful to new mothers.

The approach of implementing quiet times suggests another way in which hospital soundscapes could be positively changed. The current method of operation in hospitals is staff centric. Procedures and various checks on patients are made on a schedule that works best for each staff member. Thus, a patient might be awakened to have his blood pressure and temperature taken and fall back asleep only to have someone come in shortly afterward to change the fluids being delivered

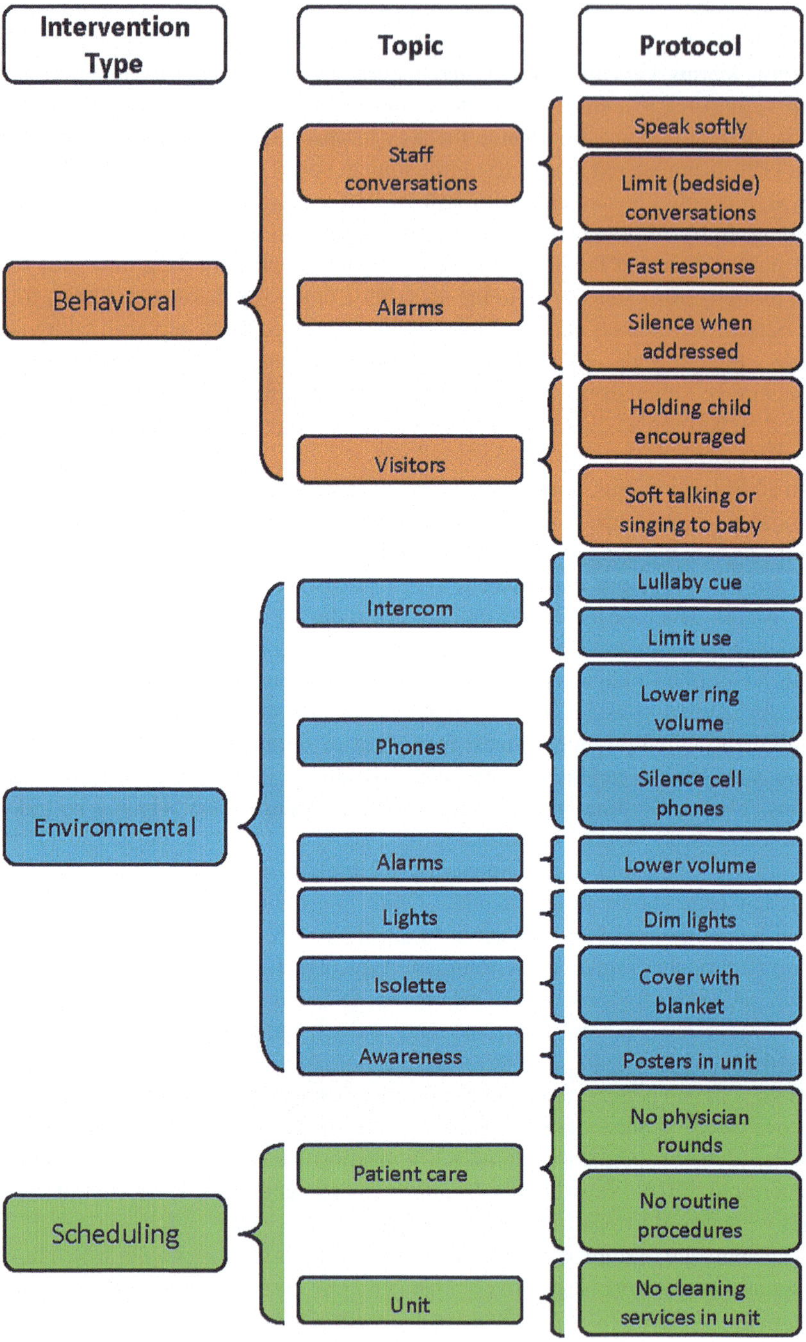


**Fig. 10.11** Example quiet time protocol components for a NICU. (Weber 2018)

intravenously, and fall back to sleep again to have someone come into the room to remove trash. A patient-centric operating schedule would cluster procedures that require entering a room in order to minimize the number of disruptions to a patient and the period of noise exposure to those in his vicinity. However, this would require a level of coordination of staff duties that is not the norm in hospitals and there is fear that as some staff members might need to wait to run procedures on a patient, this mode of operation might require more staff in hospitals.

### 10.5.2 Path Interventions

One of the most common means of accomplishing along-the-path noise control in buildings is to add acoustical absorption to surfaces. Hospitals typically have hard surfaces due to the need for their easy and regular cleaning, and such surfaces do not tend to exhibit much sound absorption. In typical office spaces, acoustical ceiling tiles are used to introduce significant sound absorption, but most of these materials aren't easily cleaned and thus they were historically used sparingly if at all in hospitals. A few of lines of acoustical materials have now been created by major manufacturers with hospitals and clinics in mind. There are also research examples of what can be done by introducing absorption into hospital spaces. MacLeod et al. (2007) quieted a unit in Johns Hopkins Hospital by introducing acoustical absorption covered with hydrophobic (and thus anti-bacterial) materials. Follow-up studies by Barnhill et al. (2010) and Hsu et al. (2010) treated cancer units at Johns Hopkins Hospital by adding absorbing materials on walls and ceilings of corridors. This improved speech intelligibility, lowered overall sound levels, and improved the staff ability to communicate and concentrate.

Private patient rooms are the norm for new hospitals, but large rooms with patient pods separated by curtains remain in many existing hospitals. The curtains are usually there for purely visual reasons—to separate one patient from another. However, thanks to new products on the market it is possible to replace thin curtains with curtains that include sound-absorbing materials in pockets sewn into the curtains. This has the impact of introducing sound absorption to the room and can dramatically reduce sound transmission from one cubical to another, even with significant gaps in the curtain at the floor and ceiling. Diminishing sound transmission has the added impact of offering greater speech privacy. Pope and Miller-Klein (2016) and Locke and Pope (2017) reported on a study in which thin curtains were replaced with a sound absorbing yet cleanable curtain and found improvements in overall sound level and speech privacy. However, Locke and Pope (2017) noted that the new curtain took as much as twice as much time to hang and longer to dry after being cleaned compared to the thin traditional curtains, so it was not immediately adopted. It is this sort of tradeoff that is a constant issue in the development of materials for hospitals. If speech privacy and improved soundscapes are to be sought, there will necessarily be compromises such as increased time to hang curtains and added costs.

Another along-the-path intervention in hospital soundscapes is the simple action of routinely closing doors. Almost all hospital rooms have doors on them—the notable exceptions being intensive care unit rooms and NICUs, which sometimes use pods or gang rooms. Just as in homes and offices, closing doors affords significant transmission loss from the room to the corridor and vice versa. In practice, busy hospitals often leave doors ajar to facilitate quicker entry and exit from the room, and to make it easier to hear patients and assure their safety. Additionally, a study by Sobieraj et al. (2006) on the impact of closed doors showed that nurses on the unit had a more difficult time hearing and localizing alarms—a potential safety issue. Closing doors, while effective, is another mitigation measure that requires a change in behavior, and thus it is unlikely to happen quickly if at all. However, in a study by Kaur et al. (2016) of intervention strategies on a pediatric intensive care unit, closing patient doors ranked at the top in effectiveness as rated by staff and patient families, with 93% of respondents saying it worked to improve the environment. Asking staff to lower voices ranked second at 88%, followed by quiet times at 82% and then reducing the number of alarms at 80%.

A third along-the-path intervention is to design hospitals while recognizing the inherent link between architectural layouts and acoustical performance. An extensive series of studies by Okcu et al. (2011) and Okcu et al. (2013) statistically investigated the links between hospital corridor layout, acoustics, and occupant response. Floor plate design features such as corridor length, number of turns and branching hallways, relative grid distance, and visual fragmentation were significantly related to reverberation time in real and simulated settings. To provide a less reverberant environment—which may in turn improve the ability of nurses to localize auditory cues—designers might consider more compact and fragmented floor plate shapes.

Finally, sound isolation properties of building partitions, floor-to-ceiling assemblies, and exterior envelope must all be considered in noise control along-the-path, though there is very little research published in this area for hospitals. One study by Pelton and Ryherd (2009) examined the acoustical remodel of a burn acute care unit (BACU), with a focus on debridement treatment areas where patients undergo the removal of dead tissue. Curtains separated the debridement stations and isolation to the rest of the unit was inadequate, resulting in patient distress sounds being heard throughout the unit. The acoustic remodel included creating sound locks, incorporating high-isolation doors and partitions, and addition of acoustic absorption. As a result, $L_1$ values (i.e., those exceeded 1% of the time) for patient distress sounds were reduced by 30 dBA and the overall soundscape was markedly improved.

### 10.5.3 Receiver Interventions

Work on hospital soundscape improvements has focused a great deal of attention on solutions at the receiver. These include adding sound locally (masking or natural sounds) and use of earplugs, earphones, or headphones.

A significant amount of work has been done on the impact of views of nature on hospital patients (see, for instance, the seminal paper by Ulrich 1984). Generally, these studies show that nature views have a strong positive influence on patients, enhancing recovery, reducing the need for pain medications, and improving moods. Based on these studies, work has also been done to examine the impact of sounds of nature on patients. A study by Annerstedt et al. (2013) found that sounds of nature reduce cardio stress markers and cortisol levels after a stressing event. A later study by Largo-Wight et al. (2016) considered the impact of nature sounds (ocean waves), classical music (Mozart), and silence on stress by monitoring muscle tension (EMG), pulse rate, and self-reported stress of subjects who listened to sounds using headphones for 15 min. Baseline measurements were taken and compared to results after the listening period. Results found that only sounds of nature had a significant impact, and these reduced stress measures.

Mackrill et al. (2013b) in their soundscape studies also looked at sounds of nature (song of a blackbird and babbling brook sound) as well as masking noise. They presented sounds with and without nature or masking sounds as part of their extended listening lab study. They found, for added nature sounds, that the ratings of hospital sounds by subjects significantly changed (improved) along the relaxation perceptual dimension. There was no change seen in the interest and understanding dimension. Further, masking noise had a much smaller impact than nature sounds. This work was expanded upon at a workshop in 2017 that compared the impact of three states (masking noise, no additional sound, and natural sounds) on the framework Mackrill et al. (2013b) developed. In this small study, the nature sounds used were falling rain and bird songs. Participants generally preferred the sound of falling rain to the bird songs, with significant individual variation.

While more work is needed on added nature sounds as a means of mitigating irritating sound sources in the hospital soundscape, it is clear that this is a potential means of improving the hospital soundscape that is relatively easy to implement. Prior to work on nature sounds added to hospital sounds, it was widely held that the soundscape in hospitals is sufficiently intense that adding sound to the mix would simply make the sound more irritating rather than less. Research to date has shown this belief to be incorrect, even if there is an irony in improving the soundscape in a loud area by adding more sound. That said, before sound is added to any hospital setting, care must be taken to ensure the existing ambient environment, delivery methods, and patient/staff interfaces are all appropriate. Additional research is warranted on optimum ways to present good sounds while also reducing unwanted sounds.

A second approach to sound control at the receiver is the use of earmuffs, earplugs, earphones, and headphones. Abou Turk et al. (2009) were early to study the impact of protecting the ears of neonates from loud noises. They used earplugs on very low weight newborns and found that this facilitated weight gain. Duran et al. (2012) looked at very low weight neonates, equipping them with earmuffs for 2 days and without for 2 days. They found that neonates with earmuffs slept more. The results on neonates with earmuffs or earplugs suggest another potential means of improving the soundscape for vulnerable individuals. However, there are issues with outfitting neonates with earplugs or earmuffs that must be considered, as their skin can be very fragile and there are concerns that posture and head shape might be affected. Further, one would anticipate that a similar approach for adult patients could improve the hospital soundscape for them as well.

In addition to simply earplugs or earmuffs, there is a growing body of work on the use of noise-canceling devices on patients. For instance, participants in the Hospital Project on Noise Sound and Sleep workshop experimented with sleep-friendly headphones and noise-canceling earphones. They concluded that both offered advantages that could be useful in the hospital environment, although a more systematic study is needed. Schlesinger et al. (2017) also looked at noise-cancellation earphones in the hospital environment. The aim of this work was to create a means of eliminating alarm noise from the soundscape for patients while passing on all other sounds with little to no distortion. Results showed significant improvement in the fraction of word scores correctly identified with the alarm canceling engaged.

These early studies using noise-canceling devices suggest a new avenue of potential improvement of the hospital soundscape for patients but there is much work yet to be done before they will be adopted by hospitals. For instance, what are the relative advantages and disadvantages of the various options: passive earplugs versus active noise-cancellation? Are there side effects to long-term wearing of such devices for patients? What conditions prevent earplugs or noise-canceling earphones or headphones from being worn and are there alternatives that accomplish essentially the same results in other means?

Taken as a whole, interventions to change the hospital soundscape have not yet taken hold on a large scale, although there is reason to be hopeful that current avenues of research might provide solutions in the future. Of particular interest are interventions that will work long term and without requiring behavioral changes. Examples of potential changes to consider are expanded implementation of quiet times, reducing audible alarms by changing the current alarm system fundamentally, developing architectural designs for hospitals that include acoustical considerations, adding sound absorption materials, piping in background sounds of nature, and using earplugs, earmuffs, earphones, or headphones on patients. All of these techniques could benefit from additional investigations.

## 10.6 Summary

The soundscape in hospitals is interesting for many reasons but paramount among them is the likelihood that soundscapes impact patient recovery and staff resilience. Current hospital soundscapes are not viewed positively by patients, their visitors, or staff.

Hospitals have been getting noticeably louder for decades, in spite of a fleet of new hospitals coming online. Key sound sources that influence perceptions in hospitals include alarms and conversations. Although alarm noise is well studied, there has been far less work to understand the extent to which the current hospital soundscape produces an environment in which speech intelligibility is marginal or poor and how to balance caregiver intelligibility with patient privacy.

Traditional acoustic measures of hospital soundscapes don't seem able to predict the impact of interventions—loudness alone does not predict human response in hospitals. Work to define newer measures, such as the occurrence rate, promises some improvement but further research is needed.

There is a significant body of literature that suggests that the hospital soundscape increases the stress felt by staff and impacts the ability of patients to sleep. Work using sophisticated techniques common in soundscape studies has found that key perceptual dimensions of hospital soundscapes are relaxation (calmness) and information.

Intervention strategies for hospital soundscapes can be divided into the typical at the source, along-the-path, and at the receiver categories. Quiet times in hospitals have been found to be effective and there are also case studies indicating the addition of sound-absorbing materials to hospital ceilings and walls can be useful. Work with earplugs, earmuffs, earphones, and headphones to control noise at the receiver is encouraging as is work using positive sounds added to the soundscape. More work is needed to introduce and promote positive sounds while reducing negative sounds.

There are many avenues of research still to be pursued to understand hospital soundscapes. These include investigations of how we might better use audible and nonaudible alarms, studies to determine whether there is a direct link between patient medical outcomes and elements of the hospital soundscape, and demonstration of interventions that can be scaled across a broad range of hospitals.

**Compliance with Ethics Requirements** Ilene Busch-Vishniac declares that she has no conflict of interest.

Erica Ryherd declares that she has no conflict of interest.

## References

Abou Turk CA, Williams A, Lasky RE (2009) A randomized clinical trial evaluating silicone earplugs for very low birth weight newborns in intensive care. J Perinatol 29(5):358–363

Adatia S, Law S, Haggerty J (2014) Room for improvement: noise on a maternity ward. BMC Health Serv Res 14:604–606

Ahmed T (2019) Spotify reveals the music surgeons love listening to in the operating room. Newsweek, 10 January 2019

Annerstedt M, Jönsson P, Wallergård M, Johansson G et al (2013) Inducing physiological stress recovery with sounds of nature in a virtual reality forest – results from a pilot study. Physiol Behav 118:240–250

Applebaum D, Fowler S (2010) The impact of environmental factors on nursing stress, job satisfaction, and turnover intention. J Nurs Adm 40:323–328

Azzahra IRN, Fela R, Sarwono J, Utami SS (2017) Hospital soundscapes: perception analysis of acoustics environment in intensive care unit. Inter-Noise, Hong Kong, 27–30 August 2017

Barnhill C, West J, Hsu T, Ryherd EE (2010) Further studies of hospital noise control at the Johns Hopkins Hospital: Part I. 159th meeting of the Acoustical Society of America, Baltimore, MD, 19–23 April 2010

Berg S (2001) Impact of reduced reverberation time on sound-induced arousals during sleep. Sleep 24(3):289–292

Berglund B., Lindvall T (eds) (1999) Guidelines for community noise. Technical Report, World Health Organization, Geneva

Bliefnick JM (2018) Evaluation of hospital soundscapes to improve patient and staff experience. PhD Dissertation, Architectural Engineering Program, University of Nebraska – Lincoln

Blomkvist V, Eriksen C, Theorell T, Ulrich R et al (2005) Acoustics and psychosocial environment in intensive coronary care. Occup Environ Med 62:1–8

Bureau of Labor Statistics (2018a) U.S. Department of Labor, https://www.bls.gov/emp/tables/employment-by-major-industry-sector.htm. Accessed 14 June 2018

Bureau of Labor Statistics (2018b), U. S. Department of Labor, Occupational Outlook Handbook. Registered Nurses. https://www.bls.gov.ooh/healthcare/registered-nurses.htm. Accessed 30 May 2018

Busch-Vishniac IJ, West JE, Barnhill C, Hunter T et al (2005) Noise levels in Johns Hopkins Hospital. J. Acoust Soc Am 118(6):3629–3645

Busch-Vishniac I (2015) A model of clinical alarm errors in hospital. Biomed Instrum Techn 49(4):280–291

Buxton OM, Ellenbogen JM, Wang W, Carballeira A et al (2012) Sleep disruption due to hospital noises: a prospective evaluation. Ann Intern Med 153:170–179

Cunha M, Silva N (2015) Hospital noise and patients' wellbeing. Procedia Soc Behav Sci 171:146–151

Cvach M (2012) Monitor alarm fatigue: an integrative review. Biomed Instrum Technol 46(4):268–277

Cvach MM, Currie A, Saperstein A, Doyle P et al (2013) Managing clinical alarms: using data to drive change. Nurs Manag 44(11):8–12

Duran R, Ciftdemir NA, Ozbek UV, Berberoğlu U et al (2012) The effects of noise reduction by earmuffs on the physiologic and behavioral responses in very low birth weight preterm infants. Int J Pediatr Otorhinol 76:1490–1493

Dyrbye LN, Shanafelt TD, Sinsky CA, Cipriano CF, et al (2017) Burnout among health care professionals: a call to explore and address this unrecognized threat to safe, high-quality care. NAME Perspectives Discussion Paper, National Academy of Medicine, Washington, DC, 5 July 2017

ECRI Institute (2013) Top ten health hazards. Health Devices 42(11):1–16

Edworthy J, Hellier E (2006) Alarms and human behavior: implications for medical alarms. Brit J Anaesth 97(1):12–17

Eurostat Data (2009). https://bit.ly/3Z8LFTC. Accessed 23 March 2023

Fick DD, Vance GL (2006) Quiet zone: reducing HVAC system noise. Hosp Fac Manag Mag 19(8):21–24

Fife D, Rappaport E (1976) Noise and hospital stay. Am J Public Health 66:680–681

Gabor JY, Cooper AB, Crombach SA, Lee B et al (2003) Contribution of the intensive care unit environment to sleep disruption in mechanically ventillated patients and healthy subjects. Am J Respir Crit Care Med 167:708–715

Godfrey RD, Feth L (2011) Benchmark measurements of noise, reverberation time, and an estimate of speech intelligibility in a representative operating room at Nationwide Children's Hospital in Columbus, OH. J Acoust Soc Am 130(4):2318

Gorges M, Markewitz BA, Westenskow DR (2009) Improving alarm performance in the medical intensive care unit using delays and clinical context. Int Anesthesia Res Soc 108(5):1546–1552

Hagerman I, Rasmanis G, Blomkvist V, Ulrich R et al (2005) Influence of intensive coronary care acoustics on the quality of care and physiological state of patients. Int J Cardiol 98:267–270

Hasegawa Y, Ryherd E (2019) Subjective and objective assessments of pediatric and neonatal hospital soundscapes. 177th meeting of the Acoustical Society of America, Louisville, KY, 13–17 May 2019

Hawksworth C, Sivalingam P, Asbury AJ (1998) The effect of music on anesthetists' psychomotor performance. Anaesthesia 53:195–197

Holmes GB Jr, Goodman KJ, Hang DW, McCorvey VM (1996) Noise levels of orthopedic instruments and their potential health risks. Orthopedics 19(1):35–37

Hsu TY, Ryherd EE, West JE, Barnhill CL (2010) Further studies of hospital noise control at the Johns Hopkins Hospital: Part II. 159th meeting of the Acoustical Society of America, Baltimore, MD, 19–23 April 2010

Hsu TY, Ryherd EE, Ackerman J (2011) Psychoacoustic measures and their relationship to patient physiology in an intensive care unit. 161st meeting of the Acoustical Society of America, Seattle, WA, 23–27 May 2011

Hsu TY (2012) Relating acoustics and human outcome measures in hospitals. PhD Dissertation, George W. Woodruff School of Mechanical Engineering, Georgia Institute of Technology 2012

Hsu T, Ryherd E, Waye KP, Ackerman J (2012) Noise pollution in hospitals: impacts on patients. J Clin Outcomes Manag 19(7):301–309

International Organization for Standardization (2014) ISO 12913-1:2014 acoustics-soundscape-part 1: definition and conceptual framework. Geneva

Iyendo TO (2016) Exploring the effect of sound and music on health in hospital settings: a narrative review. Int J Nurs Studies 63:82–100

Jha AK, Orav EJ, Zheng J, Epstein AM (2008) Patients' perception of hospital care in the United States. N Engl J Med 359(18):1921–1931

Johansson L, Bergbom I, Waye KP, Ryherd E et al (2012) The sound environment in an ICU patient room – a content analysis of sound levels and patient experiences. Int Crit Care Nurs 28(5):269–279

Kaur H, Rohlik GM, Nemergut ME, Tripathi S (2016) Comparison of staff and family perceptions of causes of noise pollution in the Pediatric Intensive Care Unit and suggested intervention strategies. Noise Health 18(81):78–84

Kracht JM, Busch-Vishniac IJ, West JE (2007) Noise in the operating rooms of Johns Hopkins Hospital. J Acoust Soc Am 121(5):2673–2680

Kwon P, Busch-Vishniac I, West J (2007) Assessing the speech intelligibility index in Johns Hopkins Hospital. 153rd Meeting of the Acoustical Society of America, Salt Lake City, UT, 4–8 June 2007

Largo-Wight E, O'Hara BK, Chen WW (2016) The efficacy of a brief nature sound intervention on muscle tension, pulse rate, and self-reported stress: nature contact micro-break in a office or waiting room. HERD J 10(1):45–51

Lawless ST (1994) Crying wolf: false alarms in a pediatric intensive care unit. Crit Care Med 22(6):981–985

Locke C, Pope DS (2017) Assessment of medical-surgical patients' perception of hospital noises and reported ability to rest. Clin Nurs Spec 31(5):261–267

Mackrill JB, Jennings PA, Cain R (2013a) Improving the hospital 'soundscape': a framework to measure individual perceptual response to hospital sounds. Ergonomics 56(11):1687–1697

Mackrill J, Cain R, Jennings P (2013b) Experiencing the hospital ward soundscape: toward a model. J Environ Psych 36:1–8

Mackrill J, Jennings P, Cain R (2014) Exploring positive hospital ward soundscape interventions. Appl Ergon 45:1454–1460

MacLeod M, Dunn J, Busch-Vishniac IJ, West JE et al (2007) Quieting Weinberg 5C: a case study in hospital noise control. J Acoust Soc Am 121:3501–3508

Madaras GS (2017) An overview of the key misperceptions that are preventing any significant improvements when addressing operational noise in acute care hospitals. Paper presented at 174th Meeting of the Acoustical Society of America, New Orleans, LA, 4–8 December 2017

Mahapatra AK (2011) Investigation of noise in hospital emergency departments. MS Thesis, George W. Woodruff School of Mechanical Engineering, Georgia Institute of Technology
McClurkin SL, Smith CD (2016) The duration of self-selected music needed to reduce preoperative anxiety. J Perianesth Nurs 31(3):196–206
Moorthy K, Munz Y, Undre S, Darzi A (2004) Objective evaluation of the effect of noise on the performance of a complex laparoscopic task. Surgery 136:25–30
Morrison W, Haas E, Shaffner DH, Garrett ES et al (2003) Noise stress and annoyance in a pediatric intensive care unit. Crit Care Med 31:113–119
Mourshed M, Zhao Y (2012) Healthcare providers' perception of design factors related to physical environments in hospitals. J Environ Psych 32:362–370
Murthy V, Malhotra S, Bala I, Raghunathan M (1995) Detrimental effects of noise on anesthetists. Can J Anaesth 42:608–611
Ng J, Man J, Fels S, Dumont G et al (2005) An evaluation of vibro-tactile display prototype for physiological monitoring. Anesth Analg 101:1719–1724
Nickl H, Nickl-Weller C (2013) Healing architecture. Braun Publishing, Salenstein
Nightingale F (1969) Notes on nursing. Dover Publishing, New York
OECD Data website. http://www.oecd.org/els/health-systems/health-data.htm. Accessed 10 Jan 2019
Okcu S, Zimring C, Ryherd EE (2008) Developments in 'aural connectivity': enhancing sound environments in critical care settings for effective nurse auditory monitoring. 155th Meeting of the Acoustical Society of America, Paris, 29 June – 4 July 2008
Okcu S (2011) Developing evidence-based design metrics and methods for improving healthcare soundscapes. PhD Dissertation, College of Architecture, Georgia Institute of Technology
Okcu S, Ryherd E, Zimring C, Samuels O (2011) Soundscape evaluations in two critical healthcare settings with different designs. J Acoust Soc Am 130(3):1348–1358
Okcu S, Ryherd E, Pelton H, Zimring C (2012) Characterizing impulsiveness of hospital sound environments. Noise Control Eng J 60(3):246–257
Okcu S, Shpuza E, Ryher E, Zimring C (2013) Linking acoustics and floor-plate shape qualities of healthcare settings. Archit Sci Rev 56(4):315–332
Park S, Song H (1994) Effect of noise on the detection of rib fractures by residents. Investig Radiol 29:54–58
Pelton H, Ryherd MM (2009) Acoustical design of a burn acute care unit for enhanced patient comfort. Noise Control Eng J 57(1):32–41
Perez-Cruz P, Nguyen L (2012) Attitudes and perceptions of patients, caregivers, and health care providers toward background music in patient care areas: an exploratory study. J Palliat Med 15(10):1130–1136
Persson Waye K, Elmenhorst EM, Croy I, Pedersen E (2013) Improvement of intensive care unit sound environment and analyses of consequences on sleep: an experimental study. Sleep Med 14(12):1334–1340
Pope DS, Miller-Klein ET (2016) Acoustic assessment of speech privacy curtains in two nursing units. Noise Health 18(80):26–35
Ryherd E, Persson Waye K et al (2008) Characterizing noise and perceived work environment in a neurological intensive care unit. J Acoust Soc Am 123(2):747–756
Ryherd E, Okcu S, Hsu T (2011) Hospital noise and occupant response, ASHRAE Winter Meeting Conference, Las Vegas, NV, 20 January – 2 February 2011
Ryherd E, Okcu J, Ackerman J, Zimring C (2012) Noise pollution in hospitals: impacts on staff. J Clin Outcomes Manag 19(11):491–500
Ryherd E, Moeller M Jr, Hsu T (2013) Speech intelligibility in hospitals. J Acoust Soc Am 134(1):586–595
Sbihi H, Hodgson M, Astrakianakis G, Ratner P (2011) Measuring how acoustical environments affect staff in health-care facilities? 161st Meeting of the Acoustical Society of America, Seattle, WA, 23–27 May 2011

Schlesinger JJ, Reynolds E, Sweyer B, Pradhan A (2017) Frequency-selection silencing device for digital filtering of audible medical alarm sounds to enhance ICU patient recovery. 23rd Int'l Conf. on Auditory Display, Penn State U, State College, PA, 20–23 June 2017
Siebein GS, Skelton R (2009) Soundscape analysis of a neonatal intensive care unit. Inter-Noise 2009, Ottawa, 23–26 August 2009
Sobieraj J, Ortega C, West I, Voepel L et al (2006) Audibility of patient clinical alarms to hospital nursing personnel. Mil Med 171(4):306–310
Stanford L, McIntyre J, Hogan JT (1985) Audible alarm signals for anaesthesia monitoring equipment. Int J Clin Monit Comput 1:251–256
Topf M (1988) Noise-induced occupational stress and health in critical care nurses. Hosp Top 66(1):30–34
Topf M, Dillon E (1988) Noise-induced stress as a predictor of burnout in critical care nurses. Heart Lung 17(5):567–574
Ulrich RS (1984) View through a window may influence recovery from surgery. Science 224(4647):420–421
Vahed N, Kabiri N (2016) The effect of music in the operating room: a systematic review. 5th International Society for Evidence-Based Healthcare Congress, Kish Island, December 2016
Wachman EM, Lahav A (2011) The effects of noise on preterm infants in the NICU. Arch Dis Childhood Fetal Neonatal Ed 96:F205–F209
Weber JR, Ryherd EE, Mahoney AD, Rolfes M, et al (2016) Evaluating hospital quiet time from engineering, medical, and nursing perspectives. 171st meeting of the Acoustical Society of America, Salt Lake City, UT, 23–27 May 2016
Weber JR (2018) The impact of quiet time on the neonatal intensive care unit soundscape and patient outcomes. PhD Dissertation, Architectural Engineering Program, University of Nebraska – Lincoln 2018
Whalen DA, Covelle PM, Piepenbrink JC, Villanova KL et al (2014) Novel approach to cardiac alarm management on telemetry units. Cardiovasc Nurs 29(5):E13–E22
Williams AL, van Drongelen W, Lasky RE (2007) Noise in a contemporary neonatal intensive care. J Acoust Soc Am 121(5):2681–2690
Willett KM (1991) Noise-induced hearing loss in orthopaedic staff. J Bone Joint Surg 73(1):113–115
Wolf JA (2013) The state of patient experience in American hospitals 2013: positive trends and opportunities for the future. The Beryl Institute
Zun LS, Downey L (2005) The effect of noise in the emergency department. Acad Emerg Med 12:663–666

# Chapter 11
# How to Put Soundscape into Practice

**André Fiebig and Brigitte Schulte-Fortkamp**

**Abstract** Over the years soundscape planning has clearly gained significance; however, it is still the case that soundscape projects and soundscape-based urban noise planning are not fully established in the fields of noise control applications and noise policy. Worldwide, numerous soundscape interventions have been implemented, indicating the value of the soundscape approach and soundscape planning that includes input from local experts. Nevertheless, there are some reservations among policy makers and planners about applying human-centered approaches and developing participatory processes with a local community for sustainable soundscape design. There are no well-established procedures for soundscape design and planning. The introduction of standardized methodology endeavors to overcome the lack of defined systematic approaches for the identification of interventions. Utilizing soundscape standards will make the decision process transparent and comprehensible and thus more acceptable for authorities and policy makers. Moreover, activities to set-up databases to document soundscape interventions and their evaluation might lead to the recognition of frequently recurring design strategies and the derivation of best practices. Therefore, an increased interest in and more frequent application of the soundscape approach in urban sound planning can be expected in the future.

**Keywords** Soundscape design · Soundscape intervention · Soundscape planning · Urban planning

---

A. Fiebig (✉)
Department of Engineering Acoustics, Technische Universität Berlin, Berlin, Germany
e-mail: andre.fiebig@tu-berlin.de

B. Schulte-Fortkamp
HEAD-Genuit Foundation, Herzogenrath, Germany
e-mail: bschulte_f@web.de



B. Schulte-Fortkamp et al. (eds.), *Soundscapes: Humans and Their Acoustic Environment*, Springer Handbook of Auditory Research 76,
https://doi.org/10.1007/978-3-031-22779-0_11

## 11.1 Introduction

In urban planning and development, visual planning plays a major role while sound is often not considered from the beginning as a relevant point of concern to be managed. Frequently, at the end of the planning stage, sound is considered as a nuisance that has to be mitigated using technical measures and interventions (de Coensel et al. 2010). Consequently, the most common approach by far to noise control is remediation after noise conflicts are identified and, most likely, the remediation focuses on reducing the sound pressure level of noise. Over time with the growing awareness that sound is an integral part of an urban environment and the recognition that innovative, site-specific solutions are best determined by participatory processes, the soundscape approach has become more and more popular. Over the decades, soundscape researchers have encouraged the integration of sound considerations at the earliest stages of urban design rather than waiting for noise problems to arise (cf., Steele et al. 2019b). The availability of international soundscape standards for data collection and analyses facilitates the integration of the soundscape approach into standard noise management and planning processes.

Soundscape approaches have been recognized by governmental organizations and national funding bodies in Europe and worldwide (cf., Kang et al. 2016). Joint efforts of an international research network to determine validated translations of the established soundscape measurement protocols into various languages have allowed worldwide national adoption of questionnaires for soundscape characterization (Aletta et al. 2020).

In the context of urban planning, it is not a matter of choosing either a method of noise control or the soundscape approach but rather choosing noise control that is complemented by soundscape planning (Brown 2012). Soundscape planning addresses aspects of quality of life and accounts for the perceptions of local citizens (Steele et al. 2019b). Such a paradigm shift toward a perception-oriented design of acoustic environments can also be observed for indoor environments. Consequently, Altomonte et al. (2020) state that indoor and outdoor environments must promote restoration, offer variation, and advance the introduction of positive stimuli for better quality of life. Merely limiting building performance standards to avoid negative sound impacts on humans, like disease or discomfort, is not sufficient anymore. Built environments must be designed to enhance positive outcomes (Altomonte et al. 2020).

### *11.1.1 Soundscape in Urban Planning*

Integrating soundscape in urban planning will continue to be a major challenge, especially when transferring the soundscape concept, with its inherent demand for a holistic approach and its interdisciplinary foundation, for real-world applications. In that sense, soundscape is sometimes perceived as an academic tool used by numerous soundscape research studies that are working on indicators and descriptors, technologies, and frameworks. This may impede the application of a soundscape approach by urban planners who must deliver sound planning schemes and implement measures and interventions.

Moreover, urban planners and soundscape planners need to have basic planning skills, to be able to conduct fieldwork, and to have sufficient knowledge of the soundscape (Xiao et al. 2018). However, considering soundscape aspects at early stages in the planning process makes it possible to evaluate sound as a resource and not just as a nuisance to be managed afterward, as Schafer (1977) recommended in the 1970s. In this sense, significant progress has been made and a large body of experiences is now available to facilitate implementation of the soundscape approach.

### *11.1.2 Soundscape Action Plan, Interventions, and Stakeholders*

In general, a large group of participating stakeholders is necessary to realize the soundscape perspective and to bring it successfully into practice. Planning and administration authorities are as important as the soundscape investigation experts and the local experts (see Sect. 11.3.1 for definition) in the communities under scrutiny (ISO 12913-2 2018). Although the soundscape approach with its interdisciplinary foundation is not yet established in urban noise projects, the increase in standards and technical specifications along with the rising number of implemented soundscape interventions are evidence of a significant change.

Related successes include the soundscape action plan (Welsh Government 2018) and the WHO guidelines for environmental noise (see Brown and van Kamp 2017) to exploit all intervention types to achieve substantially healthier acoustic environments. In Montreal, a cross-sector partnership called *Sounds in the City* was formed by university soundscape researchers, acoustic consultants, and the City of Montreal several years ago with the aim of connecting research and practice to make cities sound better (Steele et al. 2019b). Based on this collaboration, among other outcomes, a sound map of Montreal was successfully developed as a web-based soundscape project (https://www.montrealsoundmap.com) that empowered local people to make their personal soundscapes tangible and also compiled a shared database (MSM 2022). These developments are evidence of the increased attention being directed toward soundscape methods and practices. In this context, innovative interventions that are based on sound-conscious designs that go beyond conventional noise control approaches are needed to exploit all opportunities to significantly decrease the negative impact of noise, provide environmental improvements for public health, and promote individual well-being.

As Fig. 11.1 illustrates, successful urban sound planning must consider interventions at various levels to preserve and/or to improve a soundscape by following the

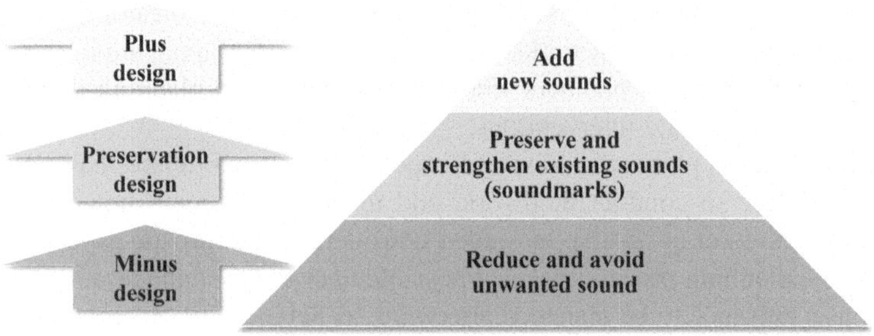


**Fig. 11.1** From minus to plus design: typical fundamental approach for soundscape design. *Minus design* refers to noise abatement measures where unwanted noise is reduced. This can be understood as the conventional way of dealing with noise in noise control. The *preservation design* step refers to the identification and preservation of sound sources that contribute to the existing acoustic environment in a positive way. This is particularly relevant for sounds that are unique for a site and allow for identification of the place by listening. Those specific sounds are often called soundmarks (Schafer 1977). The *plus design* as the final step in the design pyramid intends to improve a soundscape by deliberately introducing new sound elements to an existing acoustic environment. Schafer (1977) described this design step as "… carefully redesigning the soundscape by adding sounds that will harmonize with the environment and with each other" (Schafer 2012, p.8). For more information see Siebein and Siebein, Chap. 5

| Intervention Type/Level | Source | Path/ Infrastructure | Design | Local expert |
|---|---|---|---|---|
| Examples of measures | traffic flow<br>road surface<br>traffic composition | barriers<br>tunnels<br>greenery<br>dwelling insulation<br>traffic management | green areas<br>sound quality<br>noise masking<br>quiet façade side<br>soundmarks | communication<br>education<br>participation<br>co-creation<br>behavior |

**Fig. 11.2** Intervention framework adapted from van Kamp et al. (2019) and extended to soundscape. Interventions can be directed at sound sources or to transmission paths. Beyond those conventional measures, interventions can be implemented on a larger scale by which further aspects of the surrounding (e.g., easy access to quiet areas) or the living environment (e.g., flats with rooms facing the quiet façade side) are addressed that have an impact on well-being without directly reducing the noise exposure. Moreover, interventions can also be related to the involvement of the local experts to jointly work on solutions based on co-creation for a better soundscape. Soundscape design considers and exploits all intervention types with special emphasis on the local expert to develop a site-specific soundscape design

soundscape approach. This extended intervention framework goes beyond the established three-step soundscape design approach: (1) to reduce, buffer, or mitigate noise elements that are unwanted; (2) to determine, preserve, and enhance those sound elements that are desired; and (3) to create new sound elements that enhance the overall soundscape, as illustrated in Fig. 11.2 (see Brooks, Chap. 4; Siebein and Siebein, Chap. 5; Siebein et al. 2006). All intervention types must be equally considered simultaneously to achieve high soundscape quality. This even concerns communication, behavior, and co-creation as well as the exploitation of non-acoustical factors to modulate perception (cf., Kang et al. 2016, see Fig. 11.1). This notion corresponds to the holistic concept for urban sound planning based on the soundscape approach. For example, the specific function of the space must be determined to ensure acoustic compatibility of any future soundscape design within the site (Cerwén et al. 2017), the specific needs and requirements of the local population must be assessed (Schulte-Fortkamp and Jordan 2016), and their local culture and history must be considered (Kang et al. 2016) in addition to the (physical) impact of the acoustic environment on the residents.

The soundscape approach uses all information available on the key components of *people*, *acoustic environment*, and *context* to investigate an existing area (ISO 12913-1 2014) and to identify the interventions and actions required to preserve or improve a soundscape as much as possible (ISO/PWI TS 12913-4 2020). Focusing simply on the characteristics of sound sources and paths as the primary objective of conventional noise control studies is not sufficient to significantly improve soundscapes in a sustainable manner. Moreover, consideration of the context is important because non-acoustic factors are as important as acoustic ones in shaping the soundscape of a space (Taghipour et al. 2022).

To consider the voice of the users and local experts in the space under scrutiny, standardized soundscape data collection tools and methods (ISO/TS 12913-2 2018) are frequently applied in an acoustic consulting environment (Mitchel 2021). On the one hand, this development leads to more practical experiences with which to integrate soundscape methods and former conventional procedures and, on the other hand, facilitates discussion of the practicality of addressing issues (Heggie et al. 2019).

## 11.2 Soundscape Design and Interventions

### *11.2.1 From Soundscape Design to Soundscape Intervention*

Soundscape design indicates the plan to specifically change an existing acoustic environment or to plan a new area. A soundscape intervention is the implementation of the design plan to preserve or improve an existing soundscape. In other words, the soundscape intervention is more than just the intention to preserve or improve soundscape: it is the factual realization of the intention (i.e., the design plan). The implementation of soundscape designs, which are usually derived from soundscape investigations that determine site-specific needs, is becoming increasingly popular. At present, collections of successful soundscape design interventions are quite scarce, and practical guidelines to implement soundscape design are rarely available in landscape or urban planning and design literature. Therefore, current initiatives and projects must overcome the relative lack of documented examples of interventions and soundscape design (Moshona et al. 2022).

Platforms and websites like *Catalogue of Soundscape Intervention* (CSI 2022), *Soundscape Design* (SD 2022), or *Urban Identity* (UI 2022) list several implemented soundscape interventions and soundscape projects. Lavia et al. (2016) gave a broad range of soundscape design and intervention examples with a comprehensive discussion of the applied methods and outcomes with the effects on citizens at the heart of the work. These developments indicate the growing interest in soundscape design in urban development and also indicate the increase in requests by authorities and urban planners to learn more about soundscape projects. The availability of databases showing successful soundscape interventions allows recognition of recurring strategies that probably can be collated into design toolkits (Moshona et al. 2022). However, as the soundscape concept relates to the perceptions of a specific location and emphasizes the specific context, the determination of universal interventions and implementation strategies might not be possible.

### *11.2.2 Examples of Soundscape Interventions*

Soundscape interventions range from conventional noise mitigation combined with new approaches (e.g., installing loudspeakers or introducing natural sound features for additional masking) to sound installations and art. Examples include a dynamic system of water fountains in a public square (Kang and Hao 2013), natural organs played by sea waves (Oberman et al. 2020), the combination of a noise barrier with loudspeakers playing back natural sounds (Cerwén 2016), and the installation of audio islands as seating furniture with integrated loudspeakers for informational masking purposes (Schulte-Fortkamp and Jordan 2016).

In general, the introduction of natural elements (e.g., sounds from birds, vegetation, or water) in soundscape projects can frequently be observed as those interventions are related to higher psychological restoration (Payne 2013) but also might reduce stress, ultimately contributing positively to general health (Hägerhäll et al. 2017). Steele et al. (2019c) confirmed that naturalistic sounds increased calmness and lowered perceived sound levels; however, they also showed that these benefits extended similarly for sound art that uses added sounds with cultural themes consisting of music extracts, speech, and urban elements.

Masking techniques can play a prominent role in soundscape design to enhance or introduce preferred sounds that mask unwanted sound components or that divert attention to other more pleasant sounds (Kang et al. 2016). In the context of masking, Cerwén (2016) observed that the use of masking strategies was effective in his soundscape project in the city of Malmo, Sweden, at the given sound pressure level (i.e., 58 dB(A)). However, the implementation of adding sound designed to energetically or informationally mask unwanted noise might have limitations. Zhang and Kang (2007) estimated that additive sound design using the principle of masking might work until approximately 70 dB(A). For higher sound pressure levels, an increase of annoyance can be expected due to sensory overload that is independent from the valence level of the sound.

Kang (2010) suggests protecting the diversity of sounds in acoustic environments because they characterize a place and can be related to cultures and history (Schulte-Fortkamp and Jordan 2016). This often intersects with the attractiveness of a place with regard to tourism. Similarly, Fiebig et al. (2016) observed in a nationwide, citizen science project in Germany (initiated by the German Federal Ministry of Education and Research) that in addition to the general desire for quiet urban places, acoustically diverse and vibrant places were frequently mentioned by citizens as their favorite places in their urban environments. In this context, the term *soundmarks* became associated with specific community sounds that provide spatial and temporal orientation. Soundmarks, as defined by Schafer (1977), are unique community sounds with special qualities that are perceived by the people in that community and are associated with landmarks that are linked to a specific geographical area. In the citizen science project soundmarks like church bells or sounds from local city festivals achieved the highest number of likes (Fiebig et al. 2016).

### 11.2.3 The Social Dimension of Soundscape Interventions

Soundscape interventions are not intended to improve the acoustic environment by defining users as solely passive receivers. Interventions are also used to increase engagement with the environment, facilitate social interactions, and attract new user groups. For example, Steele et al. (2019a) observed those outcomes after providing an amenity that allowed for public music playing in a small pocket park in Montreal, Canada. This soundscape intervention, implemented as a democratic soundscape installation called Musikiosk, resulted in increased and diversified social interactions in the park while still ensuring a high level of restoration for visitors. This example illustrates how soundscape design can go beyond classical noise abatement to enrich the acoustic experience in an environment.

Soundscape design also considers the function of space, the person-environment interactions, and the support of social interactions. An in situ study on the effect of water features by Trudeau et al. (2020) suggested the value of including slightly audible misters in outdoor urban environments. Those misters support in space and design the user's activities which have a positive effect on the quality of the soundscape. Similarly, Cerwén et al. (2017) regards the location of specific functions within a place as a most important aspect of the overall planning. The location of functions should focus on the compatibility of new and existing functions, and planning should take into account how different functions affect the sound in space and time. Such considerations go beyond the aim of conventional noise control, which simply reduces the sound pressure level of unwanted noise. Urban sound must be evaluated and designed considering the specific place and contexts of use.

According to Lavia et al. (2016), typical soundscape interventions and sound management types include introducing sounds to a soundscape, utilizing sounds that already exist in a location, incorporating sonic art installations, employing noise control elements, and introducing design alterations, among other possibilities specific to a location. As no well-established framework of soundscape interventions exists so far, research projects and initiatives are underway to overcome this deficit based on data collected with the specific intent of developing a comprehensive taxonomy (Moshona et al. 2022).

## 11.3 Implementing Soundscape into Practice

The soundscape approach demands consideration of key elements that include *people*, *acoustic environment*, and *context* (ISO/TS 12913-2 2018). Understanding the specific context, inclusion of all relevant stakeholders, and successful communication across research disciplines during soundscape investigations are complex and sometimes challenging requirements. This complexity can impede the application of a soundscape approach in practice and the value of guidelines, recommendations, and examples is clear. To address these needs, further standardization efforts are being made. For example, the ISO working group 54 "Perceptual assessment of soundscape quality" works on a part 4 within the ISO 12913 series to provide information about the determination of the need for soundscape interventions and guidance on how interventions should be implemented.

### *11.3.1 Co-creation*

A further challenge concerns the participation of local communities in the development of solutions through co-creation. There is a broad consensus that the knowledge of the local experts is indispensable. Accordingly, Foth (2017) demanded that the role of the city government and the citizen must change: authorities must grow from a simple administrator role to a collaborator involved in forging a positive relationship between cities and the people living in those cities. Moreover, citizens must move from the passive role of residents to active participants and enthusiastic co-creators in an increasingly collaborative approach to city making (see Botteldooren, De Coensel, Aletta, and Kang, Chap. 8).

In the past, the creation of new urban environments used to be the responsibility of architects, urban designers, and local authorities; however, the role of the citizen has changed over time to a partner in the co-creation process (Winne et al. 2020). This kind of collaboration requires that information and participation tools are provided to the people involved, those we call the local experts. Local experts, as defined in the ISO/TS 12913-2 (2018), are persons familiar with the area under scrutiny, either living in the area or having daily activities there, who can provide valuable information about what they consider to be necessary measures (see Schulte-Fortkamp and Jordan, Chap. 3). The users of a space are the primary experts in any environment, and their feedback enables creative and responsive solutions for the acoustic design (Schulte-Fortkamp and Fiebig 2006; Lavia et al. 2016). Their participation sharply focuses the subsequent analyses of any soundscape data, as the information provided enhances the sensitivity of the research team to the particularities of the examined areas (Schulte-Fortkamp and Jordan 2016).

Accordingly, Schulte-Fortkamp (2017) states it is essential to provide advice to local participants and stakeholders on how to use the given resources as sustainable solutions, considering future generations as well. Overall, a comprehensive consideration of all socio-cultural, aesthetic, and economic effects is necessary. Moreover, a platform must be available for communication that allows all stakeholders to participate. In this regard, a variety of sources of electronic communication, such as social networks, social media, and smartphone applications, are readily available for much of the population (see Brambilla and Fiebig, Chap. 7). Those platforms provide the public with the opportunity to generate data, to provide recommendations, and to pinpoint noise conflicts. The public is empowered to report on pleasant, restorative soundscapes that should be protected (Radicchi 2019a) or to document noise conflicts (NA 2022) that should be addressed by local authorities. Residents also are encouraged to support the generation of participatory noise maps (D'Hondt et al. 2013). Thus, Radicchi et al. (2018) point out that smart digital technologies are expected to play an important role for acousticians, city planners, and policy makers in the future.

Radicchi (2019b) observed an increase in the development of mobile applications that have been deployed as environmental noise and soundscape evaluation tools to allow everyone to contribute to addressing open questions in the field of environmental noise and soundscape research. At the time of the review, she found over 20 smart applications were available that allow for the collection of mixed data, such as noise levels, audio recordings, and the collection of user feedback. Figure 11.3 displays important aspects of successful electronic participation that are needed to encourage the public to take part.

Co-creation opens a wealth of opportunities to improve public spaces and their use but, of course, a few critical pitfalls must be avoided. For example, local experts are less familiar with technical terms which challenge the communication between stakeholders (Botteldooren et al. 2020). As discussed by Brooks (Chap. 4), introducing the concept of soundscape into the urban master-plan process early is vitally important to ensure that soundscape considerations are integrated into any comprehensive plan and are part of a publicly available, visionary document. The key is to inspire proactive planning versus reactive measures regarding sound resources, their benefits, and impacts.

### *11.3.2 Training in Participation*

Sometimes the lack of technical expertise among residents limits their ability to estimate the (perceivable) impact of noise control measures as they are less familiar with the technical terms, as Botteldooren et al. (2018) observed. Therefore, training sessions are proposed for the people who are participating in the co-creation

**Fig. 11.3** Elements of successful electronic participation of the public in the context of urban sound. According to Fiebig et al. (2016), electronic participation opportunities can only have their intended effects if several requirements are met. The opportunity alone to participate electronically in a project does not automatically lead to a large turnout. Therefore, designers of electronic participatory projects need to carefully consider the conditions (as shown in the figure) to meet the desired public resonance

process. Technical terms, such as the averaged equivalent sound levels, can be explained and made more understandable. The result will be a more effective co-creation process with informed input from the residents.

Technological developments of virtual and augmented reality might help in the future to preview urban design and architecture both off site and on site. Experiencing the relative effectiveness of potential measures, at least to some extent, can facilitate discussions between different stakeholders. The soundscape experts, whether residents, regular visitors, or concerned public stakeholders, can communicate and work on solutions using different formats, such as workshops, public discussions, soundwalks, public consultations, or presentation of creative seminars.

Brooks et al. (2014) presented a case study in which they focused on co-creation with people from the downtown area in Jamestown, Rhode Island. The issue was noise from a beer garden, and they used a variety of methods for community participation, including physical sound surveys and stakeholder workshops (see Brooks, Chap. 4). The outcome of this participatory procedure was the unexpected decision by the residents to oppose limitations of the sound emissions from the beer garden as the activity is important for community identity (Schulte-Fortkamp et al. 2007; Brooks et al. 2014).

Table 11.1 describes how different tools and methods can be applied to involve local experts successfully as co-creators. Steele et al. (2019b) support increasing the connections between soundscape evaluations and emerging models of participatory design and planning that will facilitate the dissemination of knowledge and, ultimately, make cities sound better. Local experts need to be empowered to take part in urban sound planning by approved participatory methods to represent their interests and needs.

**Table 11.1** Forms of cooperation and communication in soundscape projects. Summary of different forms of co-creation

| Measure of cooperation and communication | Description |
|---|---|
| Workshop | A group of local experts together with few other stakeholders develop soundscape design solutions by means of a collaborative process without addressing any technical requirements and boundary conditions. Outcomes of the workshops are evaluated with respect to criteria including feasibility, regulatory requirements, costs, and sustainability in subsequent steps |
| Public consultation | Process by which the public opinion about a project is collected and discussed. Sometimes the public consultation is also used to inform the public about potential measures and developments, which increases transparency and acceptance |
| Soundwalk | Site inspection to raise awareness of site-specific demands with soundscape project stakeholders and/or for data collection |
| Creative seminar | A group of local experts together with a few additional stakeholders develop creative soundscape design solutions for inspirational purposes. Ideas and proposals can be based on artistic approaches |
| Focus group discussion | Problems and objectives are identified through interactive and directed discussions with stakeholders and local people |

Figure 11.4 illustrates a model of a flexible participatory, urban soundscape-planning process suggested by Xiao et al. (2018). Different participation and engagement methods are proposed for different urban scales: from street level, community level, and city level. Various methods are needed to successfully engage stakeholders at these different stages of soundscape planning and the design process. In this context, it is important to mention that at each planning phase, the strategies and design plans should be re-evaluated to allow maximum freedom in the next stages and to avoid missing opportunities to create the most pleasant and healthy living environments (van Renterghem et al. 2020).

Engaging the local experts is critical. Achieving a good understanding of soundscape concepts by the stakeholders is an essential step to be completed before conducting further work. Greater understanding among all participants guarantees due consideration of regulatory requirements and further aspects of the project.

### *11.3.3 Integrating the Virtual Experience*

Several examples are available that illustrate how the soundscape approach can be extended by using virtual reality (VR) for non-existent or future environments (see Brambilla and Fiebig, Chap 7). Vorländer (2020) acknowledged that tools of acoustic virtual reality offer ground-breaking opportunities to advance sound assessment both in preliminary consulting and in actual practice.

| PLANNING STAGE | OBJECTIVES | METHODS | OUTCOMES |
|---|---|---|---|
| Start-up | Definition of soundscape goals | Panel meeting | Soundscape planning scheme |
| Preparation | Soundscape objectives, action plan | Focus groups, workshops, conferences, interviews | Awareness, identification of soundscape features and approaches |
| Forecasting and Implementation | Generation and implementation of soundscape design | Site assessments, soundwalks, onsite experiments | Design and experiments for a better soundscape |
| Evaluation | Evaluation of soundscape design and outcomes | Panel meeting, survey, interviews | Establishment of evaluation criteria |
| Future Developments | Establishment of new goals for soundscape planning in a city | Panel meeting, survey, interviews | Generation of sustainable planning process |

**Fig. 11.4** Proposal of an agile participatory process for urban soundscape planning by Xiao et al. (2018) shown in a condensed version. The soundscape planning process is understood as an iterative, agile, and circular process to manage interactions with local experts as co-specifiers. The process is flexible: each stage can be referred back to the last stage and re-developed to meet the objectives. As shown in the figure, different methods like panel meetings, surveys, soundwalks, or interviews can be performed at different stages to achieve the respective objectives, to support the co-creation process, and to consider sufficiently location-specific aspects. (Adapted from Xiao et al. 2018)

A practical example is presented by Sajeev et al. (2022). They applied the soundscape approach for a co-housing development project in combination with a conventional noise impact assessment. They performed in situ soundwalks with future residents during daytime and night-time periods, covering locations with different sounds and different future uses. Beyond that, they created virtual soundwalks based on auralizations to provide an actual experience of the future soundscape. They concluded that their procedure helped the residents to recognize design improvements for the sound environment of their future homes, fostered inclusivity, and facilitated co-creation (Sajeev et al. 2022).

Oberman et al. (2020) used virtual techniques to investigate the value of soundscape interventions with musical features that were introduced to public spaces as permanent sound art. They used second-order ambisonics (a special surround sound format) that reproduced the three-dimensional spatial relationships of both static and dynamic sound sources and combined the ambisonics with panoramic photographs to determine the effect of sound installations at specific sites (see Brambilla and Fiebig, Chap.7). They showed based on the virtual scenarios the potential which public sound art has when applied within urban design; adding sound art to a site can influence not only pleasantness but also appropriateness of the overall acoustic environment (Oberman et al. 2020).

Although a gap between on-site and virtual experiences of soundscape may still exist, studies suggest that there are similarities in sound source recognition and sound assessments under those different conditions (e.g., Hong et al. 2019). Once a certain level of immersivity is reached, soundscape evaluations obtained in VR-based experiences seem to be like those obtained directly at the same location, at least when analyzing the results from standard questionnaire surveys (Rajguru et al. 2020). Thus, the application of VR for experiencing non-existing acoustic environments appears basically conceivable.

Fraisse et al. (2022) applied a higher order ambisonics soundscape simulation tool to design a permanent sound installation in an urban public space in Paris. Lugten et al. (2018) used virtual reality experiments to evaluate the benefit of introduced water features in soundscapes affected by aircraft noise. They observed the reduction of the saliency of aircraft flyovers with the presence of moving water sound features, which clearly indicated that soundscape strategies can complement noise abatement in areas prone to aircraft noise. These examples show that simulation techniques allow perceptual assessments of various options for soundscape interventions before they are implemented, and modifications of less successful interventions as indicated by the virtual experiences.

As shown by the projects and case studies presented here, using sound reproduction techniques facilitates co-creation and allows participatory approaches during project development (see Schulte-Fortkamp and Jordan, Chap. 3). Although some challenges remain for successfully producing virtual soundwalks that allow a future soundscape to be experienced, there are no real obstacles to providing a holistic approach and putting human perception in the center of soundscape considerations.

## 11.4 Summary

There is no doubt that there is increasing interest in the soundscape approach and application of soundscape techniques when dealing with environmental noise (Aletta and Xiao 2018); however, integration of the soundscape approach with standard community noise procedures is not mandatory at present. Nevertheless, there is evidence for an ongoing paradigm shift from noise control to soundscaping and soundscape approaches are increasingly applied in noise management projects and everyday practice (cf., Jiang et al. 2022).

Indeed, further research is required on soundscape design to reduce the gap between theory and practice (Carvalho et al. 2019). In particular, knowledge gained about best practices must be shared among academic researchers, urban planners, and designers (Aletta and Xiao 2018). The main issue is to bring the soundscape approach successfully into practice for all kinds of environmental noise projects. Goals for the future of soundscaping can be summarized as follows:

- Raising awareness of all stakeholders for the need to integrate the soundscape approach in all stages of urban sound planning through public activities
- Promote education for students, professionals, and practitioners in soundscape methods to enable consistent and proper use of soundscape data collection and analysis methods according to national or international specifications (e.g., ISO/TS 12913-2 2018; ISO/TS 12913-3 2019)
- Integrate soundscape methods in legal frameworks to promote their regular use in everyday practice and to stimulate future developments based on the gained experiences in practice
- Improve virtual reality technologies to allow the perceptual assessment of non-existing environments in a multidimensional way with the help of local experts
- Provide more scientific evidence regarding the sustainable design of soundscapes and impacts on public health
- Disseminate current developments in conferences and workshops

While some challenges remain, inclusion of the soundscape approach in a variety of noise management and development projects is becoming more common. Increasingly, soundscape projects are overcoming imagined obstacles in noise management and development and providing successful temporary or permanent interventions (Oberman et al. 2020). In those projects, the expertise of locals was considered and acknowledged at each stage of the planning process, which produced effective outcomes and sustainable solutions.

The soundscape interventions that were implemented in various successful projects and studies have been documented and have demonstrated the benefit of the soundscape approach, validating these practical applications beyond academic interests. Understanding the performance of previous soundscape investigations leads to more established best practices that include the involvement of local experts in co-creative processes and the consideration of location-dependent particularities. In addition, ongoing research activities will provide further information about soundscape interventions and how to achieve the intended purpose, for example, the detailed design of waterfalls, fountains with upward jets, and the configuration of flowing streams to effectively promote peacefulness and relaxation in the presence of road traffic noise (e.g., Galbrun and Ali 2013; Calarco 2015) or the effect of various water features and vegetation on the perceived levels of aircraft noise (Lugten et al. 2018).

In general, soundscape planning could be successfully introduced to large urban (re)development projects and multi-stage development projects (van Renterghem et al. 2020) because the technical, economic, and organizational feasibility of integrating soundscape design actions has been demonstrated. Guidance on how to set-up a communication framework to bring all stakeholders successfully on board and how to implement innovative soundscape interventions will additionally increase the interest in, and applicability of, the soundscape approach in urban sound planning.

The ISO working group 54 (ISO/PWI TS 12913-4 2020) provided standardized guidance on how to design soundscapes to preserve or improve a soundscape. This guidance intends to encourage the further rise of the soundscape concept within urban planning processes and establish new ways of managing urban acoustic environments beyond noise regulations that focus on designating noise level limits.

Finally, the built environment should not be viewed as a monolithic construction. The history and current purpose of a place must be considered in urban planning, particularly with respect to comfort, engagement, and community connections. Over time, an acoustic environment is also constructed by the various people who use it and their interactions within the space. Thus, the strategies used to understand any particular location must be adapted to these singularities. Considering how to most effectively balance acoustic measurements, architectural planning, and input based on the expertise of local experts will lead to a new understanding of co-creation methods in urban planning.

**Compliance with Ethics Requirements** André Fiebig declares that he has no conflict of interest.

Brigitte Schulte-Fortkamp declares that she has no conflict of interest.

## References


Aletta F, Xiao J (2018) What are the current priorities and challenges for (urban) soundscape research? Challenges 9(1):16. https://www.mdpi.com/2078-1547/9/1/16

Aletta F, Oberman T, Axelsson Ö, Xie H et al. (2020) Soundscape assessment: towards a validated translation of perceptual attributes in different languages. In: Internoise 2020, Proc., Seoul, pp 3137–3146

Altomonte S, Allen J, Bluyssen PM, Brager G et al (2020) Ten questions concerning well-being in the built environment. Build Environ 180. https://doi.org/10.1016/j.buildenv.2020.106949, 1–13

Botteldooren D, Dekoninck L, Meeussen C, Van Renterghem T (2018) Early-stage sound planning in urban re-development: the Antwerp case study. In: Internoise 2018, Proc., Chicago

Botteldooren D, De Pessemier T, Filipan K, Sun K et al (2020) Modifying and co-creating the urban soundscape through digital technologies, December 2020. In: Smaniotto Costa C, Mačiulienė M, Menezes M, Goličnik Marušić B (eds) Co-creation of public open places. Practice - Reflection – Learning, C3Place, pp 185–200. https://doi.org/10.24140/2020-sct-vol.4-2.2

Brooks B, Schulte-Fortkamp B, Voigt KS, Case AU (2014) Exploring our sonic environment through soundscape research and theory. Acoust Today 10(1):30–40

Brown AL (2012) A review of progress in soundscapes and an approach to soundscape planning. Int J Acoust Vib 17(2):73–81

Brown AL, van Kamp I (2017) WHO environmental noise guidelines for the European region: a systematic review of transport noise interventions and their impacts on health. Int J Environ Res Public Health 14(8):873, 1–44

Calarco FMA (2015) Soundscape design of water features used in outdoor spaces where road traffic noise is audible, doctoral thesis. Heriot-Watt University, Edinburgh

Carvalho ML, Davies WJ, Fazenda B (2019) Investigation of emotional states in different urban soundscapes through laboratory reproductions of 3D audiovisual samples. In: Proceedings of the 14th international postgraduate research conference in the built environment: contemporary and future directions in the built environment, Salford, pp 327–339

Cerwén G (2016) Urban soundscapes: a quasi experiment in landscape architecture. Landsc Res 41(5):481–494

Cerwén G, Wingren C, Qviström M (2017) Evaluating soundscape intentions in landscape architecture: a study of competition entries for a new cemetery in Järva, Stockholm. J Environ Plan Manag 60(7):1253–1275

CSI (2022) Catalogue of soundscape intervention. https://soundscape-intervention.org. Last accessed on 31 Aug 2022

D'Hondt E, Stevens M, Jacobs A (2013) Participatory noise mapping works! An evaluation of participatory sensing as an alternative to standard techniques for environmental monitoring. J Pervasive Mob Comput 9(5):681–694. https://doi.org/10.1016/j.pmcj.2012.09.002

De Coensel B, Bockstael A, Dekoninck L, Botteldooren D et al (2010) The soundscape approach for early-stage urban planning: a case study. In: Internoise 2010, Proc., 2333–2342, Lisbon

Fiebig A, Schulte-Fortkamp, Genuit K (2016) The city of the future - noise initiative in the "year of science 2015". J Acoust Soc Am 139:1982. https://doi.org/10.1121/1.4949784

Foth M (2017) Participation, co-creation, and public space. J Public Space 2(4):21–36. https://doi.org/10.5204/jps.v2i4.139

Fraisse V, Schütz N, Wanderley MM, Guastavino C et al (2022) Informing sound art in public space through soundscape stimulations. In: Internoise 2022, Proc., Glasgow

Galbrun L, Ali TT (2013) Acoustical and perceptual assessment of water sounds and their use over road traffic noise. J Acoust Soc Am 133(1):227–237. https://doi.org/10.1121/1.4770242

Hägerhäll C, Taylor R, Cerwén G, Watts G et al (2017) Biological mechanisms and neurophysiological responses to sensory impact from nature. In: Van den Bosch M, Bird W (eds) Oxford textbook of nature and public health. Oxford University Press, Oxford

Heggie C, Smyrnova J, Allen M, Klein A (2019) The practicalities of soundscape data collection by systematic approach according to ISO 12913-2. In: Internoise 2019, Proc., Madrid, Spain

Hong JY, Lam B, Ong ZT, Ooi K et al (2019) Quality assessment of acoustic environment reproduction methods for cinematic virtual reality in soundscape applications. Build Environ 149:1–14

ISO 12913-1:2014 Acoustics-Soundscape-Part 1: definition and conceptual framework. Geneva. https://www.iso.org/standard/52161.html

ISO/PWI TS 12913-4:2020 Acoustics — Soundscape — Part 4: Guidelines for the assessment of soundscape investigation results. Geneva. https://genorma.com/en/project/show/iso:proj:81507

ISO/TS 12913-2:2018 Acoustics-Soundscape-Part 2: data collection and reporting requirements. Geneva. https://www.iso.org/standard/75267.html

ISO/TS 12913-3:2019: Acoustics-Soundscape Part 3: data analysis. Geneva. https://www.iso.org/standard/69864.html

Jiang L, Bristow A, Kang J, Aletta F et al (2022) Ten questions concerning soundscape valuation. Build Environ 219:1–10. https://doi.org/10.1016/j.buildenv.2022.109231

Kang J (2010) From understanding to designing soundscapes. Front Archit Civ Eng China 4(4):403–417. https://doi.org/10.1007/s11709-010-0091-5

Kang J, Hao Y (2013) Soundscape of waterscape and square on the Sheffield Gold Route. Soundscape Travelling Exhibition, Meran 2013

Kang J, Aletta F, Gjestland TT, Brown LA et al (2016) Ten questions on the soundscapes of the built environment. Build Environ 108(2016):284–294. https://doi.org/10.1016/j.buildenv.2016.08.011

Lavia L, Dixon M, Witchel HJ, Goldsmith M (2016) Applied soundscape practices. In: Kang J, Schulte-Fortkamp B (eds) Soundscape and the built environment. CRC Press, Taylor & Francis Group, Boca Raton

Lugten M, Karacaoglu M, White K, Steemers K (2018) Improving the soundscape quality of urban areas exposed to aircraft noise by adding moving water and vegetation. J Acoust Soc Am 144(5):2906–2917. https://doi.org/10.1121/1.5079310

Mitchel A (2021) The implementation of a soundscape approach and methodologies in an acoustic consulting context. J Acoust Soc Am 149:A72. https://doi.org/10.1121/10.0004553

Moshona CC, Aletta F, Henze H, Chen X et al (2022) hat is a soundscape intervention? Exploring definitions and identification criteria and a platform to gather real-world examples. In: Internoise 2022, Proc., Glasgow

MSM (2022) Montreal sound map. https://www.montrealsoundmap.com. Last accessed on 31 Aug 2022

NA (2022) The Noise App. Simplify noise reporting and investigation. https://www.thenoiseapp.com/. Last accessed on 31 Aug 2022

Oberman T, Jambrošić K, Horvat M, Bojanić Obad Šćitaroci B (2020) Using virtual soundwalk approach for assessing sound art soundscape interventions in public spaces. Appl Sci 10(6):2102. https://doi.org/10.3390/app10062102

Payne S (2013) The production of a perceived restorativeness soundscape scale. Appl Acoust 74(2):255–263. https://doi.org/10.1016/j.apacoust.2011.11.005

Radicchi A (2019a) Hush City. From crowdsourced data to open source planning of quieter and healthier cities. In: Besters M, Marrades Sempere R, Kahne J (eds) Our city? Countering exclusion in public space. Placemaking Europe Publications, pp 367–370

Radicchi (2019b) Mobile applications for environmental noise and soundscape evaluation. In: Internoise 2019, Proc., Madrid

Radicchi A, Henckel D, Memmel M (2018) Citizens as smart, active sensors for a quiet and just city. The case of the "open source soundscapes" approach to identify, assess and plan "everyday quiet areas" in cities. Noise Mapp 4:104–123. https://doi.org/10.1515/noise-2018-0001

Rajguru C, Obrist M, Memoli G (2020) Spatial soundscapes and virtual worlds: challenges and opportunities. Front Psychol 11:569056. https://doi.org/10.3389/fpsyg.2020.569056

Sajeev S, Pitone RR, Nabialek D, Batten J et al (2022) Soundscape design for a residential development: a case study. In: Internoise 2022, Proc., Glasgow

Schafer RM (1977) The tuning of the world. Alfred a. Knopf, Inc, New York

Schafer RM (2012) Soundscape studies: the early days and the future. Soundscape: J Acoust Ecol 12(1):6–8

Schulte-Fortkamp B (2017) Soundscape and noise management – a matter of life quality. In: Keynote, Inter-noise 2017, Proc., Hong Kong

Schulte-Fortkamp B, Fiebig A (2006) Soundscape analysis in a residential area: an evaluation of noise and people's mind. Acta Acust united with Acust, Special Issue on Soundscapes 92(6):875–880

Schulte-Fortkamp B, Jordan P (2016) When soundscape meets architecture. Noise Mapp 3(1):216–231

Schulte-Fortkamp B, Brooks B, Bray W (2007) Soundscape- an approach to rely on human perception and expertise in the post-modern community noise era. Acoust Today 3(1):7–15

SD (2022) Soundscape design. https://soundscapedesign.info. Last accessed on 31 Aug 2022

Siebein GW, Kwon YM, Smitthakorn P (2006) An acoustical palette for urban design. 13. In: International congress on sound and vibration, Vienna

Steele D, Bild E, Tarlao C, Guastavino C (2019a) Soundtracking the public space: outcomes of the Musikiosk soundscape intervention. Int J Environ Res Public Health 16:1865

Steele D, Kerrigan C, Guastavino C (2019b) Sounds in the city: bridging the gaps from research to practice through soundscape workshops. J Urban Des 25(5):646–664. https://doi.org/10.1080/13574809.2019.1699399

Steele D, Legast E, Trudeau C, Fraisse V et al (2019c) Sounds in the city: improving the soundscape of a public square through sound art. In: International Congress on Sound and Vibration, Proc., Montreal

Taghipour A, Gisladottir A, Aletta F, Bürgin M et al (2022) Improved acoustics for semi-enclosed spaces in the proximity of residential buildings. In: Internoise 2022, Proc., Glasgow

Trudeau C, Steele D, Guastavino C (2020) A tale of three misters: the effect of water features on soundscape assessments in a Montreal public space. Front Psychol 11(570797):1–13. https://doi.org/10.3389/fpsyg.2020.570797

UD (2022) Urban identity. https://urbanidentity.info. Last accessed on 31 Aug 2022 van Kamp I, Brown AL, Schreckenberg D (2019) Soundscape approaches in urban planning: implications for an intervention framework. In: 23rd International Congress on Acoustic, Proc., 405–410, Aachen. https://doi.org/10.18154/RWTH-CONV-238792

Van Renterghem T, Dekoninck L, Botteldooren D (2020) Multi-stage sound planning methodology for urban redevelopment. Sustain Cities Soc 62(102362):1–12. https://doi.org/10.1016/j.scs.2020.102362

Vorländer M (2020) Are virtual sounds real? Acoust Today 16(1):46–54

Welsh Government (2018) Noise and soundscape action plan 2018–2023, December 2018. Crown Copyright, Cardiff

Winne J, Filipan K, Moens B, Devos P et al (2020) The soundscape hackathon as a methodology to accelerate co-creation of the urban public space. Appl Sci 10(6):1932(1–19). https://doi.org/10.3390/app10061932

Xiao J, Lavia L, Kang J (2018) Towards an agile participatory urban soundscape planning framework. J Environ Plan Manag 61(4):677–698. https://doi.org/10.1080/09640568.2017.1331843

Zhang M, Kang J (2007) Towards the evaluation, description, and creation of soundscape in urban open spaces. Environ Plann B Plann Des 34:68–86. https://doi.org/10.1068/b31162

The manufacturer's authorised representative in the EU is Springer Nature Customer Service Centre GmbH, Europaplatz 3, 69115 Heidelberg, Germany. If you have any concerns regarding our products, please contact ProductSafety@springernature.com

Printed and bound by CPI Group (UK) Ltd, Croydon, CR0 4YY
01/10/2026
02225892-0010